Groundwater Geochemistry and Isotopes

Groundwater Geochemistry and Isotopes provides the theoretical understanding and interpretive methods for the use of geochemistry and isotopes in the practice of hydrogeology. This text simplifies the basics of aqueous geochemistry and the complementary use of environmental isotopes, using case studies to illustrate principles. This second edition includes new methods and case studies since the first publication. Accessible to consultants and practitioners as well as undergraduates and graduates, this new edition is supported by a workbook of example problems and test problems with solutions in a separate document for instructors. Additional material in the workbook is included on groundwater sampling, field measurements, testing alkalinity and routine analytical methods.

- Covers the basics of solutes, gases, and isotopes in water, and concentration-activity relationships and reactions.
- Describes tracing the water cycle, weathering, and the geochemical evolution of water quality.
- Explores dating groundwater as young as a few years to over hundreds of millions of years.
- Uses case studies to demonstrate the application of geochemistry and isotopes for contaminated groundwaters.

A *Groundwater Geochemistry and Isotopes* workbook, including example problems, test questions with solutions and sampling and analysis guide, is freely available to instructors by request from idclark@uottawa.ca and at the website for this textbook at the link: www.routledge.com/9781032405933.

Groundwater Geochemistry and Isotopes

Second Edition

Ian Clark

CRC Press
Taylor & Francis Group
Boca Raton London New York

CRC Press is an imprint of the
Taylor & Francis Group, an **informa** business

Cover: The blue springs of Oman discharge hyperalkaline Ca-OH groundwater with pH 12 and H_2 gas following low temperature serpentinization of ophiolites. The uptake of atmospheric CO_2 forms calcite that lines these incredible pools. Photo credit: Balz Abplanalp, former Swiss Ambassador to Oman.

Second edition published 2026
by CRC Press
2385 NW Executive Center Drive, Suite 320, Boca Raton FL 33431

and by CRC Press
4 Park Square, Milton Park, Abingdon, Oxon, OX14 4RN

CRC Press is an imprint of Taylor & Francis Group, LLC

ISBN: 978-1-032-40593-3 (hbk)
ISBN: 978-1-032-40597-1 (pbk)
ISBN: 978-1-003-35383-6 (ebk)

DOI: 10.1201/9781003353836

Typeset in Times
by KnowledgeWorks Global Ltd.

Access the Support Material: www.routledge.com/9781032405933

To the skilled and dedicated isotope and geochemical technologists in our laboratories who strive to produce the high quality data used in this book and supporting publications, and to the students they train in this science.

Contents

About the Author

Ian Clark is emeritus professor in the Department of Earth and Environmental Sciences at the University of Ottawa. He completed a bachelor of science in earth sciences and a master of science in hydrogeology at the University of Waterloo, followed by his doctoral degree at the Université de Paris-Sud (Orsay) in isotope hydrogeology and paleoclimatology. Since his earliest work on geothermal systems in western Canada, Dr. Clark's research has focused on the integration of geochemistry and isotopes to address questions on the origin, age, and geochemical history of groundwater and solutes in natural and contaminated settings. He and his colleagues recently established geoscience labs for accelerator mass spectrometry, stable isotopes, noble gases, and geochemistry in the Advanced Research Complex at the University of Ottawa.

1 Water Rocks

THE NATURE OF WATER

Water is a remarkable compound. From the mantle to the atmosphere, it plays a leading role in how the earth functions. Below the crust, high-pressure magnesium silicates in the mantle contain oceans of water that affect mantle rheology and convection. Within the crust, hydrated minerals reduce friction and allow plate tectonics, and water lowers the melting temperature of rock to generate magma that builds the continents. As chloride-rich brine, deep crustal waters carry gold, copper, and other metals to form ore deposits. Water of the ancient oceans provided the crucible for early life to evolve. In the terrestrial environment, rainwater is the agent for weathering rocks to release nutrients and then carry them osmotically through the roots and stems to nourish plants as it cycles back to the atmosphere. As vapor, water provides the earth with some 30°C of greenhouse warming and makes the planet habitable. Condensed as clouds, it reflects solar radiation and regulates our climate. Its high heat capacity and enthalpy fuel hurricanes and typhoons. Crystallized as ice, it has a remarkable open-tetrahedral framework that is less dense than liquid water, and so, unlike most other compounds, its solid phase is buoyant. Fortunately so, as aquatic habitats would freeze from the bottom up, and there would be no surface ice to regulate seasonal albedo. Even the search for life on Mars and establishing outposts on the moon follow the water. All who have plunged their face into a fountain of cold, clear water to slake their thirst on a hot day feel its quality for sustaining life.

In geochemistry, it is water's capacity to dissolve solids and transport solutes that makes H_2O so remarkable. The water molecule's asymmetrical charge distribution is central to this property. The two hydrogen atoms and single oxygen atom share four tetrahedrally oriented electron orbits, which give rise to a dipolar configuration. Although the overall molecule is electrically neutral, the surface charge is unevenly distributed, making H_2O a polar molecule (Figure 1.1).

The positive and negative poles of neighboring water molecules attract to form a weak hydrogen bond such that loose clusters of molecules are continually breaking apart and reforming as a loosely structured liquid at earth surface temperatures. At 25°C, about 80% of the water molecules are so structured. Near the boiling point, this is slightly less, and as water freezes, it becomes fully structured. Hydrogen bonds are weak and easily break, allowing water molecules to evaporate as vapor. When humidity exceeds the dew point of the air, hydrogen bonds reform and water vapor condenses—key steps in the hydrological cycle.

The dipolar nature gives water a second important property. These polar molecules orient themselves around ions, thus isolating them and holding them in solution. This makes water such a good solvent. The crystal-clear water of the Dead Sea, for example, holds about 30% of its mass as Mg^{2+} and Cl^- ions. The brines of the Atacama Desert in Chile are so laden with lithium ions leached from the Andes that they are mined.

DOI: 10.1201/9781003353836-1

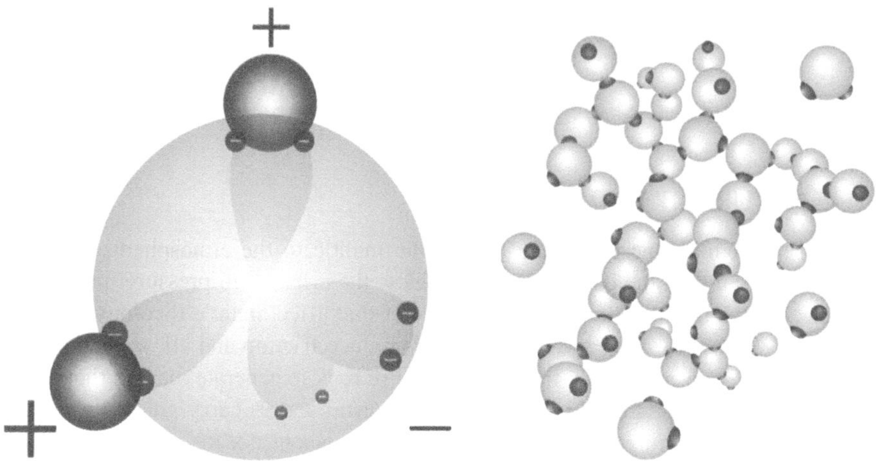

FIGURE 1.1 The dipolar molecule of H_2O with two hydrogen atoms configured within the oxygen atom at an angle of 105° (left) provides the tetrahedral structure of water (right). Clusters of water molecules held by weak hydrogen bonds give structure to liquid water.

A more arcane property of water is how the hydrogen bond is stronger for the heavier isotopes of water. This affects their evaporation and condensation rates such that these isotopes are partitioned throughout the hydrological cycle and can be used to trace the origin and movement of water.

ELEMENTS AND ISOTOPES

There are 92 naturally occurring elements, with atomic number, *Z*, from 1 to 92 (Table 1.1). Most are stable or very long-lived and formed some five billion years ago when the red giant precursor to our Sun was in its final stage. These primordial elements comprise some 286 isotopes or nuclides (neutron number, *n*, from 0 for hydrogen to 146 for uranium-238). These stable and radio-nuclides were produced by stellar nucleosynthesis, whereby hydrogen fusion in the earlier sun evolved to helium fusion in a red giant. In its final years, helium depletion led to burning higher masses to form the most abundant terrestrial elements, including C, O, Na, Al, Mg, Si, Ca, and Fe. Fusion of nuclei heavier than Fe is endothermic, and so the elements from iron up to uranium were produced by neutron and proton activation reactions on the growing inventory of higher-mass nuclides. As the dying red giant ran out of fuel, it exploded as a supernova, and the scattered mass orbiting a residual star, our Sun, gradually coalesced into the planets, comets, and asteroids of our solar system today.

The chemistry of the elements arises from the configuration of the electrons (e^-) into shells. Those with unfilled outer shells will shed or acquire electrons to stabilize their outer valence shell, producing a negative or positive charge, or oxidation state. Although most elements can have a variety of stable or quasi-stable oxidation states, most have only one or two common states in the environment. These are shown for each element in Table 1.1.

TABLE 1.1
Periodic Table of the Elements, with Atomic Mass Units (amu), Atomic Number (*z*) and Common Redox States

z amu **H** redox

IA	IIA	IIIA	IVA	VA	VIA	VIIA	← VIIIA		→	IB	IIB	IIIB	IVB	VB	VIB	VIIB	VIIIB
1 1.01 **H** 0 1																	2 4.00 **He** 0
3 6.94 **Li** 1	4 9.01 **Be** 2											5 10.8 **B** 3	6 12.0 **C** -4 -2 0 4	7 14.0 **N** -3 0 3 5	8 16.0 **O** -2 0	9 19.0 **F** -1	10 20.2 **Ne** 0
11 23.0 **Na** 1	12 24.3 **Mg** 2											13 27.0 **Al** 3	14 28.1 **Si** 4	15 30.1 **P** -3 5	16 32.1 **S** -2 0 4 6	17 35.5 **Cl** -1	18 40.0 **Ar** 0
19 39.1 **K** 1	20 40.1 **Ca** 2	21 45.0 **Sc** 3	22 47.9 **Ti** 4	23 50.9 **V** 5	24 52.0 **Cr** 3 6	25 54.9 **Mn** 2 4	26 55.8 **Fe** 2 3	27 58.9 **Co** 2 3	28 58.7 **Ni** 2	29 63.5 **Cu** 1 2	30 65.4 **Zn** 2	31 69.7 **Ga** 3	32 72.6 **Ge** 4	33 74.9 **As** -3 3 5	34 79.0 **Se** -2 0 4 6	35 79.9 **Br** -1	36 83.8 **Kr** 0
37 85.5 **Rb** 1	38 87.6 **Sr** 2	39 88.9 **Y** 3	40 91.2 **Zr** 4	41 92.9 **Nb** 3 5	42 95.9 **Mo** 4 6	43 98 **Tc** 7	44 101 **Ru** 3 4	45 103 **Rh** 2 3 4	46 106 **Pd** 2 4	47 108 **Ag** 1	48 112 **Cd** 2	49 115 **In** 3	50 119 **Sn** 4,2	51 122 **Sb** 3,5	52 128 **Te** -2 0 4 6	53 127 **I** -1 7	54 131 **Xe** 0
55 133 **Cs** 1	56 137 **Ba** 2	57 139 **La** 3	72 178 **Hf** 4	73 181 **Ta** 5	74 184 **W** 4 6	75 186 **Re** 7	76 190 **Os** 3 4	77 192 **Ir** 2 4 6	78 195 **Pt** 2 4	79 197 **Au** 1 3	80 200 **Hg** 2	81 204 **Ti** 1 3	82 207 **Pb** 2	83 209 **Bi** 3 5	84 209 **Po** 2 4	85 210 **At** -1	86 222 **Rn** 0
87 223 **Fr** 1	88 226 **Ra** 2	89 227 **Ac** 3															

58 140 **Ce** 3	59 141 **Pr** 3	60 141 **Nd** 3	61 145 **Pm** 3	62 150 **Sm** 3	63 152 **Eu** 3	64 157 **Gd** 3	65 159 **Tb** 3	66 163 **Dy** 3	67 165 **Ho** 3	68 167 **Er** 3	69 169 **Tm** 3	70 173 **Yb** 3	71 175 **Lu** 3
90 232 **Th** 4	91 231 **Pa** 5	92 238 **U** 4 6	93 237 **Np** 4 5	94 244 **Pu** 4	95 243 **Am** 3	96 247 **Cm** 3	97 247 **Bk**	98 251 **Cf** 3	99 252 **Es**	100 257 **Fm**	101 258 **Md**	102 259 **No**	103 260 **Lr**

Each of the 92 natural elements so produced, from hydrogen to uranium, is defined by its atomic number, *Z*, for the number of positively charged protons, *p*, in the nucleus (Table 1.1). The isotopes of the elements are defined by the number of neutrons, *N*. Table 1.2 shows this for the isotopes of elements up to oxygen. The stable isotopes of these elements are highlighted in gray with their abundance given in percent. Most elements have more than one stable isotope, although some, like beryllium and fluorine, have only one. The fractional atomic masses of the elements (amu in Table 1.1) are accounted for by the abundance of their rare isotopes.

With addition or loss of neutrons, the nuclide becomes unstable and will decay with a given half-life. The radioisotopes of a given element typically become increasingly unstable (shorter half-life) with increasing departure from its stable isotopes (greater or fewer neutrons). Carbon, as an example, is stable with the addition of one neutron, ^{13}C, but with two becomes radiocarbon with a half-life of 5730 years (Table 1.2). A third

TABLE 1.2

Stable Isotopes of the Light Elements, in Gray, for the First Two Rows of the Periodic Table (H to Ne), Showing Their Percent Abundances of the Stable Isotopes

^{2}H 0.015	Stable Isotope Percent abundance
^{3}H 12.32 a β^-	Long-lived Radioisotope Half-life Decay mode
^{15}C 2.45 s β^-	Short-lived Radioisotope Half-life Decay mode

10											^{20}Ne 90.48	^{21}Ne 0.27	^{22}Ne 9.25
9									^{17}F 64.5s ec	^{18}F 110m ec	^{19}F 100	^{20}F 64.5s β^-	^{21}F 4.16s β^-
8								^{14}O 70.6s ec	^{15}O 122s ec	^{16}O 99.76	^{17}O 0.04	^{18}O 0.20	^{19}O 26.9s β^-	^{20}O 13.5s β^-
7							^{12}N 11ms ec	^{13}N 9.97m ec	^{14}N 99.63	^{15}N 0.37	^{16}N 7.13s β^-	^{17}N 4.17s β^-	^{18}N 0.62s β^-	^{19}N 0.3s β^-
6						^{10}C 19.3s ec	^{11}C 20.3m ec	^{12}C 98.9	^{13}C 1.1	^{14}C 5730a β^-	^{15}C 2.45s β^-	^{16}C 0.75s β^-	^{17}C 20ms β^-	12
5					^{8}B 770ms ec 2α	^{9}B 10^{-3} fs 2p α	^{10}B 19.9	^{11}B 80.1	^{12}B 12ms β^-	^{13}B 17ms β^-	^{14}B 14ms β^-	10	11	
4				^{6}Be 10^{-5}fs 2p	^{7}Be 53.3d ec	^{8}Be 0.07fs 2α	^{9}Be 100	^{10}Be 1.6Ma β^-	^{11}Be 13.8s β^-	^{12}Be 24ms β^-	9			
3				^{5}Li 10^{-7} fs p α	^{6}Li 7.5	^{7}Li 92.5	^{8}Li 0.84s β^- 2α	^{9}Li 0.18s β^- α	7	8				
2			^{3}He 0.000	^{4}He 99.99	^{5}He 10^{-6} fs n α	^{6}He 807ms β^-	^{7}He 10^{-6} fs n	6						
1		^{1}H 99.98	^{2}H 0.015	^{3}H 12.32 a β^-	3	4	5							
Z↑ N→			1	2										

Source: From Chart of the Nuclides, 16th ed., General Electric Co. and KAPL, Inc. 2002. This *n–z* chart of the isotopes extends up to the trans-uranic elements and is available through the International Atomic Energy Agency (IAEA; https://www-nds.iaea.org/relnsd/vcharthtml/VChartHTML.html).

neutron makes ^{15}C with a half-life of only 2.45 seconds. Removing one neutron, ^{11}C has a half-life of only 12.3 minutes.

AQUIFERS AND AQUITARDS

Aquifers are formally defined as a water-saturated geological medium capable of storing and transmitting groundwater. They are commonly composed of fine- to coarse-grained porous media or are fractured and dissolution-fissured (karst) bedrock. Aquitards, by contrast, are hydrogeological units that have variable water content but do not readily transmit water. Aquitards are typically low-permeability materials such as clay-rich sediments and shales or relatively unfractured rock masses. Unconfined or phreatic aquifers are the shallowest hydrostratigraphic unit and comprise a saturated zone where porosity is water-filled with a water table and an overlying unsaturated or phreatic zone between the water table and ground surface that has porosity filled with both water and air. Unconfined aquifers are generally recharged by direct infiltration of meteoric waters through the overlying soil and unsaturated zone or by infiltration from adjacent surface water bodies. Atmospheric gases and soil CO_2 are dissolved in groundwaters in this unsaturated zone. Confined or artesian aquifers are overlain by lower-permeability aquitards. Recharge occurs either by flow from a region where the aquifer "outcrops" and has unconfined conditions or by leakage of groundwater from the adjacent confining aquitard.

The geochemical evolution of groundwater substantially develops during recharge, where chemical interaction with the soil and geologic medium begins. The nature and extent of geochemical evolution are very much determined by both soil conditions and the aquifer and aquitard materials. The crust is dominated by two principal rock types: (i) the crystalline terrains of igneous and metamorphic silicate rocks and (ii) the sedimentary formations, dominantly carbonates, sandstones and shales with far less frequent evaporitic units. Unconsolidated surface sediments add an important veneer that often hosts the best aquifers.

Because they are more resistant to physical abrasion during transport, unconsolidated sediments are primarily siliciclastic and are largely the weathering products of crystalline terrains. Unconsolidated sands and gravels, deposited by rivers, glacial outwash, shorelines, and wind make some of the best aquifers due to their highly permeable intergranular porosity (Figure 1.2). Increased silt and clay content reduces permeability and forms aquitards.

Siliciclastics are the product of crystalline rock weathering and so are less reactive to further weathering during groundwater recharge. The minerals that dominate siliciclastic terrains are largely quartz and feldspars, together with varying amounts of clay minerals. Calcite [$CaCO_3$] can be present as a secondary mineral coating. Other secondary mineralization often includes ferrihydrite [$Fe(OH)_3$] from oxidation of ferrous iron–bearing minerals. Because of the sedimentary nature of unconsolidated aquifers, they can contain buried sedimentary organic matter, which has important considerations for redox reactions in groundwater. This is particularly so for clay-rich aquitards, which typically form in lacustrine or marine settings, with significant input of sedimentary organic debris.

Accumulations of coastal sands along continental margins are the precursor to sandstone formations generated during marine transgressions and diagenesis, which represent major aquifers in many parts of the world (Figure 1.3). Highly abraded and

FIGURE 1.2 Siliciclastic sand and gravel aquifer formed by glacial outwash during deglaciation in the Ottawa region, Canada. The pit face is 12 m high.

sorted quartz sand provides uniform grain size with high connected porosity, making sandstones excellent aquifers.

Fractures in crystalline and carbonate rocks represent important aquifers, particularly for rural wells, although porosity in fracture aquifers represents only 1–2% by volume. Fractured shales and sandstones typically have a secondary intergranular or dual porosity. Intact granites and metamorphic rocks also have minor porosity as microfractures and intergranular discontinuities. Movement of water and solutes through this porosity is dominated by diffusion but can be a source of salinity to groundwater moving through adjacent open fractures. The dominant mineralogy of fractured crystalline aquifers is feldspar. Granites are composed of potassium and sodium feldspars (alkali feldspars), which are sources of low but significant concentrations of Na, Ca, and K in groundwaters through weathering. Rocks of intermediate to mafic composition are dominated by calcium feldspars and by increasing amounts of Mg-Fe minerals (Figure 1.4).

The fracture and fissure networks in carbonate formations form karst aquifers and are among the most exploited freshwater resources in the world (Figure 1.5). The principal mineral, calcite [$CaCO_3$], is soluble under the acidic conditions generated in soils with carbonic acid. This leads to a continual opening of porosity through fissure dissolution. Although dolomite [$CaMg(CO_3)_2$] is less soluble, it too provides good aquifer conditions. Karst groundwaters are dominated by Ca^{2+}, Mg^{2+}, and HCO_3^-. As carbonate aquifers are most often found within sedimentary basins,

FIGURE 1.3 Nepean sandstone near Ottawa is a significant aquifer with intergranular and fracture porosity supplying high-quality groundwater for much of the rural region. The upper inset shows a sandstone outcrop surface with glacial striations; the lower inset is a 2-mm-wide thin section showing sand grains with overgrowths of quartz cement and connected porosity in blue.

FIGURE 1.4 Granitic gneiss is one of the most common rock units of the crust, composed largely of feldspars (pink and white), quartz and ferromagnesian biotite, and amphiboles (inset). Fractures provide permeability for significant yields of high-quality groundwater, evident here from the iron staining from discharge points.

FIGURE 1.5 Karst development in limestone formations provides highly permeable aquifers with significant storage. The inset shows dissolution fissures. Geochemical weathering of carbonate rocks contributes significant Ca^{2+} and Mg^{2+} hardness to karst groundwaters.

additional solutes, including Na^+, Cl^-, and SO_4^{2-}, may be derived from saline pore waters and minor gypsum in the carbonates and/or shaley units.

ROCK-FORMING MINERALS

As the earth coalesced during its early bombardment phase some 5 billion years ago, accumulating mass began a process of geochemical differentiation. Liquid iron and nickel, the most abundant terrestrial components, sank to form a metallic core, leaving the lighter oxides of Si, Mg, Al, and other metals to form a cooling mantle. Further differentiation of the elements occurred during crust formation through volcanism and tectonism, enriching the growing continents in lighter aluminosilicate minerals, making oxygen, silica, and aluminum the three most abundant crustal elements. Iron is the fourth most abundant, followed by the major alkali and alkaline-earth elements (Na, K, Ca, Mg) (Figure 1.6).

The distribution of elements was addressed by the pioneer of modern geochemistry, V.M. Goldschmidt, who classified the elements by their affinities to different geological environments and processes as an extension of Mendeleev's periodic table. Based on Goldschmidt's classification, the elements can be divided according to their affinity to:

i. native iron of the core (siderophile; e.g., Fe, Co, Ni, Pt, Au, C, and P),
ii. the silicate phases of the earth's crust (lithophile; e.g., O, Ca, Mg, Na, K, Cl, Mn, Cr, and U),
iii. the sulfides of base metal deposits (chalcophile; e.g., S, Cu, Zn, Pb, and Hg),
iv. and a final group that includes elements with a tendency to volatilize (atmophile; e.g., H_2, N_2, and the noble gases).

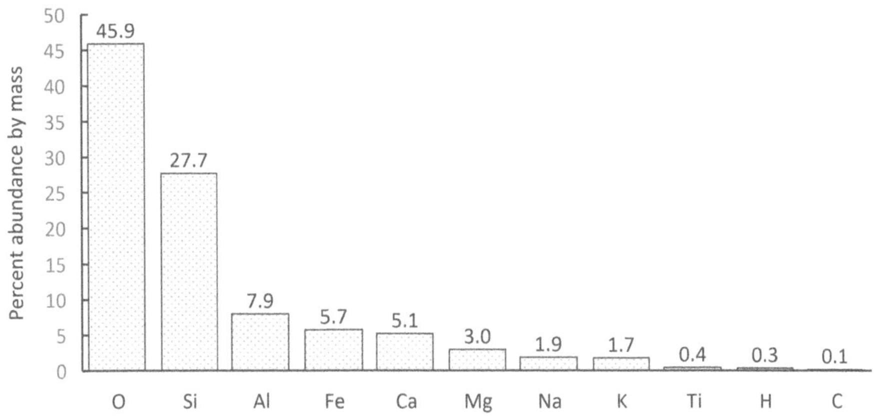

FIGURE 1.6 Chart of the most abundant elements in the earth's crust. (Data from Faure 1998.)

The earth is differentiated according to Goldschmidt's classification with a siderophilic core, a lithophilic crust, and the volatile (atmophile) phases of the hydrosphere and atmosphere. Chalcophile phases are largely present as sulfide deposits within the lithosphere. The major and minor rock-forming minerals are described below.

Silicates

The two most abundant elements, oxygen (–II) and silicon (+IV), coordinate to form the most abundant minerals in the crust—the silicates. Substitution of the third most abundant element, aluminum (+III), for silica produces the aluminosilicates, which are dominated by feldspars. Feldspars constitute over half of the minerals in the crust. Silicates are hard, durable minerals with low solubilities and weather slowly to clay minerals, contributing low concentrations of silica and cations to solution. Bowen's reaction series for a cooling melt summarizes the major aluminosilicate minerals found in igneous and metamorphic crystalline terrains.

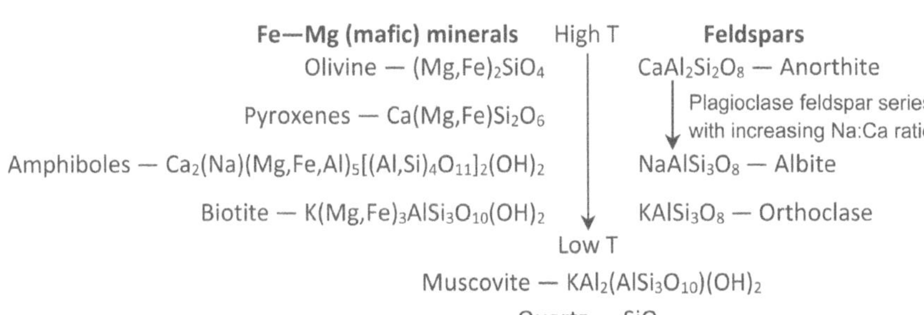

Granites and granitic gneiss (Figure 1.4) are formed by the lower-temperature alkali feldspars and amphiboles down to quartz, whereas more mafic basalts comprise higher-temperature anorthites and mafic minerals. Mechanical weathering of crystalline terrains favors quartz, which is the dominant component of siliciclastic sediments.

Clay Minerals

Clay minerals, the phyllosilicates, are the weathering product of silicates and form the class of minerals known for their sheet structure and very fine grain size. This is not to be confused with the geotechnical definition, where mineral particles that are less than 2 μm in diameter are clay-sized but may be quartz or other minerals. An important feature of clay minerals is their sheet structure, which has negatively charged boundaries and faces onto which cations can be fixed and exchanged. They are of considerable importance given this affinity for cation exchange, which can affect the geochemistry of groundwater.

Clay minerals carry hydroxyl groups (OH^-) as a product of low temperature weathering. Serpentine [$Mg_3Si_2O_5(OH)_4$] and chlorite [$(Mg,Fe,Al)_6(Si,Al)_4O_{10}(OH)_8$] are commonly produced by hydrothermal alteration of mafic minerals (olivine and pyroxene), whereas kaolinite [$Al_2Si_2O_5(OH)_4$] and illite [$KAl_2(AlSi_3)O_{10}(OH)_2$] form by meteoric weathering of feldspar-rich rocks. Smectites [$Na(Al,Mg)_2Si_4O_{10}(OH)_2$] are expanding clays, bringing water into their inter-sheet positions.

Carbonates

Carbonate rocks are the greatest reservoir of carbon on earth, comprising over 100,000 times the living biomass equivalent of C. They are dominated by calcite [$CaCO_3$] and dolomite [$CaMg(CO_3)_2$], both considerably more soluble than the aluminosilicate minerals. Marine limestone and dolostones can host significant aquifers due to their higher solubility and so have an impact on the geochemistry of drainage from such regions. Carbonate minerals are also common fracture-filling minerals in silicate rocks and are more readily dissolved and so create permeability in crystalline rocks. Because calcite readily forms at low temperatures, it is important to the geochemistry of natural waters.

Sulfides

Sulfide (S^{2-}) is the reduced form of sulfur in the environment, usually present in groundwaters as hydrogen sulfide gas (H_2S) or ionized as HS^-. Reduced sulfur is readily complexed by most base metals (Fe, Cu, Zn, Pb, Ni) as sulfide minerals. The most common is pyrite [FeS_2]. It is an uneconomical accessory mineral in many mining camps that is concentrated in mine tailings and waste rock. Pyrite is ubiquitous in organic-rich sediments such as shales and some limestones. Glacial till often contains a small percentage of pyrite exposed by erosion of bedrock. Mackinawite [FeS] is an amorphous iron sulfide mineral that forms in low-temperature aqueous settings. With their capacity for oxidation, iron sulfides are important controls on redox, sulfur, and iron geochemistry in natural waters.

Evaporites

Evaporite minerals are the highly soluble salts generally formed by the evaporation of seawater or groundwater in restricted basins. These are most commonly found in certain sedimentary rocks with limestone, dolomite, or shale, formed in Paleozoic seas, although modern evaporites occur, such as the Bonneville Salt Flats in Utah and the salt cliffs and caves of the northern Atacama Desert in Chile. The common evaporite minerals most relevant to groundwater geochemistry include halite [$NaCl$] and gypsum [$CaSO_4 \cdot 2H_2O$]. Rarer is sodium nitrate [$NaNO_3$], which collects in hyperarid settings.

Oxides and Hydroxides

Oxides and hydroxides of iron and manganese form during weathering of pyrite and mafic minerals. High-temperature oxides include magnetite [Fe_3O_4], hematite [Fe_2O_3], and pyrolucite [MnO_2]. At lower temperatures, these are hydrated as hydroxides, forming secondary minerals such as ferrihydrite [$Fe(OH)_3$] and pyrochroite [$Mn(OH)_2$], often as cement in sandstones or as lake and ocean bed concretions. They can also be found as rock-forming minerals, particularly in metamorphic rocks.

Phosphates

Phosphate, PO_4, is a minor species in aqueous systems, yet is often the rate-limiting nutrient for photosynthetic activity. The most common minerals are fluorapatite [$Ca_5(PO_4)_3F$] as an accessory mineral in many igneous rocks and rarer vivianite [$Fe_3(PO_4)_2 \cdot 2H_2O$] found in anoxic sediments.

2 Aqueous Geochemistry

DISSOLVED SOLIDS IN GROUNDWATER

The natural geochemistry of fresh groundwater resources most often begins with infiltration through soils, dissolution of organics, and CO_2, which allows the weathering of primary minerals and generation of dissolved solids in groundwater. It comprises a range of solutes that comprise the load of dissolved solids. The concentrations of dissolved species evolve through mass transfers among solid, gaseous, and aqueous phases in the system. The solid phase includes minerals in the subsurface as well as organic material in soils and aquifers from degradation of vegetation in soils or buried in clastic aquifers. The gaseous phase is largely restricted to the phreatic zone, which communicates with the open atmosphere. The aqueous phase includes the dissolved ions and uncharged species, dissolved gases, soluble organic complexes and molecules, as well as colloids. The transfer of mass between these phases, whether involving the dissolution or precipitation of minerals, the degradation of organics, and/or the gain and loss of gases, occurs as the system moves towards a state of equilibrium with a minimum of free energy, thus thermodynamics.

The major cations include the abundant alkali and alkaline-earth metals Na^+, K^+, Ca^{2+}, and Mg^{2+}, which are highly soluble, mineral-forming ions. The major anions include bicarbonate (HCO_3^-) from CO_2 uptake during recharge and from carbonate dissolution, chloride (Cl^-) from saline sources, and sulfate (SO_4^{2-}) from oxidation of sulfides, dissolution of evaporates, and other saline sources. These major ions make up most of the total dissolved solids (TDS) load of groundwater. Neutral species such as silicic acid (H_4SiO_4) are present in minor concentrations. Minor and trace elements make up only a small percentage of dissolved solids. The total mass of solids in water comprises a range of sizes (Figure 2.1), from dissolved ions to suspended solids (total suspended solids, TSS). The size limit for what is dissolved is operationally defined by the filter used to remove the suspended mass. For geochemical analysis, filtering to 0.45 μm (micron or 0.00045 mm) is considered best practice, although membranes with smaller pore throats such as 0.2 or 0.1 μm, can be used for research on metal speciation and colloids.

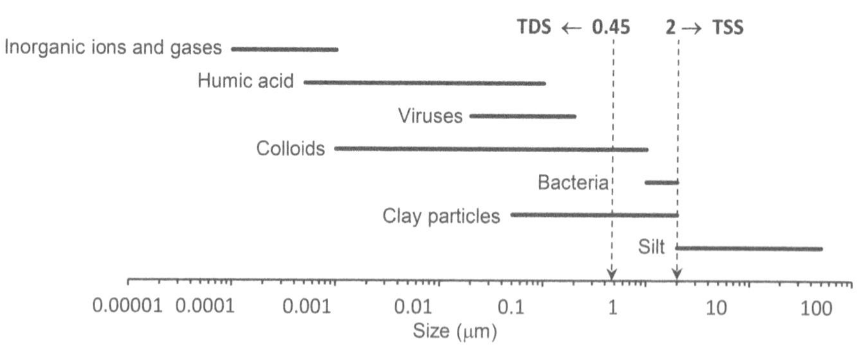

FIGURE 2.1 Size range of solutes and compounds dissolved in natural water.

DOI: 10.1201/9781003353836-2

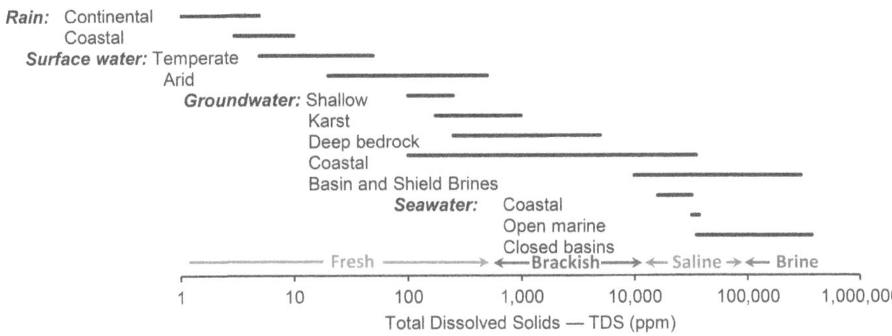

FIGURE 2.2 Typical range for total dissolved solids in different water types.

TDS is measured gravimetrically by drying a given mass of the filtered water sample and so is expressed as mass of dried solutes per mass of solution, in units of milligrams per kilogram or parts per million (ppm). It can be approximated by summing the mass (ppm) of dissolved ions and uncharged species that have been analyzed. A measurement of electrical conductivity (EC), which is proportional to the concentration of charged ions in solution, can also approximate the TDS.

TDS can vary enormously in natural waters, from concentrations below 10 ppm in rainwater to values exceeding 250,000 ppm (25%) in brines (Figure 2.2). Surface waters typically have low TDS (less than 500 ppm), including mainly dissolved inorganic solutes derived during runoff and shallow groundwater discharge, plus organics from biological activity in soils and surface waters. Groundwaters, in contrast, have higher TDS derived through weathering reactions during recharge. Coastal groundwater may be brackish to saline due to infiltration of seawater, while deep bedrock formations typically host high-density brines derived from the infiltration of ancient evaporated seawater from continental seas.

Solute concentrations are typically reported as mass of a dissolved solute in a mass (ppm) or as mass solute per volume (mg/L) of the solution, depending on the analytical protocols and preparation of standard solution in the laboratory. The sum of these concentrations can be used to calculate TDS as an alternative to a gravimetric analysis of TDS by drying.

$$\text{TDS}(\text{ppm}) = \Sigma \,\text{cations}(\text{ppm}) + \text{anions}(\text{ppm})$$

ppm = *parts per million*, milligrams of solute per kilogram of solution, mg/kg_s.
mg/L = *milligrams per liter,* milligrams of solute per liter of solution.

Solute concentrations measured as ppm (mg/kg_s) and mg/L differ as salinity increases due to the change in density (and therefore the volume of 1 kg_s). They can then be equated through the measured density, d, of the solution.

$$d = \frac{\text{mass}(\text{kg}_s)}{\text{volume}(\text{L})}$$

$$\text{mg/L} = \text{ppm} \times d$$

For infinitely dilute waters where 1 L of solution weighs 1 kg, density is equal to 1, and these units become equivalent. For low-salinity water, milligrams per liter and parts per million can therefore be used interchangeably. In highly saline waters, with density greater than 1, 1 kg of the solution will occupy a volume that is less than 1 L. Accordingly, ion concentrations measured as parts per million are lower than milligrams per liter for the same solution. This difference only becomes relevant for salinities greater than seawater ($d > 1.03$), where it begins to exceed the analytical error.

The use of geochemical data in calculations to investigate mineral solubility and geochemical processes involves solute concentrations in moles rather than mass. Recall that one mole of an element is equal to Avogadro's number (6.022×10^{23}) of atoms of that element and has a mass equal to its molar mass, M, in grams. One mole of carbon has a molar mass $M_C = 12g$. One mole of a molecule has a molar mass equal to the sum of the molar masses of its composite elements. Water [H_2O] then has molar mass $M_{H_2O} = 2 \times 1 + 16 = 18$. One mole of $CaCO_3$ has a molar mass $M_{CaCO_3} = 40 + 12 + 3 \times 16 = 100g$.

The number of moles of a solute is usually normalized to the mass of water (kg_w) in the solution (molality, $m = mol/kg_w$) for geochemical calculations. This requires conversion of analytical data measured as parts per million or milligrams per liter to molality. Conversion of parts per million to molality requires the molar mass of the solute. It also requires the TDS of the solution (Figure 2.2). This is expressed as kg/kg_s (TDS_{kg}). TDS_{kg} is simply TDS divided by 1,000,000 to convert milligrams to kilograms ($TDS_{kg} = TDS_{ppm}/10^6$).

$$\text{molality, } m = \frac{\text{ppm}}{(\text{M} \times 1000)\left(1 - \text{TDS}_{\text{kg}}\right)}$$

Converting concentrations measured as milligrams per liter to molality requires M, TDS_{kg} and density, d.

$$\text{molality, } m = \frac{\text{mg/L}}{(\text{M} \times 1000)\left(1 - \text{TDS}_{\text{kg}}\right) \times d}$$

For most groundwaters, TDS is sufficiently low such that density is essentially equal to 1, so solute concentrations expressed as either parts per million or milligrams per liter have no significant difference. Molality can then be approximated by dividing by ($M \times 1000$), which converts milligrams to grams and grams to moles.

$$\text{for low salinity waters, molality, } m \cong \frac{\text{ppm or mg/L}}{(\text{M} \times 1000)}$$

In most reactions, it is necessary to take into account the valence (z) of the ions in solution. For example, it takes twice the concentration of Na^+ ($z = 1$) to balance with Ca^{2+} ($z = 2$) in cation exchange reactions. In this case, units of equivalents, eq, are used. In mass-based concentrations (molal), this is expressed as milliequivalents per kilogram (or equivalents per million, epm):

$$\text{epm} = \frac{\text{ppm}}{\text{M}} \times z \qquad \text{Ca}^{2+} = \frac{74.6 \text{ ppm}}{40.1} \times 2 = 3.72 \text{ meq}$$

The accuracy of a geochemical analysis can be tested by comparing the total of the negative charges in solution with the total of positive charges. The laboratory analyses are converted to equivalents or milliequivalent units (meq/L = ppm × valence ÷ molar mass) for the major and minor species. The difference between the sum of all major cations (Σcat) and anions (Σan) is then compared with the sum of all major ions to give the charge balance error:

$$\text{Charge balance error}(\%) = \frac{\sum \text{cat}_{\text{eq}} - \sum \text{an}_{\text{eq}}}{\sum \text{cat}_{\text{eq}} + \sum \text{an}_{\text{eq}}} \times 100$$

In solutions with a very low or high pH, the concentrations of H^+ or OH^- should be used in the calculation. A charge balance error of less than 5% is acceptable. When errors are greater than 5%, a major contributing species may have been overlooked or not accurately analyzed. Achieving a good charge balance in low-salinity waters such as surface waters is difficult due to the inherently larger percent error as concentrations approach the detection limit.

GEOCHEMICAL REACTIONS

Geochemical reactions are driven by an excess of *Gibbs free energy* (ΔG kJ/mol) in the system. A geochemical system having a minimum of free energy, where $\Delta G = 0$, is considered to be in thermodynamic equilibrium with no net transfers of mass between phases. All reactions move towards a state of equilibrium where concentrations are unchanging until new conditions are encountered. Determining the state of equilibrium for geochemical reactions helps understand what reactions have occurred and how they will evolve as conditions change. The condition of equilibrium for geochemical reactions is defined by the activities, a, of the reactants and products, and their reaction constant (K).

SOLUTE ACTIVITY, *a*, AND THE REACTION CONSTANT, *K*

The concentration of a solute will have an analytical concentration (m). It will also have a thermodynamic concentration, or activity (a), which differs from m due to electrostatic interactions with other ions in solution. These interactions increase with the abundance of ions or ionic strength (I) of the solution. A reaction constant, K, defines the activities of products and reactants in a given reaction at equilibrium.

$$K_{\text{reaction}} = \frac{a_{\text{products}}}{a_{\text{reactants}}}$$

Consider the example of halite [NaCl] dissolving into pure water to produce an $Na^{+} - Cl^{-}$ solution.

$$NaCl \Leftrightarrow Na^{+} + Cl^{-}$$

The concentrations of Na^+ and Cl^- will increase as the reaction proceeds. As Na^+ and Cl^- concentrations increase, these solutes interact and reform halite in a reverse reaction. Provided there is excess halite, the concentrations of Na^+ and Cl^- will increase until the reverse reaction rate for halite precipitation equals that of halite dissolution. Here the system is at equilibrium, and there is no net flux of mass. At equilibrium, the activities of Na^+ and Cl^- are defined by the reaction constant, K_{NaCl}.

$$K_{NaCl} = \frac{a_{Na^+} \times a_{Cl^-}}{a_{NaCl}}$$

When the product/reactant activity ratio for a reaction is less than K, then the reaction has not reached equilibrium and will proceed. If the activity ratio is greater than K, the reverse reaction will dominate. Accordingly, K can be used to establish the state of equilibrium for a reaction. It is required for any geochemical assessments and is the basis of geochemical models such as the United States Geological Survey (USGS) program PHREEQC (Parkhurst and Appelo 1999). Its value can be determined for any given reaction from the standard Gibbs Free Energy ($\Delta G°$) of the reactants and products.

Ionic Strength and Ion Activity

The potential for reaction for a given ion increases with its concentration, m, but decreases with increased concentration of ions in solution due to electrostatic interactions. This effect is quantified through the *ionic strength, I*, of the solution, which is used to determine ion activities from their analytical concentrations. Ionic strength is calculated from the sum of the molar concentrations of all major ions in solution multiplied by the square of their valence,

$$I = \frac{1}{2}\sum m_i z_i^2$$

where z is the charge of the ith ion and m is the molality of the ith ion.

Ion activity, a (unitless), is related to ion concentration, m (mol/kg), through the activity coefficient, γ, which has units of inverse concentration, or kg/mol.

$$a = m \times \gamma$$

In low-salinity water with low ionic strength, activity coefficients, γ, are close to 1, and so activities are close to concentrations. However, with increasing ionic strength, activity coefficients decrease, and ion activities become a fraction of the ion concentration. The values for γ and a decrease not only with increasing salinity, but also they are lower for ions with higher charge (Figure 2.3).

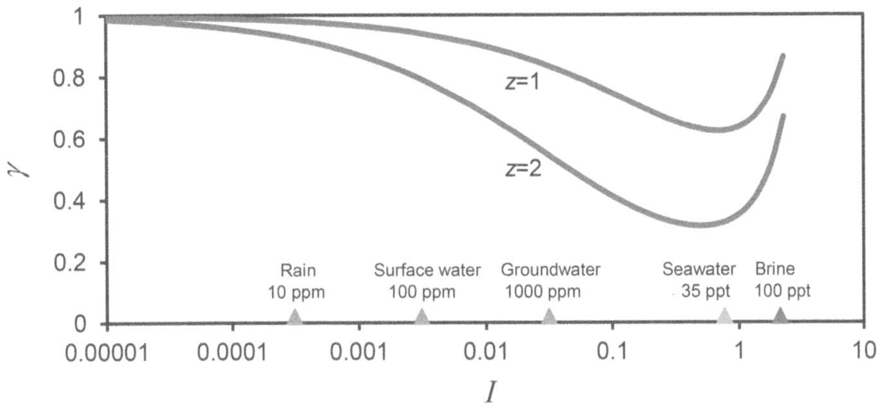

FIGURE 2.3 Activity coefficient γ for monovalent ($z = 1$) and divalent ($z = 2$) ions (å = 5), as a function of ionic strength. For uncharged species (e.g., $H_4SiO_4^o$ or $CaSO_4^o$) and dissolved gases, $z = 0$, and so activity coefficients are equal to 1. The values for the hydrated diameter of major ions, å, are NH_4^+ å = 2.5 | K^+, Cl^- å = 3.5 | Na^+ å = 4.0 | Ca^{2+}, SO_4^{2-}, Sr^{2+} Ba^{2+} å = 5.0 | Mg^{2+}, HCO_3^-, CO_3^{2-} å = 5.5.

Debye and Hückel (1923) developed the fundamental relationship between activity coefficients and ionic strength, showing γ to be related to z and I. The formulation of their relationship takes into account the hydrated radius of the ion (å in angstrom units or = 10^{-8} cm). A refinement by Davies (1962) adds a coefficient ($0.3I$) to resolve departures of the Debye–Hückel model from measurements at high ionic strength, giving the generalized equation,

$$\log \gamma = \frac{-0.51\, z^2 \sqrt{I}}{1 + 0.33\, \text{å} \sqrt{I}} + 0.3I$$

Note that the activity of pure minerals such as calcite [$CaCO_3$] is defined as 1 but is less than 1 as the concentration of substituting elements increases, such as calcite with 5% Mg^{2+} [$Ca_{0.95}Mg_{0.05}CO_3$], for which its activity would be 0.95. Similarly, the activity of pure water is 1 but decreases with increasing ionic strength.

Determining K from ΔG

All minerals, gases, geochemical species, and the various phases of water itself have a standard Gibbs free energy of formation ΔG°, measured in kiloJoules per mole. This is the change in free energy when forming one mole of the compound or ion from its component elements in their elemental states. By convention, all elements in their elemental state (e.g., O_2, S, or zero-valence metals, such as Fe°), as well as protons, H^+, and electrons, e^-, have $\Delta G^\circ = 0$. Compounds not in their elemental state will have positive or negative ΔG° values, depending on whether they gain (positive ΔG°) or release energy (negative ΔG°) in their formation from an elemental state. These values are determined for most compounds through detailed calorimetric

TABLE 2.1
Standard Gibbs Free Energy Data for a Few Common Aqueous Species and Minerals (kJ/mol)

Species	$\Delta G°$	Species	$\Delta G°$	Species	$\Delta G°$
$CO_{2\,(gas)}$	−394.4	Fe^{3+}	−8.56	Mg^{2+}	−455.4
H_2CO_3	−623.14	Fe^{2+}	−82.88	Mn^{2+}	−228.1
HCO_3^-	−586.8	$Fe(OH)^{2+}$	−233.2	MnO_2	−465.14
CO_3^{2-}	−527.9	$Fe(OH)_2^+$	−450.5	NH_4^+	−79.31
$CH_2O_{carbohydrate\ (OM)}$	−129.7	$Fe(OH)_3^0$	−648.3	NO_3^-	−108.74
CH_4	−34.45	$Fe(OH)_4^-$	−833.83	NO_2^-	−32.2
Ca^{2+}	−552.8	$Fe(OH)_{2\ amorphous}$	−486.5	$N_2O_{(g)}$	104.3
$CaCO_{3\ calcite}$	−1129.07	$Fe(OH)_{3\ ferrihydrite}$	−692.07	Na^+	−262.0
$CaF_{2\ fluorite}$	−1176.3	$FeS_{2\ pyrite}$	−155.9	$NaCl_{halite}$	−384.18
$CaMg(CO_3)_{2\ dolomite}$	−2161.7	$H_2O_{\ liquid}$	−237.14	S^{2-}	85.8
$CaSO_{4\ anhydrite}$	−1321.98	OH^-	−157.2	HS^-	12.08
$CaSO_4{\cdot}2H_2O_{\ gypsum}$	−1797.36	H^+	0	$H_2S_{(aq)}$	−27.83
Cl^-	−131.2	K^+	−282.5	$SO_{2(g)}$	−300.1
F^-	−281.5	$KCl_{sylvite}$	−408.6	SO_4^{2-}	−744.0

Sources: Drever 1997; Langmuir 1997; Faure 1997.
Some databases use kcal/mol, where 1kcal = 4.184kJ.
$\Delta G° = 0$ kJ/mol for species in their elemental (zero valence) state, and for H^+.

experiments. Values of $\Delta G°$ for some common geochemical species used in this book are given in Table 2.1.

From the standard Gibbs free energies of formation for reacting compounds, we can calculate the standard Gibbs free energy of any reaction, ΔG_r° which is used to determine the reaction constant, K. The Gibbs free energy of reaction, ΔG_r°, is equal to the sum of $\Delta G°$ for all products minus the sum of all reactants.

$$\Delta G_r^\circ = \sum \Delta G_{products}^\circ - \sum \Delta G_{reactants}^\circ$$

Using the dissolution of fluorite [CaF_2] to demonstrate and ΔG values from Table 2.1:

$$CaF_2 \Leftrightarrow Ca^{2+} + 2F^-$$

The Gibb's free energy of the reaction:

$$\begin{aligned}\Delta G_{fluorite}^\circ &= \Delta G_{Ca^{2+}}^\circ + 2\Delta G_{F^-}^\circ - \Delta G_{CaF_2}^\circ \\ &= (-552.8) + (2 \times (-281.5)) - (-1176.3) \\ &= 60.5 \text{ kJ/mol}\end{aligned}$$

The Gibbs free energy of the reaction is related to the activities of the reacting components through the gas constant R (8.3143 J/mol·K or 0.0083143 kJ/mol·K) and absolute temperature, T (298.16 K),

$$\Delta G^\circ_r = -RT \ln K$$

Substituting values for R and T and converting from natural logarithm to $\log_{10}$ provides the simple formulation to calculate the equilibrium reaction constant, K, from standard Gibbs free energy data for any reaction,

$$\log K = -\frac{\Delta G^\circ_r}{5.708}\left(\Delta G \text{ in kJ/mol}\right)$$

and for the dissolution of fluorite, K_{fluorite} can then be calculated from $\Delta G^\circ_{\text{fluorite}}$:

$$\Delta G^\circ_{\text{fluorite}} = 60.5 \text{ kJ/mol}$$

$$\Delta G^\circ_{\text{fluorite}} = -RT \ln K_{\text{fluorite}}$$

$$\log K_{\text{fluorite}} = -\frac{\Delta G^\circ_{\text{fluorite}}}{5.708} = -\frac{60.5}{5.708} = -10.6$$

$$K_{\text{fluorite}} = 10^{-10.6} = 2.52 \times 10^{-11}$$

The value of K can be calculated for all relevant geochemical reactions using ΔG° for all relevant reactants and products. This makes it possible to calculate mineral solubility, ion activities and ion concentrations, and speciation for equilibrium conditions.

EFFECT OF TEMPERATURE ON *K*

The calculation of K in these examples is specified for the standard temperature of 25°C. However, as it is thermodynamically derived from the Gibbs free energy of reaction, it will vary with temperature. Adjustments to K for temperatures different from 25°C are made using standard enthalpy measurements for the reaction ΔH°_r and the Van't Hoff equation. The approach is based on the standard enthalpy H and entropy S of reaction (ΔH°_r in J/mol and ΔS°_r in J/mol·K) and temperature T (K), which relate to the standard Gibbs free energy of reaction according to

$$\Delta G^\circ_r = \Delta H^\circ_r - T\Delta S^\circ_r$$

$$-RT \ln K = \Delta H^\circ_r - T\Delta S^\circ_r$$

The Van't Hoff equation differentiates with respect to T and is integrated over the range from 298 K (25°C) to the temperature of interest,

$$\log K_{T_2} = \log K_{298} + \frac{\Delta H_r}{2.303 R}\left(\frac{1}{298} - \frac{1}{T_{\text{kelvin}}}\right) \qquad \text{Van't Hoff equation}$$

TABLE 2.2
Standard Enthalpy of Reaction for Some Common Mineral Dissociation Reactions

Mineral	Dissociation Reaction	ΔH_r° (kJ/mol)
Calcite	$CaCO_3 \Leftrightarrow Ca^{2+} + CO_3^{2-}$	–9.610
Dolomite	$CaMg(CO_3)_2 \Leftrightarrow Ca^{2+} + Mg^{2+} + 2CO_3^{2-}$	–39.4425
Gypsum	$CaSO_4 \cdot 2H_2O \Leftrightarrow Ca^{2+} + SO_4^{2-} + 2H_2O$	–0.109
Anhydrite	$CaSO_4 \Leftrightarrow Ca^{2+} + SO_4^{2-}$	–1.71
Fluorite	$CaF_2 \Leftrightarrow Ca^{2+} + 2F^-$	19.6042

Source: Parkhust and Appelo 1999.

This integration is valid for only a limited temperature range of about ± 25°C (i.e., from 0°C to 50°C). Beyond this range, the change in enthalpy and entropy from standard values at 25°C to the new temperature must also be taken into account. Standard enthalpy of reaction data, ΔH_r°, for some common reactions are given in Table 2.2. The gas constant, R, in joules, is used ($R = 0.0083143$ kJ/mol/K).

K and Mineral Saturation Index

For mineral dissolution reactions, K defines the activities of the reaction products. Comparing K with the measured activities of reaction products provides an indication of whether the mineral is dissolving or precipitating or is at equilibrium. This is the mineral saturation index, SI, defined as the log of the ratio of the ion activity product (IAP) to the solubility constant, K.

$$\mathrm{SI} = \log\left(\frac{\mathrm{IAP}}{K}\right)$$

For example, a solution at saturation for gypsum [$CaSO_4 \cdot 2H_2O$] will have dissolved Ca^{2+} and SO_4^{2-} activities defined by the equilibrium constant for gypsum,

$$CaSO_4 \cdot 2H_2O \Leftrightarrow Ca^{2+} + SO_4^{2-} + 2H_2O$$

$$\begin{aligned} K_{\text{gypsum}} &= \frac{a_{\text{products}}}{a_{\text{reactants}}} = \frac{a_{Ca^{2+}} \times a_{SO_4^{2-}} \times a_{H_2O}^2}{a_{CaSO_4 \cdot 2H_2O}} \\ &= a_{Ca^{2+}} \times a_{SO_4^{2-}} \quad \left(\text{recall that } a_{\text{mineral}} = 1 \text{ and } a_{H_2O} = 1\right) \\ &= 10^{-4.6} \end{aligned}$$

From the measured concentrations, $m_{Ca^{2+}}$ and $m_{SO_4^{2-}}$, the activities of these ions can be calculated using the Debye–Hückel activity coefficients, $\gamma_{Ca^{2+}}$ and $\gamma_{SO_4^{2-}}$, where

$a = m \times \gamma$. The gypsum saturation index for a geochemical analysis is then determined by comparing the IAP with K_{gypsum},

$$IAP_{gypsum} = a_{Ca^{2+}} \times a_{SO_4^{2-}}$$

$$K_{gypsum} = 10^{-4.60}$$

$$SI = \log\left(\frac{a_{Ca^{2+}} \times a_{SO_4^{2-}}}{10^{-4.60}}\right) \text{or} \qquad SI = \log a_{Ca^{2+}} + \log a_{SO_4^{2-}} - (-4.60)$$

The SI can then provide a qualitative assessment of the degree of saturation of any mineral for a given solution as follows:

$SI > 0$ The mineral is supersaturated and can precipitate.
$SI = 0$ The mineral is in equilibrium with the solution.
$SI < 0$ The mineral is undersaturated and will dissolve if present.

The greater the deviation from 0, the greater the degree of disequilibrium, and mineral saturations can vary by a few orders of magnitude.

DISSOCIATION REACTIONS

A range of aqueous reactions involves the dissociation of ionic bonds and the formation and exchange of ions in solution. Although often classified as acid–base reactions, they are more than this, ranging from simple acid dissociation reactions to mineral dissolution, ion hydration, and formation of complex ions.

pH and Dissociation of Acids

The most fundamental of all aqueous geochemical reactions is the dissociation of water:

$$H_2O \Leftrightarrow H^+ + OH^-$$

$$K_{H_2O} = a_{H^+} \times a_{OH^-} = 10^{-14.0} \text{ at } 25°C$$

This reaction proceeds rapidly in both forward and reverse directions, such that in pure water at 25°C, the activity of H^+ and OH^- is 10^{-7} and because pH is the negative log of the hydrogen ion activity, the pH of pure water at 25°C is equal to 7.

$$pH = -\log(a_{H^+})$$

The dissociation constant for water changes with temperature and therefore changes the pH scale. At 0°C, $K_{H_2O} = 10^{-14.9}$ and the pH is 7.45; that is, less water is dissociated. For water at 100°C, $K_{H_2O} = 10^{-13.0}$ and the pH is 6.50, indicating higher activities for H^+ and OH^- due to greater dissociation of H_2O.

TABLE 2.3
Common Acids in Groundwater and Their First Dissociation Constant

Acid	Dissociation Constant, *K*	pH at 50% Dissociated
Bisulfate (HSO_4^-)	$K_{HSO_4^-} = \frac{a_{SO_4^{2-}} \times a_{H^+}}{a_{HSO_4^-}} = 10^{-1.99}$	1.99
Phosphoric acid (H_3PO_4)	$K_{H_3PO_4} = \frac{a_{H_2PO_4^-} \times a_{H^+}}{a_{H_3PO_4}} = 10^{-2.15}$	2.15
Humic acid (HA)	$K_{HA} = \frac{a_{HA^-} \times a_{H^+}}{a_{HA}} \approx 10^{-2} \text{ to } 10^{-4}$	~2–4
Carbonic acid (H_2CO_3)	$K_{H_2CO_3} = \frac{a_{HCO_3^-} \times a_{H^+}}{a_{H_2CO_3}} = 10^{-6.35}$	6.35
Hydrogen sulfide (H_2S)	$K_{H_2S} = \frac{a_{HS^-} \times a_{H^+}}{a_{H_2S}} = 10^{-6.99}$	6.99
Boric acid (H_3BO_3)	$K_{H_3BO_3} = \frac{a_{H_2BO_3^-} \times a_{H^+}}{a_{H_3BO_3}} = 10^{-9.24}$	9.24
Silicic acid (H_4SiO_4)	$K_{H_4SiO_4} = \frac{a_{H_3SiO_4^-} \times a_{H^+}}{a_{H_4SiO_4}} = 10^{-9.83}$	9.83
Bicarbonate (HCO_3^-)	$K_{HCO_3^-} = \frac{a_{CO_3^{2-}} \times a_{H^+}}{a_{HCO_3^-}} = 10^{-10.33}$	10.33

A variety of acids are present and dissociate in groundwater over a wide pH range, contributing to the buffering of acidity and alkalinity (Table 2.3). The major sources of acidity in groundwater are derived from carbonic acid (H_2CO_3), from the dissolution of CO_2 in soils produced by the oxidation of organics and from sulfuric acid (H_2SO_4) produced by the oxidation of sulfide minerals such as pyrite [FeS_2]. Organic acids such as humic and fulvic acids are produced by the degradation of vegetation and contribute to the dissolved organic carbon (DOC) content of soil and groundwater and dissociate at low pH, between 2 and 4. The dissolution of carbonic acid produces bicarbonate, HCO_3^-, which is the dominant anion in fresh groundwater. Silicic acid is an even weaker acid, produced by the weathering of many silicate minerals. Dissolved silica, $H_4SiO_4^\circ$, is stable at most pH values. Only at high pH does dissolved silicic acid dissociate to the anion $H_3SiO_4^-$ and become very soluble (Table 2.3).

Mineral Dissolution and Ion Hydration

The evaporite minerals, such as gypsum [$CaSO_4{\cdot}2H_2O$] and halite [NaCl], are examples of simple mineral dissolution, producing an anion and cation. Others, such as the dissolution of feldspars, are more complicated, producing secondary minerals through incongruent dissolution, discussed below. Mineral dissolution reactions are facilitated by the nature of the bipolar water molecule to coordinate around ions, creating a hydration sphere with a diameter several times larger than the ion's

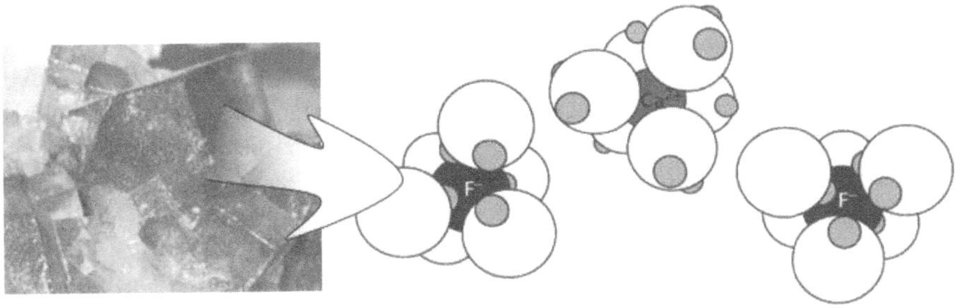

FIGURE 2.4 Dissolution of fluorite (CaF_2) and hydration of Ca^{2+} and F^-.

crystallographic or non-hydrated diameter. The dissolution of a salt, such as fluorite, CaF_2, is a good example (Figure 2.4).

$$CaF_2 \Leftrightarrow Ca^{2+} + 2F^-$$

In writing this reaction for the dissolution of fluorite, the hydration role played by water is not explicitly written, but the hydration reactions and their aqueous complex can be approximated by

$$Ca^{2+} + 6H_2O \rightarrow Ca(H_2O)_6^{2+}$$
$$F^- + 6H_2O \rightarrow F(H_2O)_6^-$$

The number of water molecules fixed in a hydration sheath depends on the valence of the ion and its ionic radius. The extent of ion hydration is proportional to the ionic potential or charge-to-radius ratio, z/r. Smaller ions, such as Li^+ or Be^{2+}, have a higher net surface charge than K^+ and Ca^{2+} and so have more coordinated water molecules. This leads to the relationship where smaller ions have a larger hydrated radius than larger ions (shown in Table 2.4) and so have higher solubilities.

TABLE 2.4
Cystallographic (Rc) and Hydrated (Rh) Radii (Angstrom Units, å = 10-10 m) for major Alkalis and Alkaline-Earth Elements

Element	*Rc*	*Rh*	Element	*Rc*	*Rh*
H^+	—	4.5	Be^{2+}	0.31	4.6
Li^+	0.94	3.8	Mg^{2+}	0.72	4.0
Na^+	1.17	3.6	Ca^{2+}	1.00	3.3
K^+	1.49	3.3	Sr^{2+}	1.26	2.5

Source: Shannon 1976; Parkhurst 1995.

Ion Hydrolysis

When the surface charge of an ion is extremely high, as in the case of trivalent Al^{3+} or Fe^{3+}, the strong positive charge will force the hydrated ion to shed one or more H^+ ions from its hydration sheath into the bulk solution. The result is a complex with hydroxide, OH^-, generating a new ion with a lower net charge. The consequence of hydrolysis of metals dissolved in water is a decrease in pH, as hydrogen ions are expelled into solution. At the same time, the pH of the solution will affect the stability of metal hydroxide complexes in solution. The following series of hydrolysis reactions for ferric iron, Fe^{3+}, is a good example, which, through the common variable of pH, produces the ferric iron speciation diagram of Figure 2.3.

$$Fe^{3+} + H_2O \Leftrightarrow Fe(OH)^{2+} + H^+ \qquad K = \frac{a_{Fe(OH)^{2+}} \times a_{H^+}}{a_{Fe^{3+}}} = 10^{-2.19}$$

$$Fe(OH)^{2+} + H_2O \Leftrightarrow Fe(OH)_2^+ + H^+ \qquad K = \frac{a_{Fe(OH)_2^+} \times a_{H^+}}{a_{Fe(OH)^{2+}}} = 10^{-3.48}$$

$$Fe(OH)_2^+ + H_2O \Leftrightarrow Fe(OH)_3^\circ + H^+ \qquad K = \frac{a_{Fe(OH)_3^\circ} \times a_{H^+}}{a_{Fe(OH)_2^+}} = 10^{-6.89}$$

$$Fe(OH)_3^\circ + H_2O \Leftrightarrow Fe(OH)_4^- + H^+ \qquad K = \frac{a_{Fe(OH)_4^-} \times a_{H^+}}{a_{Fe(OH)_3^\circ}} = 10^{-9.04}$$

Aluminum, the third most abundant element in the earth's crust, is another good example of ion hydrolysis and the effect on pH. Like ferric iron, Al^{3+} has a high surface charge and forms a series of aluminum hydroxide species that range from uncomplexed Al^{3+} to an aluminum anion that is coordinated with four hydroxide ions ($Al(OH)_4^-$) (Figure 2.5).

Formation of Ion Pairs

In aqueous solutions, ions can have electrostatic interactions. The product is an aqueous species with reduced or neutral charge. The dissociation constants of the

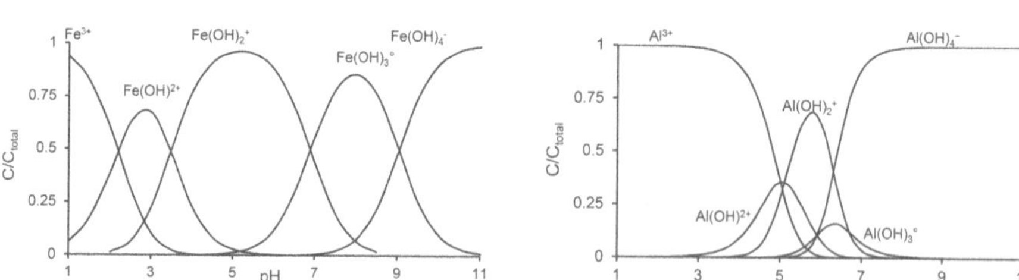

FIGURE 2.5 The relative distribution of iron hydroxide species (left) and aluminum hydroxide species (right) with pH, given as the concentration of each species over total dissolved Fe or Al, C/Ctotal.

TABLE 2.5
Dissociation Constants for Common Ion Pairs

Ion Pair	Log K_{diss}	Ion Pair	Log K_{diss}	Ion Pair	Log K_{diss}	Ion Pair	Log K_{diss}
$MgOH^+$	−2.21	$NaHCO_3^-$	+0.25	$NaCO_3^-$	−1.27	$NaSO_4^-$	−0.70
$CaOH^+$	−1.40	$MgHCO_3^+$	−1.07	$MgCO_3^\circ$	−2.02	$MgSO_4^\circ$	−2.25
		$CaHCO_3^+$	−1.11	$CaCO_3^\circ$	−3.23	$CaSO_4^\circ$	−2.31

common ion pairs that form in natural waters are given in Table 2.5. Consider, for example, gypsum-saturated groundwater. Gypsum dissolution is established by

$$K_{gypsum} = 10^{-4.60}$$

and with ion activities $\quad a_{Ca^{2+}} = a_{SO_4^{2-}} = 10^{-2.30}$

These ions also participate in an additional reaction with constant K:

$$CaSO_4^\circ \Leftrightarrow Ca^{2+} + SO_4^{2-}$$

$$K_{CaSO_4^\circ} = \frac{a_{Ca^{2+}} \times a_{SO_4^{2-}}}{a_{CaSO_4^\circ}} = 10^{-2.31}$$

This dissociation constant shows that the activities of Ca^{2+} and SO_4^{2-} in a gypsum-saturated solution will support a high activity for $CaSO_4^\circ$, which then contributes to greater dissolution of gypsum than would be calculated without consideration of such ion pairs. So too will other minerals have greater solubility due to the formation of ion pairs, such as those given in Table 2.5.

The formation of ion pairs increases the solubility of minerals. The total analytical concentration of a given major ion, say Ca^{2+}, will be the sum of all solutes and ion pairs that contain it:

$$m_{Ca-total} = m_{Ca^{2+}} + m_{CaSO_4^\circ} + m_{CaOH^+} + m_{CaHCO_3^+} + m_{CaCO_3^\circ} + m_{CaCl^+} + \cdots$$

Such ion pairs are not analyzed. Their activities are determined by iterative calculation using their reaction constant and a mass balance equation for each major ion.

Incongruent Dissolution

The final group of dissolution reactions to be introduced is that where minerals dissolve incongruently, producing both solutes and a secondary mineral phase. Weathering of silicate minerals generally occurs incongruently. For example, feldspar weathering involves the dissolution of a primary aluminosilicate and the formation of a secondary clay mineral.

$$CaAl_2Si_2O_{8\,anorthite} + H_2O + 2H^+ \rightarrow Ca^{2+} + Al_2Si_2O_5(OH)_{4\,kaolinite}$$

Clay minerals, such as kaolinite, can be further leached of their silica, producing more highly insoluble aluminum oxide and dissolved silica:

$$Al_2Si_2O_5(OH)_{4\,\text{kaolinite}} + 5H_2O \rightarrow 2Al(OH)_{3\,\text{gibbsite}} + 2H_4SiO_4^{\circ}$$

Another common example involves the dissolution of dolomite by Ca^{2+}-bearing waters, resulting in the precipitation of calcite:

$$Ca^{2+} + CaMg(CO_3)_{2\,\text{dolomite}} \rightarrow Mg^{2+} + 2CaCO_{3\,\text{calcite}}$$

REDOX REACTIONS

Although dissociation reactions involve largely electrostatic interactions between ionic components of minerals and solutions, a second class of reactions involves the exchange of electrons between key components. These are oxidation and reduction, or redox, reactions. Free electrons do not exist in solution and so are exchanged between paired redox couples, specifically an electron donor and an electron acceptor. Organic matter is the most common electron donor in natural (and contaminated) systems. Elemental oxygen is the most common electron acceptor, which supports aerobic biodegradation or respiration in soils and shallow groundwaters. It takes place as a pair of reactions involving an electron donor and an electron acceptor:

Respiration	$CH_2O + O_2 \rightarrow CO_2 + H_2O$	$\Delta G^{\circ}_r = -501.84\,\text{kJ/mol}$
Electron donor	$CH_2O + H_2O \rightarrow CO_2 + 4e^- + 4H^+$	$C^0 \rightarrow C^{IV} + 4e^-$
Electron acceptor	$O_2 + 4H^+ + 4e^- \rightarrow 2H_2O$	$O^0 + 2e^- \rightarrow O^{-II}$
Redox couples	CO_2/CH_2O and O_2/H_2O	

Redox regimes in groundwater are defined by their abundance of certain electron donors and acceptors. Shallow, phreatic systems with abundant O_2 and other electron acceptors from the unsaturated zone are *oxidizing*, whereas groundwaters with abundant organic carbon from buried peat or organic contaminants are *reducing* with little to no O_2 or other electron acceptors.

REDOX COUPLES AND HALF-CELL REACTIONS

In natural and contaminated waters, several redox couples exist that can exchange electrons in paired half-cell reactions, which include an electron donor and an electron acceptor. The overall redox reaction proceeds with an output of energy that can be exploited by bacteria and archaea for metabolism and growth. The redox reaction will proceed until either the electron-acceptor reservoir or the electron-donor substrate is exhausted. For this reason, redox reactions in aqueous systems are generally unidirectional rather than forward and backward, as required for equilibrium. Such systems are very much like a charged battery, which dissipates its energy by a flow of electrons from the cathode (electron donor) to the anode (electron acceptor). Many groundwater systems have abundant concentrations of electron acceptors but

TABLE 2.6
Common Redox-Sensitive Elements in Groundwaters with Their Species and Valence Redox State; below Are Examples of the Common Redox Couples

H	C	N	O	S	Fe
H_2—H^0	CO_2—C^{IV}	NO_3^- – N^V	O_2—O^0	SO_4^{2-} – S^{+VI}	$Fe(OH)_3$—Fe^{III}
H^+—H^I	HCO_3^-—C^{IV}	NO_2^- – N^{III}	H_2O—O^{-II}	S_{native}—S^0	Fe^{3+}—Fe^{III}
H_2O—H^I	CH_2O—C^0	N_2O—N^I	CH_2O—O^{-II}	$FeS_{2\ pyr}$—S^{-I}	Fe^{2+}—Fe^{II}
CH_2O—H^I	CH_4—C^{-IV}	N_2—N^0		H_2S—S^{-II}	$Fe(OH)_2$—Fe^{II}
		NH_4^+ – N^{-III}		FeS—S^{-II}	$FeS_{2\ pyr}$—Fe^{II}
$\frac{H_2O}{H_2}$	$\frac{CO_2}{CH_4}$	$\frac{NO_3^-}{N_2}$	$\frac{O_2}{H_2O}$	$\frac{SO_4^{2-}}{H_2S}$	$\frac{Fe(OH)_3}{Fe^{2+}}$

no viable electron donors, or vice versa, and so have no redox reactions taking place. The most important redox-sensitive elements are given in Table 2.6 with their common redox species.

Redox reactions are inherently out of thermodynamic equilibrium and require heat or cellular enzyme activation. Consider the example of respiration:

$$CH_2O + O_2 \rightarrow CO_2 + H_2O \qquad \Delta G_r^\circ = -501.84 \text{ kJ/mol}$$

$$\log K = \Delta G_r^\circ / 5.708 = 87.9$$

$$K_{\text{respiration}} = \frac{P_{CO_2} \times a_{H_2O}}{a_{CH_2O} \times P_{O_2}} = \frac{P_{CO_2}}{P_{O_2}} = 10^{87.9} \qquad \text{for thermodynamic equilibrium}$$

For earth surface conditions, with a partial pressure for CO_2 and $P_{CO_2} = 420$ ppm (abundance ratio of 0.00042 or $10^{-3.4}$), the partial pressure of O_2 in air would need to be $10^{-91.3}$ for thermodynamic equilibrium. As the atmosphere is 21% O_2 ($P_{O_2} = 0.21$ atm) it is far from thermodynamic equilibrium, which makes forests at extreme risk of wildfire.

Electron Activity and Electromotive Potential

The relative abundance of the reduced species (electron donors) of a redox couple imparts an electron activity in groundwater. A solution with a high concentration of reduced species, such as Fe^{2+} or H_2S, will have a high electron activity. Notation for electron activity adopts the pH convention, and so is expressed as the negative log of the electron activity, p*e*:

$$pe = -\log a_{e^-}$$

As pH is a parameter fundamental to dissociation reactions in groundwater, p*e*, is fundamental to redox reactions. It is determined from the activity ratio of a given redox couple, such as Fe^{2+}/Fe^{3+} or H_2S/SO_4^{2-} and the redox half-cell reaction constant, *K*. Using a generic reaction and the example of the iron redox couple, p*e* is related to the activity ratio as follows:

$$\text{Ox} + ne^- \Leftrightarrow \text{Red} \qquad\qquad Fe^{3+} + e^- \Leftrightarrow Fe^{2+}$$

$$\Delta G_r^\circ = \Delta G_{\text{Red}}^\circ - \Delta G_{\text{Ox}}^\circ \qquad\qquad \Delta G_r^\circ = (-82.88) - (-8.56) = -74.32$$

$$\log K = -\frac{\Delta G_r^\circ}{5.708} \qquad\qquad \log K = -\frac{-74.32}{5.708} = 13.02$$

$$K = \frac{a_{\text{Red}}}{a_{\text{Ox}} \times a_{e^-}^n} \qquad\qquad K = \frac{a_{Fe^{2+}}}{a_{Fe^{3+}} \times a_{e^-}} = 10^{13.02}$$

$$a_{e^-} = \left(\frac{a_{\text{Red}}}{a_{\text{Ox}} \times K}\right)^{1/n} \qquad\qquad a_{e^-} = \frac{a_{Fe^{2+}}}{a_{Fe^{3+}} \times K}$$

$$\log a_{e^-} = 1/n\left[-\log K + \log\left(\frac{a_{\text{Red}}}{a_{\text{Ox}}}\right)\right] \qquad\qquad \log a_{e^-} = -\log K + \log\left(\frac{a_{Fe^{2+}}}{a_{Fe^{3+}}}\right)$$

For single-electron transfers, $n = 1$:

$$pe = \log K - \log\left(\frac{a_{\text{Red}}}{a_{\text{Ox}}}\right) \qquad\qquad pe = 13.02 - \log\left(\frac{a_{Fe^{2+}}}{a_{Fe^{3+}}}\right)$$

A given redox couple in solution will exert an electromotive potential, *E*, which is related to p*e* through the activity ratio of the redox species in the half-cell and the standard electromotive potential of a redox half-cell, *E*°. The standard potential is calculated from Gibb's free energy for the half-cell reaction and Faraday's constant, $\Im$ (96.485 coulombs per mole):

$$E_{\text{half-cell}}^\circ = -\frac{\Delta G_{\text{half-cell}}^\circ}{n\Im} \qquad\qquad E_{Fe^{2+/3+}}^\circ = -\frac{-82.88 + 8.56}{96.485} = 0.77\text{ V}$$

The electromotive potential of a redox couple in solution is then related to its activity ratio in solution according to the Nernst equation:

$$E_{\text{half-cell}} = E_{\text{half-cell}}^\circ - \frac{RT\ln\left(\frac{a_{\text{Red}}}{a_{\text{Ox}}}\right)}{n\Im} \qquad\qquad E_{Fe^{3+/2+}} = 0.77 - \frac{RT\ln\left(\frac{a_{Fe^{2+}}}{a_{Fe^{3+}}}\right)}{n\Im}$$

$$= E_{\text{half-cell}}^\circ - 0.059\log\left(\frac{a_{\text{Red}}}{a_{\text{Ox}}}\right) \qquad\qquad = 0.77 - 0.059\log\left(\frac{a_{Fe^{2+}}}{a_{Fe^{3+}}}\right)$$

Recalling that $\Delta G_r^\circ = -5.708 \log K$ with substitution for $E°$ above ($E° = 0.059 \log K$), and expressing $E_{\text{half-cell}}$ against the standard hydrogen electrode (with $E° = 0$ volts), as Eh (or Eh):

$$\text{Eh} = 0.059 \log K - 0.059 \log\left(\frac{a_{\text{Red}}}{a_{\text{Ox}}}\right)$$

$$= 0.059\left(\log K - \log\left(\frac{a_{\text{Red}}}{a_{\text{Ox}}}\right)\right)$$

In this we recognize the pe relationship from above,

$$pe = \left(\log K - \log\frac{a_{\text{Red}}}{a_{\text{Ox}}}\right)$$

Substitution then gives the fundamental relationship between pe, needed for thermodynamic calculations, and Eh, which can be measured in the field:

$$\text{Eh} = 0.059\ pe$$

Redox Evolution in Groundwater

Each redox couple considered in groundwaters, and indeed in any chemical system, has an intrinsic electromotive potential or energy level required to donate or accept electrons. All batteries exploit this chemical energy by pairing an electron acceptor with a high electromotive potential with an electron donor having a lower electromotive potential to achieve energy output. So too in groundwater, where bacteria and archaea derive energy through electron exchanges from low-E electron donor species such as CH_4 to high-E electron acceptors such as O_2. Those commonly found in groundwater are shown in Table 2.7. Accordingly, the electron acceptor in a redox couple with higher Eh (higher on the scale in Table 2.7) can oxidize an electron donor from a couple with lower Eh.

The arrangement of the redox couples in Table 2.7 according to Eh is often referred to as the redox tower or chain, which defines the evolution of redox along a groundwater flow path. Each half-cell reaction involves both electrons and protons, which allows the illustration of the redox tower in a pe–pH diagram. Given that Eh = 0.059 pe, an ordinate axis for Eh is also used. As all reactions are aqueous, the bounds to such a diagram are the stability limits for water itself, as defined by the following two redox half-cell reactions. The upper stability field for water is defined by the redox half-cell reaction for elemental oxygen conversion to water.

$$\tfrac{1}{2}O_2 + 2e^- + 2H^+ \rightarrow H_2O \quad \Delta G^\circ_r = -237.14\,\text{kJ/mol}$$

TABLE 2.7
Major Redox Couples in Groundwater, Electrons Transferred (*n*), and Their Electromotive Potential at Neutral pH

Couple	*n*	*E*°
O_2/H_2O	2	0.81
NO_3^-/N_2	5	0.71
MnO_2/Mn^{2+}	2	0.58
$Fe(OH)_3/Fe^{2+}$	1	0.17
SO_4^{2-}/H_2S	8	−0.22
CO_2/CH_4	8	−0.24
H_2O/H_2	1	−0.41

The lower limit of the stability field for water is established by the redox couple by which water is reduced to hydrogen gas, H_2.

Upper Stability Field:

$$½O_2 + 2e^- + 2H^+ \rightarrow H_2O$$

$$\Delta G_r^\circ = -237.14 \text{ kJ/mol}$$

$$\log K = -\frac{\Delta G_r^\circ}{5.708} = -\frac{-237.14}{5.708} = 41.55$$

$$= \log a_{H_2O} - ½\log P_{O_2} - 2\log a_{e^-} - 2\log a_{H^+}$$

$$pe = 20.78 - \text{pH} + ½\log P_{O_2}$$

Lower Stability Field:

$$H_2O + e^- \rightarrow ½H_2 + OH^-$$

$$\Delta G_r^\circ = 79.9 \text{ kJ/mol}$$

$$\log K = -\frac{\Delta G_r^\circ}{5.708} = -\frac{79.9}{5.708} = 14$$

$$= ½\log P_{H_2} + \log a_{OH^-} - \log a_{e^-}$$

$$pe = -14 - ½\log P_{H_2} - \left(-14 - \log a_{H^+}\right)$$

$$= -½\mathit{log}P_{H_2} - \text{pH}$$

For near-surface conditions with P_{H_2} or P_{O_2} of 1 atm, where O_2 and H_2 gas bubbles would occur in shallow waters, these relationships defining the stability field for water become as follows:

$$pe = 20.78 - \text{pH} \quad \text{For } O_2/H_2O$$

$$pe = -\text{pH} \quad \text{For } H_2O/H_2$$

Using the electromotive scale, where Eh = 0.059 *pe*, gives us the following:

$$\text{Eh} = 1.23 - 0.059 \text{ pH} \quad \text{For } O_2/H_2O$$

$$\text{Eh} = -0.059 \text{ pH} \quad \text{For } H_2O/H_2$$

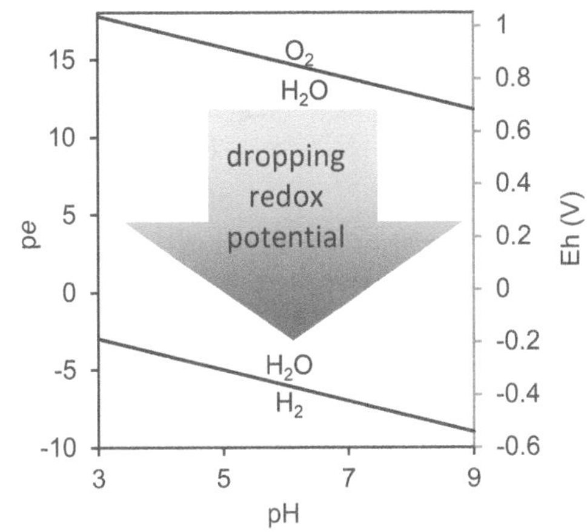

FIGURE 2.6 pe–pH diagram showing the stability field of water.

Plotted in Figure 2.6, these relationships define the H_2O stability field. At electron activities lower than the upper line (more positive p*e*), the oxygen in H_2O is oxidized to O_2. This would not occur in groundwater as it requires a redox couple with even higher electromotive potential, but can occur in nature from lightning strikes. At electron activities greater than those along the lower line (more negative p*e*), the H^+ in H_2O is reduced to H_2 gas.

The common redox reactions in Table 2.7 take place within this p*e*-pH stability field for water. As an electron acceptor with a higher electromotive potential is depleted, the p*e* of the system drops to that of next available redox couple. Accordingly, each successive redox reaction takes place at decreasing p*e* or Eh. These are not only p*e*-dependent reactions but also pH dependent. Using the approach shown for the water stability field, the p*e*–pH relationship for these redox reactions can be determined, defining gas pressures at atmospheric pressure and ion activities at environmental levels.

$$NO_3^- + 5e^- + 6H^+ \Leftrightarrow \tfrac{1}{2}N_2 + 3H_2O^\circ \quad \Delta G_r^\circ = -602.68 \text{ kJ/mol} \quad pe = 20.5 - 6/5\,\text{pH}$$

$$MnO_2 + 2e^- + 4H^+ \Leftrightarrow Mn^{2+} + 2H_2O \quad \Delta G_r^\circ = -237.45 \quad pe = 23.8 - 2\text{pH}$$

$$Fe(OH)_3 + e^- + 3H^+ \Leftrightarrow Fe^{2+} + 3H_2O \quad \Delta G_r^\circ = -102.23 \quad pe = 23.9 - 3\text{pH}$$

$$SO_4^{2-} + 8e^- + 10H^+ \rightarrow H_2S + 4H_2O \quad \Delta G_r^\circ = -232.26 \quad pe = 5.09 - 5/4\,\text{pH}$$

$$CO_2 + 8e^- + 8H^+ \rightarrow CH_4 + 2H_2O \quad \Delta G_r^\circ = -130.58 \quad pe = 2.86 - \text{pH}$$

Plotting each of these redox half-reactions on a p*e*–pH diagram shows the evolution of redox conditions in natural waters (Figure 2.7).

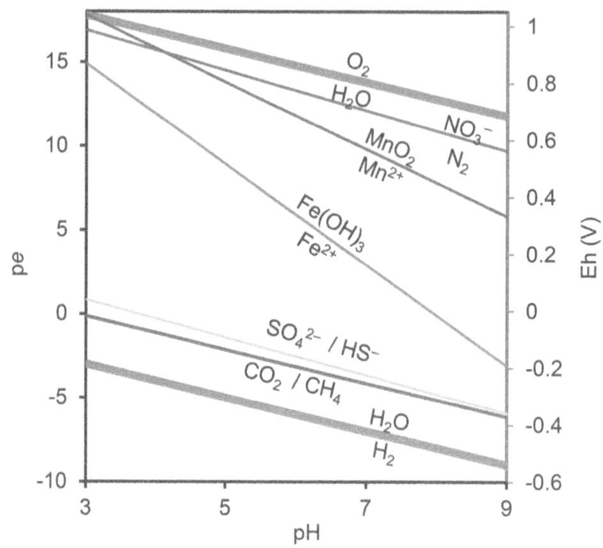

FIGURE 2.7 *pe*–pH diagram for principal redox buffering pairs in water.

Considering the electron acceptor in each couple, each will react in a paired redox reaction, providing an electron donor in a couple from deeper down the ladder is present. Often the electron donor is some form of carbohydrate [CH_2O] derived from biodegradation of vegetation. The following are common reactions in groundwater. As they proceed with a net energy yield, they are mediated by bacteria adapted for each particular reaction. The electron acceptor (ea) and electron donor (ed) in each are indicated.

$O_{2(ea)} + CH_2O_{(ed)} \rightarrow CO_2 + H_2O$	Respiration
$½O_{2(ea)} + 2Fe^{2+}_{(ed)} + 5H_2O \rightarrow 2Fe(OH)_3 + 4H^+$	Iron oxidation
$4NO^-_{3(ea)} + 5CH_2O_{(ed)} \rightarrow 5HCO^-_3 + 2N_2 + 2H_2O + H^+$	Denitrification
$4MnO_{2(ea)} + H_2S_{(ed)} + 6H^+ \rightarrow 4Mn^{2+} + SO^{2-}_4 + 4H_2O$	Manganese reduction
$4Fe(OH)_{3(ea)} + CH_2O_{(ed)} + 7H^+ \rightarrow HCO^-_3 + 4Fe^{2+} + 10H_2O$	Iron reduction
$SO^{2-}_{4(ea)} + CH_{4(ed)} + 2H^+ \rightarrow H_2S + CO_2 + 2H_2O$	Sulfate reduction
$CO_{2(ea)} + 4H_{2(ed)} \rightarrow CH_4 + 2H_2O$	CO_2 reduction

Balancing Redox Reactions

In writing and balancing redox reactions, it is useful to write them as half-reactions first, balance the electron transfer, then add them together to balance oxygen and hydrogen. Take, for example, the oxidation of organic carbon, CH_2O, using sulfate as an electron acceptor. In this case, methane is oxidized to CO_2, and sulfate is reduced to hydrogen sulfide.

	Oxidation	**Reduction**
	$CH_2O \rightarrow CO_2$	$SO_4^{2-} \rightarrow H_2S$
Half-cell reactions	$CH_2O \rightarrow CO_2 + 4e^-$	$SO_4^{2-} + 8e^- \rightarrow H_2S$
Balancing electrons	$2CH_2O \rightarrow 2CO_2 + 8e^-$	$SO_4^{2-} + 8e^- \rightarrow H_2S$
Add half-reactions	$2CH_2O + SO_4^{2-} \rightarrow 2CO_2 + H_2S$	
Balance O with H_2O	$2CH_2O + SO_4^{2-} \rightarrow 2CO_2 + H_2S + 2H_2O$	
Balance H with H^+	$2CH_2O + SO_4^{2-} + 2H^+ \rightarrow 2CO_2 + H_2S + 2H_2O$	
Resolving	$2CH_2O + SO_4^{2-} \rightarrow 2HCO_3^- + H_2S$	

A final check on the charge balance shows it to be electrically neutral and so the reaction is balanced.

GAS REACTIONS

Dissolved gases in groundwater are important components of geochemical reactions. All groundwaters have some component of dissolved gas, which has been incorporated by dissolution from soil air during recharge and may have some component of geogenic gas from subsurface redox reactions and radioactivity. Typically, the dominant gases are the three abundant atmospheric gases, N_2, O_2 and Ar, although O_2 is reactive and decreases after recharge through the oxidation of organics and other electron donors. Noble gases are derived from atmospheric dissolution, while some radiogenic noble gas isotopes like He and Ar can accumulate in the subsurface. CO_2 generated in the soil by biodegradation is incorporated into groundwaters during infiltration and provides acidity for weathering. Tectonic and metamorphic zones can release abundant CO_2 to groundwaters. CH_4 in groundwater can be generated by anaerobic degradation of buried organic carbon or migrate from deeper sources in coal and hydrocarbon formations. Nitrogen gases, N_2 and N_2O are produced by bacterial degradation of NO_3^- (denitrification and anammox). H_2S is produced under highly reducing conditions by bacterial reduction of sulfate, usually when significant sources of organic carbon are available.

Provided that the water depth is sufficient to maintain high confining pressure (hydrostatic pressure), a considerable amount of gas from geogenic sources can be retained in solution. Such high gas solubilities can then be exceeded when groundwater circulates to surface pressure, causing effervescence. Similarly, effervescence occurs in hot springs where geothermal heating has lowered the gas solubilities, causing gas exsolution upon discharge.

Unlike ions from mineral dissolution, gases are dissolved in water as neutral species and therefore have no charge. In the recharge environment at atmospheric pressure, the dissolution of gases is controlled by their atmospheric partial pressure and

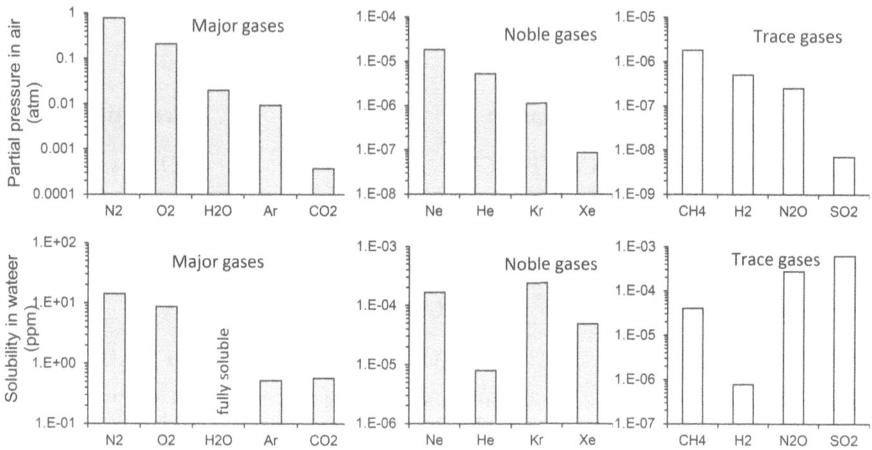

FIGURE 2.8 Partial pressure of the major, noble, and trace gases (atm) in air (upper) and their solubilities (ppm) in water (lower).

their relative solubility. The atmospheric abundances of the major and trace gases are shown here in Figure 2.8.

Partial Pressure of Gases

An ideal gas in a gas phase is present as a molar fraction of all gases present. Gases of low to trace concentration are usually expressed as parts per million by *volume* (ppmv). They also have a partial pressure. The partial pressure of a gas, P_{gas}, is its fractional contribution to the total gas pressure, P_{total}, and is calculated as its molar fraction (not mass fraction) multiplied by the total gas pressure:

$$P_{gas} = \frac{\text{moles}_{gas}}{\text{moles}_{total}} \times P_{total}\ (\text{atm})$$

Note that some publications express partial pressures of gases as pO_2 or pO_2, despite the fact that small "p" represents the negative log (i.e., pH, pK, and p*e*). Pressure, however, is represented by *P*, and so partial pressures are better expressed as P_{gas}. Partial pressures for atmospheric gases and their solubilities are shown in Figure 2.8 and Table 2.8.

Gas pressure is commonly expressed in units of atmospheres (atm), although bars and kilopascals can also be used (1 atm = 1.01325 bar = 101.325 kPa). For example, O_2 in air has a molar fraction of 0.209, or about 21%, and at the earth's surface has a partial pressure, $P_{O_2} = 0.21$ atm. At an altitude of 10 km, where air pressure is only 0.27 atm, however, O_2 has the same molar fraction (0.209) but a partial pressure of only 0.06 atm.

TABLE 2.8
Common Atmospheric Gases, Their Henry's Law Constants and Bunsen Coefficients at 25°C, and Their Aqueous Solubility When at Equilibrium with the Atmosphere

Gas	M	Partial Pressure *P* (atm)	Molar Fraction in Air	K_H (mol/L/atm)	Air-Equilibrated Water (mg/L)	Bunsen Coefficient (cc_{STP}/cc_{H_2O}/atm)[a]	Atmospheric Equilibrium (cc_{STP}/cc_{H_2O})[a]
N_2	28	0.78084	78.10%	0.00065	14.2	0.01457	1.14E-02
O_2	32	0.209476	20.90%	0.0013	8.71	0.02914	6.10E-03
H_2O[b]	18	~0.02	~2%	—	—	—	—
Ar	39.9	0.00934	0.93 %	0.0014	0.522	0.03138	2.92E-04
CO_2	44	0.00042	420 ppmv	0.034	0.63	0.78449	3.30E-04
Ne	20.2	1.82E-05	18.2 ppmv	0.00045	1.65E-04	0.01024	1.86E-07
He	4	5.24E-06	5.24 ppmv	0.00038	7.96E-06	0.00876	4.59E-08
CH_4	16	1.80E-06	1.8 ppmv	0.0014	4.03E-05	0.03138	5.65E-08
Kr	83.8	1.14E-06	1.14 ppmv	0.0025	2.39E-04	0.05620	6.41E-08
H_2	2	5.00E-07	0.5 ppmv	0.00078	7.80E-07	0.01748	8.74E-09
N_2O	44	2.50E-07	0.25 ppmv	0.025	2.75E-04	0.56035	1.40E-07
Xe	131.3	8.60E-08	0.086 ppmv	0.0043	4.86E-05	0.09810	8.44E-09
SO_2	64	7.00E-09	0.007 ppmv	1.4	6.28E-04	31.37980	2.20E-07
H_2S	34	trace	trace	0.0001	trace	0.00224	trace

Source: Sander 1999; Smith 1983.

[a] cc gas at STP per cc water. STP = standard temperature and pressure (273.15K or 0°C and 1 atm pressure).

[b] Water vapor is the third most abundant gas in the atmosphere but has high variability.

Gas Solubility and Henry's Law

When gas dissolves from an adjacent free-gas phase, its solubility is proportional to its partial pressure in the gas phase. In the case of gas production below the water table, gas can accumulate to the point where its concentration exceeds its solubility for the pressure (depth) conditions and then will form a free-gas phase. In both cases, the equilibrium saturation point is defined by Henry's Law, which states that the concentration of a gas dissolved in a liquid in equilibrium with an adjacent gas phase is proportional to its partial pressure in the gas phase.

The proportionality constant (K_H) for Henry's Law defines the solubility of a gas, expressed as molarity, *M*—moles of gas per liter solution (not to be confused with molar mass, M, used in Chapter 1):

$$K_{gas} = \frac{M_{gas}}{P_{gas}} \qquad \text{Henry's Law}$$

As the partial pressure for the gas, P_{gas}, is typically in atmospheres, the units of K_{gas} are mol/L/atm. Because gases are not ionized, their activity coefficient is equal to 1 for most low-salinity waters, and so aqueous concentrations, M_{gas}, are the same as activities. For example, the solubility of O_2 is defined by the Henry's Law constant (at 25°C):

$$K_{O_2} = \frac{M_{O_2(aq)}}{P_{O_2}} = 0.0013$$

The concentration of dissolved oxygen (DO or $O_{2(aq)}$) in air-equilibrated freshwater ($P_{O_2} = 0.21$ atm) at 25°C would be equal to 0.27 mmol/kg or 8.7 mg/L. Note that for aqueous concentrations, the standard mass units of mg/L or ppm are used rather than volumetric units of ppmv for gases in air. Henry's constants for gas solubility are given above in Table 2.8.

Partial Pressure of Dissolved Gas

The increase in pressure with depth below the water table is such that free-gas phases seldom exist in most aquifers. Nonetheless, dissolved gases still have a partial pressure. This is the partial pressure, in atmospheres, that the dissolved gas would maintain in an adjacent gas phase were it to be present. It can be calculated from Henry's Law, using K_H for the in situ temperature and the dissolved concentration to give P_{gas}.

$$P_{gas} = \frac{M_{gas}}{K_{gas}}$$

It is useful to express dissolved gas concentrations as partial pressures, as this allows direct comparison with their partial pressure of the original free-gas phase during recharge. If groundwater at some point of observation has a dissolved gas with a higher partial pressure than it does in the soil air, then it must have gained concentration by subsurface reactions, whereas if its concentration is lower, then some must have been lost by reaction along the flow path. For example, say groundwater at a depth of 10 m has a dissolved concentration of O_2 of 2.1 mg/L. Its partial pressure is then

$$M_{O_2} = 2.1/32000 = 0.0000656 \text{ mol/L}$$

$$K_{O_2} = 0.0013 \text{ mol/L/atm (from Table 2.8)}$$

$$P_{O_2} = \frac{M_{O_2(aq)}}{K_{O_2}} = \frac{0.0000656}{0.0013} = 0.05 \text{ atm}$$

Because this is much less than 0.209 atm in air (Table 2.8), the groundwater has lost DO by reaction, typically by oxidation of organics.

The partial pressure exerted by dissolved gas in groundwater can be demonstrated with a device that uses this principle to sample dissolved gases for analysis

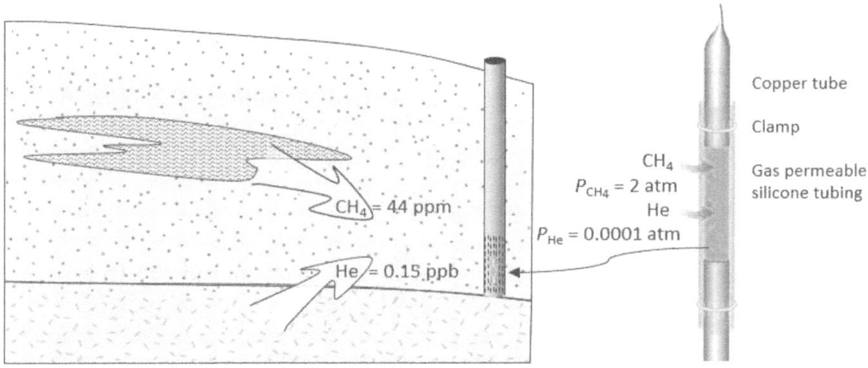

FIGURE 2.9 Dissolved gases in an aquifer sampled using a passive diffusion gas sampler with gas-permeable silicon tubing attached to air-filled copper tubes sealed at the ends to keep out water. Dissolved gases diffuse into the sampler to their respective groundwater partial pressure as established by Henry's constant for each gas. Volume changes in the silicone tubing due to over- or underpressuring when redeployed to the surface are mitigated with rigid internal and external shielding (not shown in the illustration).

(Gardner and Solomon 2009). A passive diffusion gas sampler typically involves a copper tube sealed with a gas-permeable silicon rubber membrane that can be set at some depth below the water level in a piezometer or well (Figure 2.9). Dissolved gases in the groundwater diffuse in and exchange with the original air in the sampler. When left for a few days to equilibrate, the sampler will contain the gases dissolved in the groundwater at concentrations equal to their respective partial pressures. Upon retrieval, the copper tube is crimp-sealed, and the sampled gases are later measured for their partial pressures and isotopes.

Gas Solubility in Volumetric Units: Bunsen Coefficient

Although concentrations of gases in water can be expressed in mass units of mg/L, they can also be expressed volumetrically as a volume of gas per volume of water. This is most common with the noble gases, which are measured volumetrically. At standard temperature and pressure (STP = 273.16 K or 0°C and 1 atm or 101.3 kPa), one mole of gas occupies 22.414 L. This derives from the gas constant, R, and temperature, T, which define the pressure–temperature–volume relationship for a mole of an ideal gas:

$$PV = nRT \quad \text{or} \quad n = \frac{PV}{RT}$$

where n is the number of moles of the gas, P is the pressure (atm), and V is the volume (L or cc, cm^3). The gas constant, R, is simply the pressure change with temperature for a unit volume of a mole of gas ($R = 0.082058$ atm L/mol/K), and T is the absolute temperature (kelvin, K). For one mole of gas at atmospheric pressure and 273.15 K (STP), its volume $V = RT = 22.414$ L, or 22,414 cc.

This leads to the expression of dissolved gas concentration in volumetric units, cc_{gas}, at STP per cubic centimeter of water or solution (cc or mL). The solubility constant is the Bunsen coefficient, *B*. From the gas law, the Bunsen coefficient is simply the Henry's Law constant multiplied by *RT*, derived according to the following:

Henry's Law constant $K_H = \text{mol/L} (\text{units} = n/\text{L})$

and so $n = K_H \times \text{L}$

Gas law $n = \frac{PV}{RT} = \frac{\text{L}_{gas}}{RT}$ at STP(1 atm and 273.16 K)

Equating Henry's Law with the gas law through *n* gives

$$\frac{\text{L}_{gas}}{RT} = K_H \times \text{L}_{H_2O}$$

and so

$$\frac{\text{L}_{gas}}{\text{L}_{water}} = \frac{cc_{gas}}{cc_{water}} = K_H\text{RT}$$

This expression is Bunsen's coefficient (Table 2.8), which expresses the solubility of a gas as its volume at STP.

$$\text{Bunsen coefficient:} \quad B = K_H RT \quad (cc_{STP}/\text{cc/atm})$$

Thus, the solubility of any atmospheric gas in cc_{STP}/cc_{H_2O} is equal to B_{gas} multiplied by the molar fraction of that gas in the atmosphere (e.g., 0.78 for N_2).

Gas Solubility with Temperature

Gases become less soluble at higher temperatures. The Henry's Law constant for gas solubility at nonstandard temperatures is calculated from the Van't Hoff equation, above, using enthalpy data, $\Delta H°$ from Table 2.9 and the gas constant in joules (0.0083143 kJ/mol/K).

$$\log K_{T_2} = \log K_{298} + \frac{\Delta H_r^o}{2.303R}\left(\frac{1}{298} - \frac{1}{T_2}\right)$$

The decreased solubilities of the major atmospheric gases with increasing temperature are shown in Figure 2.10 for a partial pressure, P_{gas}, of 1 atm. The right-hand chart shows their solubilities in air-equilibrated water in parts per million. The temperature equations on which these graphs are based are also given. The temperature

TABLE 2.9
Standard Enthalpy of Reaction for Some Gas Dissolution Reactions

Gas	log K_H 25°C	ΔH_r° (kJ/mol)
N_2	−3.23	−11.36
O_2	−2.89	−15.43
Ar	−2.85	−12.47
CO_2	−1.47	−19.95
CH_4	−2.85	−13.3

Source: Parkhurst and Appelo 1999.

control of gas solubility during recharge allows calculation of recharge temperature for old groundwaters based on concentrations of nonreactive, atmospherically derived noble gases. Applications of this tool are presented in the chapter on groundwater dating.

Geothermal waters are an example of the effect of temperature on gas solubility. As air-saturated recharge water becomes heated, gas solubility decreases, although the great hydrostatic pressure precludes the formation of bubbles at depth. As the geothermal water circulates back to the low hydrostatic pressure at the surface, the N_2 gained at recharge will be oversaturated and effervesce at the discharge vent.

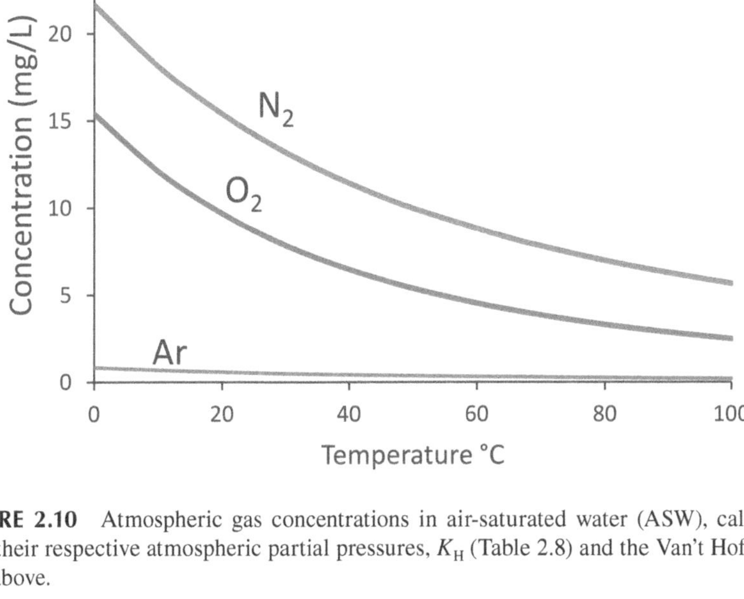

FIGURE 2.10 Atmospheric gas concentrations in air-saturated water (ASW), calculated from their respective atmospheric partial pressures, K_H (Table 2.8) and the Van't Hoff equation, above.

Gas Solubility with Pressure

Hydrostatic pressure (depth) plays an important role in gas solubility. Groundwater typically becomes saturated with air during recharge, but as groundwaters circulate more deeply, the greater pressure increases the equilibrium concentration. The water then becomes undersaturated with its dissolved gases, and concentrations are lower than would be for equilibrium solubility at greater hydrostatic pressure (Figure 2.11). In the case of groundwater recharge through an unsaturated zone in porous media and fractured rock, microbubbles of air can be entrained and carried below the water table. At higher hydrostatic pressure with depth below the water table, they then become dissolved, contributing an "excess air" component to the dissolved gas concentration. Excess air can increase atmospheric gas concentrations above equilibrium by a few to several tens of percent.

At depth during subsurface circulation, the excess hydrostatic pressure keeps the gas in solution. For this reason, geogenic gases like CH_4 and CO_2 can gain partial pressures well above atmospheric levels. With the pressure drop as the groundwater discharges at the surface, the dissolved gas rapidly becomes oversaturated for the temperature–pressure conditions and effervesces (Figure 2.11).

Gas Solubility with Salinity

Gases also have decreased solubility with increased salinity. The salting-out effect is well known, where the addition of salt to water can cause effervescence. This has to do with the restructuring of water by solutes and the reduced activity that lowers the gas solubility for a given partial pressure. The effect is shown in Figure 2.12 for argon. Groundwater can therefore become oversaturated with dissolved atmospheric gases through increases in salinity along the flow path.

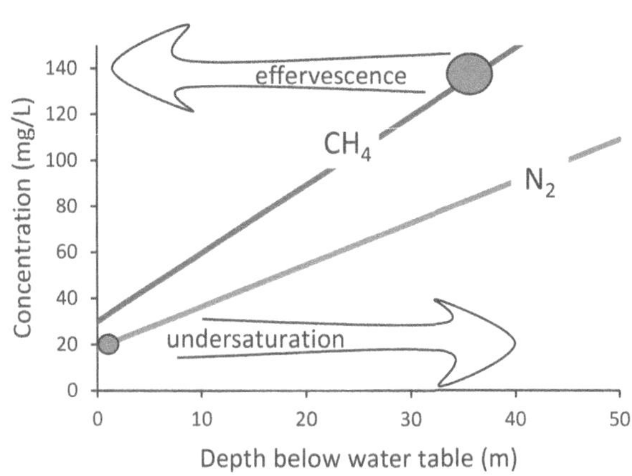

FIGURE 2.11 Gas solubility with pressure for N_2 and CH_4 at 25°C, showing a linear increase in equilibrium solubility with increased pressure (pressure as depth below the water table, where 10 m is 2 atm). A decrease in water pressure by pumping or discharge to a spring can cause effervescence, as shown for CH_4. By contrast, groundwater saturated with atmospheric N_2 in this example becomes undersaturated at greater depth.

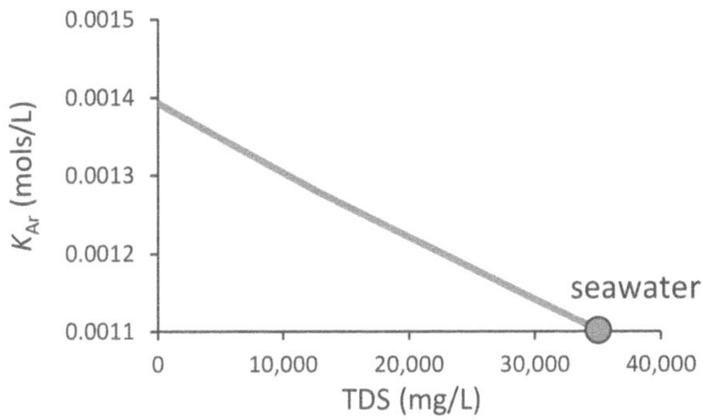

FIGURE 2.12 Effect of salinity on the solubility of Ar at 20°C. Modified from Kipfer et al. 2002.

Biogenic and Geogenic Gases in the Subsurface

Below the surface, a variety of biological and abiotic reactions contribute dissolved gases to groundwater. These range from the soil environment to deep crustal settings. Soil atmospheres differ from the open atmosphere due mainly to aerobic biodegradation and CO_2 production. Below the water table, where redox conditions become anoxic, anaerobic microbial activity in the subsurface mediates biogeochemical reactions that can produce *biogenic* gases:

$$CH_2O + O_2 \rightarrow CO_2 + H_2O \qquad \text{Respiration}$$

$$2NO_3^- + 2\tfrac{1}{2}CH_2O \rightarrow 2\tfrac{1}{2}HCO_3^- + \tfrac{1}{2}H^+ + H_2O + N_{2(aq)} \qquad \text{Denitrification}$$

$$CH_4 + SO_4^{2-} + 2H^+ \rightarrow CO_2 + H_2S + 2H_2O \qquad \text{Sulfate reduction}$$

$$CH_3COOH \rightarrow CO_2 + CH_{4(aq)} \qquad \text{Acetate fermentation}$$

$$CO_2 + 4H_2 \rightarrow CH_{4(aq)} + 2H_2O \qquad CO_2 \text{ reduction}$$

Geogenic gases are produced by various processes including radioactive decay. The principle reactions are the decay of the two major radioisotopes in the crust, ^{238}U and ^{40}K:

$$^{238}U \rightarrow {}^{234}U + {}^{4}He \rightarrow {}^{206}Pb + 8\,{}^{4}He_{(aq)} \qquad \text{alpha decay of uranium}$$

$$^{40}K \rightarrow {}^{40}Ar_{(aq)} \qquad \text{beta decay of potassium}$$

Geogenic gases are also produced in certain tectonic settings by high-temperature metamorphism of carbonates:

$$CaCO_3 + SiO_2 \rightarrow CaSiO_{3(wollastonite)} + CO_2 \qquad \text{decarbonation of calcite}$$

or through mantle and crustal degassing of CO_2, He, and other volatile species.

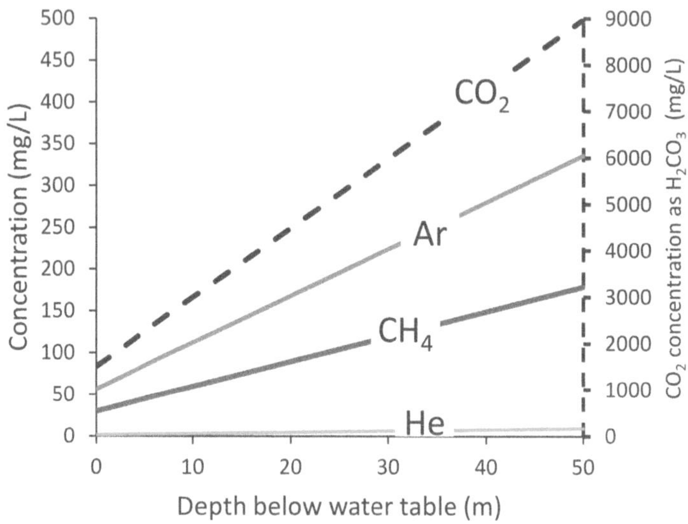

FIGURE 2.13 The solubility of the major geogenic gases that can occur in groundwaters. Equilibrium solubilities with increasing pressure (in depth below the water table) at 25°C calculated from K_H values.

The subsurface accumulation of dissolved biogenic and geogenic gases is greatly enhanced by the effect of pressure on solubility. Below the water table, hydrostatic pressure increases with depth. According to Henry's Law, this increases equilibrium solubility and so allows greater concentrations of gas to remain in a dissolved state. This has important implications for settings with high biogenic and geogenic gas production. Figure 2.13 shows the solubility of a selection of biogenic and geogenic gases with depth (pressure). The linear increase of equilibrium solubilities demonstrates the high gas concentrations that can exist in the subsurface in a dissolved state. Note the very low solubility for He, which has high diffusivity and preferentially partitions into the free-gas phase.

Subsurface production can then exceed the solubility limit for a given temperature and pressure and form a free-gas phase in the aquifer. Further, the production of groundwaters with gas partial pressures greater than 1 atm can be hazardous when pumped into a water distribution system, where the lower pressure can allow buildup of a free-gas phase, commonly methane, in the plumbing.

CARBONATE REACTIONS

Carbon cycles from atmospheric CO_2 through the biosphere and soils to groundwater as dissolved inorganic carbon (DIC), where it plays a central role in aqueous geochemistry and weathering. Inorganic or mineralized carbon carries a valence of +IV and coordinates with oxygen to form CO_2. Dissolved in water, CO_2 hydrates to form carbonic acid, H_2CO_3, and its two dissociation products, bicarbonate [HCO_3^-] and carbonate [CO_3^{2-}]. Collectively, these species comprise DIC:

$$m_{DIC} = m_{H_2CO_3} + m_{HCO_3^-} + m_{CO_3^{2-}}$$

DIC can originate from a variety of reactions, but most are related to soil respiration, carbonate dissolution and anaerobic fermentation of buried organics:

$$\text{Soil respiration:}\quad CH_2O + O_2 \rightarrow CO_2 + H_2O \rightarrow H_2CO_3$$

$$\text{Calcite dissolution:}\quad CaCO_3 + H^+ \rightarrow Ca^{2+} + HCO_3^-$$

$$\text{Fermentation:}\quad C_6H_{12}O_6 \rightarrow CH_3CH_2CH_2COOH + 2CO_2 + 2H_2$$

DIC, in most groundwater, is dominated by bicarbonate. Most low-salinity groundwaters are dominated by a $Ca^{2+} - HCO_3^-$ geochemical facies, driven by the dissolution of CO_2 during recharge and subsequent mineral weathering:

$$H_2CO_3 + CaCO_3 \rightarrow Ca^{2+} + 2HCO_3^-$$

$$2H_2CO_3 + H_2O + CaAl_2Si_2O_{8\ \text{anorthite}} \rightarrow Al_2Si_2O_5OH_{4\ \text{kaolinite}} + Ca^{2+} + 2HCO_3^-$$

The source of CO_2 and DIC in these reactions is of interest for reconstructing geochemical reactions and carbon cycling with groundwaters. Here, ^{13}C and radiocarbon ^{14}C are invaluable tracers of carbon sources and reactions. The application of these isotopes is presented later, in Chapter 5.

pH and the Distribution of DIC Species

For most natural waters, it is the production of CO_2 from the degradation of plant debris in soils that contributes most of the DIC. In unsaturated, aerobic soils, CO_2 accumulates in the gas phase, sustaining high CO_2 partial pressures. The sequence of reactions begins with the dissolution and hydration of CO_2 in water to produce carbonic acid:

$$CO_{2\,(g)} \Leftrightarrow CO_{2\,(aq)} + H_2O \Leftrightarrow H_2CO_3$$

Although unhydrated $CO_{2(aq)}$ is an important species in solution, the hydration step is usually included in the overall reaction for CO_2 dissolution. By convention, all dissolved CO_2, whether hydrated or not, is taken to be in the form H_2CO_3.

Carbonic acid is a weak acid and readily dissociates to form bicarbonate and a hydrogen ion. Bicarbonate then dissociates to form carbonate, and carbonate can precipitate calcite and other carbonate minerals. The equilibrium constants for these reactions are as follows:

$$CO_{2\,(g)} + H_2O \Leftrightarrow H_2CO_3 \quad K_{CO_2} = \frac{a_{H_2CO_3}}{P_{CO_2}} = 10^{-1.47}$$

$$H_2CO_3 \Leftrightarrow H^+ + HCO_3^- \quad K_1 = \frac{a_{HCO_3^-} \times a_{H^+}}{a_{H_2CO_3}} = 10^{-6.35}$$

$$HCO_3^- \Leftrightarrow H^+ + CO_3^{2-} \quad K_2 = \frac{a_{CO_3^{2-}} \times a_{H^+}}{a_{HCO_3^-}} = 10^{-10.33}$$

TABLE 2.10
Solubility and Dissociation Constants for the Carbonate System

T (°C)	$\log K_{CO_2}$	$\log K_{1(H_2CO_3)}$	$\log K_{2(HCO_3^-)}$	$\log K_{CaCO_3}$
0	−1.11	−6.58	−10.62	−8.38
5	−1.19	−6.52	−10.55	−8.39
10	−1.27	−6.47	−10.49	−8.41
15	−1.34	−6.42	−10.43	−8.43
20	−1.41	−6.38	−10.38	−8.45
25	−1.47	−6.35	−10.33	−8.48
50	−1.72	−6.28	−10.16	−8.65

Equations: (T in °C)
$\log K_{CO_2} = 0.00009\,T^2 - 0.0167\,T - 1.11$
$\log K_{H_2CO_3} = -0.00013\,T^2 + 0.0125\,T - 6.58$
$\log K_{HCO_3^-} = -0.00009\,T^2 + 0.0139\,T - 10.62$
$\log K_{CaCO_3} = 0.00006\,T^2 - 0.00252\,T - 8.38$

Precipitation of calcite:

$$CaCO_3 \Leftrightarrow Ca^{2+} + CO_3^{2-} \quad K_{CaCO_3} = a_{Ca^{2+}} \times a_{CO_3^{2-}} = 10^{-8.48}$$

Note that K_1 and K_2 are the first and second dissociation constants for carbonic acid (i.e. $K_{H_2CO_3}$ and $K_{HCO_3^-}$). Values for these equilibrium constants are given in Table 2.10 for different temperatures. The hydrogen ion activity, a_{H^+}, is the common species in all these equations, and so the relative concentrations of H_2CO_3, HCO_3^-, and CO_3^{2-} are a function of pH (Figure 2.14). From this figure, it is apparent that a decrease in pH will increase the activity of carbonic acid and therefore P_{CO2} which would provoke CO_2 degassing. Increasing the pH of the solution would increase CO_3^{2-} activity and provoke calcite precipitation.

Concentrations of DIC Species

Knowing the concentrations of these species is required to determine their activities for calculations of carbonate mineral saturation and CO_2 partial pressures for degassing. Total DIC is measured with a filtered sample by one of two methods, and then the activities of the individual species can then be calculated from pH. Their concentrations are measured either as total DIC by acidification to pH 2 to convert all species to CO_2 followed by purging and measurement:

$$\text{Total DIC (ppm)} = \left(m_{H_2CO_3} + m_{HCO_3^-} + m_{CO_3^{2-}}\right) \times 12{,}000$$

or by alkalinity titration. The alkalinity of a solution is a measure of its capacity to buffer acidity. Bicarbonate, HCO_3^-, is the dominant alkalinity species in most

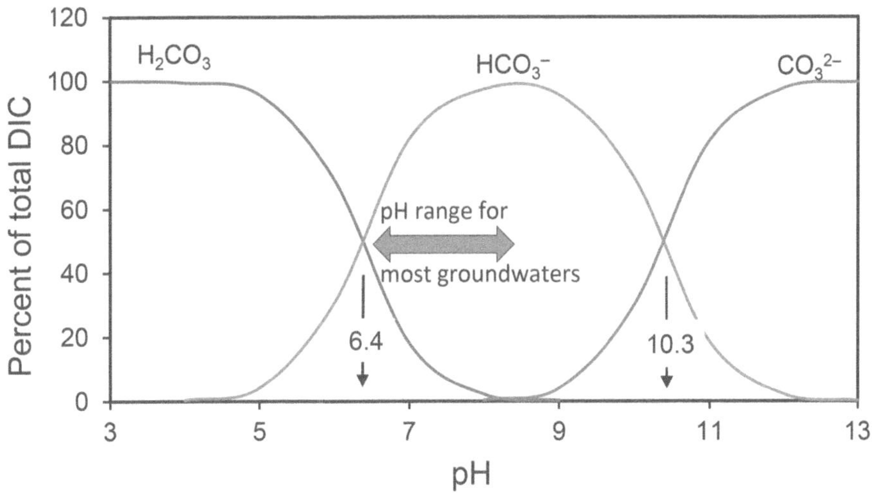

FIGURE 2.14 Relative distribution of dissolved inorganic carbon species in pure water as a function of pH at 25°C.

waters. It can buffer the addition of acid by consuming protons to convert to carbonic acid. Carbonate, CO_3^{2-}, is a minor contributor to solution alkalinity given its typically low concentration in most waters. The alkalinity of a solution is the sum of carbonate and bicarbonate and any other proton-accepting species present. It is expressed in units of equivalence, meq/L:

$$\text{Total alkalinity, alk}_T = m_{HCO_3^-} + 2m_{CO_3^{2-}} + m_{H_3SiO_4^-} + m_{H_2BO_3^-} + m_{HS^-} + m_{OH^-}$$

$$\text{Carbonate alkalinity, alk}_c = m_{HCO_3^-} + 2m_{CO_3^{2-}} \ (\text{meq/kg})$$

Carbonate alkalinity is calculated by correcting a total alkalinity measurement for the concentrations of non-carbonate species. In most cases, these concentrations are negligible, and so $\text{alk}_c \cong \text{alk}_T$. The minor alkaline species only become important at higher pH.

With a measurement of total DIC or carbonate alkalinity, together with pH, the concentrations of all DIC species can be calculated. The calculations can be simplified by considering that DIC is dominated by one major and one minor carbonate species at any given pH: H_2CO_3 plus HCO_3^- below pH 8.4 and HCO_3^- plus CO_3^{2-} above pH 8.4 (Figure 2.14).

The following equations allow calculation of the relative concentrations or fraction, f, of these species in either pH range. Note that although the third component (e.g., CO_3^{2-} at low pH or carbonic acid at high pH) is insignificant in such a mass balance calculation, it is important in thermodynamic calculations. CO_3^{2-} may contribute insignificantly to the total DIC at neutral pH, but it will have an activity that can be readily calculated from the concentration of DIC and that will play a vital role in the dissolution and precipitation of carbonate minerals. Thus, although DIC is almost entirely composed of HCO_3^- at neutral pH, it is important to calculate the activities of the minor DIC species, H_2CO_3 and CO_3^{2-}, for their role in CO_2 degassing and carbonate mineral precipitation reactions. The following derives generic formulae for calculating the concentrations of DIC species from a measured DIC value. Note that these are based on the activity relationships for the DIC species and therefore include activity coefficients to determine species concentrations.

Calculating concentrations of DIC species from measured DIC (mol/kg) (note that $\gamma_{H_2CO_3} = 1$):

Acidic—below pH 8.4

$$m_{DIC} = m_{H_2CO_3} + m_{HCO_3^-}$$

$$K_1 = \frac{a_{H^+} \times a_{HCO_3^-}}{a_{H_2CO_3}}$$

$$= a_{H^+} \times \frac{m_{HCO_3^-} \times \gamma_{HCO_3^-}}{\left(m_{DIC} - m_{HCO_3^-}\right)}$$

$$K_1 \times m_{DIC} = a_{H^+} \times m_{HCO_3^-} \times \gamma_{HCO_3^-} + K_1 \times m_{HCO_3^-}$$

$$m_{HCO_3^-} = \frac{K_1 \times m_{DIC}}{a_{H^+} \times \gamma_{HCO_3^-} + K_1}$$

Alkaline—above pH 8.4

$$m_{DIC} = m_{HCO_3^-} + m_{CO_3^{2-}}$$

$$K_2 = \frac{a_{H^+} \times a_{CO_3^{2-}}}{a_{HCO_3^-}}$$

$$= a_{H^+} \times \frac{m_{CO_3^{2-}} \times \gamma_{CO_3^{2-}}}{\left(m_{DIC} - m_{CO_3^{2-}}\right)\gamma_{HCO_3^-}}$$

$$K_2 \times m_{DIC} \times \gamma_{HCO_3^-} = a_{H^+} \times m_{CO_3^{2-}} \times \gamma_{CO_3^{2-}} + K_2 \times m_{CO_3^{2-}} \times \gamma_{HCO_3^-}$$

$$m_{CO_3^{2-}} = \frac{K_2 \times m_{DIC} \times \gamma_{HCO_3^-}}{a_{H^+} \times \gamma_{CO_3^{2-}} + K_2 \times \gamma_{HCO_3^-}}$$

Often, a geochemical analysis will include a measurement of bicarbonate, HCO_3^-, measured by an alkalinity titration rather than DIC as measured by a dissolved carbon analyzer. HCO_3^- is most often the major component of DIC and together with CO_3^{2-}, participates in carbonate reactions. Calculating the distribution of the remaining DIC species from $m_{HCO_3^-}$ and pH is simply a matter of substitution into the appropriate equations for K_1 and K_2.

Partial Pressure of CO_2 and Dissolved Inorganic Carbon

Any water that has a free surface (i.e., surface water or soil water in the unsaturated zone of an aquifer) will dissolve CO_2 from the adjacent gas phase according to the partial pressure of CO_2 (P_{CO_2}) present in that gas. In soils, where active aerobic decay of vegetation is producing CO_2, the P_{CO_2} of the soil air is usually between $10^{-2.5}$ and $10^{-1.5}$ atm (0.003 and 0.03 atm). In the open atmosphere, it is much lower, close to $10^{-3.4}$ atm (0.0004 atm). The concentration of carbonic acid in groundwater that is generated by dissolution of CO_2 in the soil air is governed by the Henry's Law constant, K_{CO_2}, which at 25°C, is $10^{-1.47}$ or 0.034 (Table 2.10). Accordingly, a P_{CO_2} of 0.03 atm supports 0.001 mol/kg or 63 ppm of H_2CO_3. This provides groundwater with the acidity required for mineral weathering.

Groundwaters below the water table (i.e., with no proximity to the free surface) may retain CO_2 acquired during recharge or may have accumulated CO_2 from subsurface reactions, such as oxidation of buried organics in the aquifer. From measurements of pH and alkalinity or DIC, we can calculate a value for P_{CO_2}. Although there is no separate gas phase, it represents the partial pressure of CO_2 that would be present in an adjacent gas phase to maintain the concentration of carbonic acid observed in the water. The importance of calculating the P_{CO_2} for groundwater is in the interpretation of its geochemical evolution.

Such calculations of P_{CO_2} are routine for geochemical interpretations. In surface water, P_{CO_2} values may be significantly greater than atmospheric, most commonly inherited from groundwater discharging to rivers and lakes. In groundwater, high P_{CO_2} values are gained during recharge through organic soils, whereas lower values are more typical of closed system weathering below the water table. Such trends are discussed further in Chapter 5. Calculations of P_{CO_2} are given in the Supplementary Material.

Problems

Example problems and test problems with solutions are available in the supplementary Workbook for Groundwater Geochemistry and Isotopes available on request from idclark@uottawa.ca and at the website for this textbook found at the link: www.routledge.com/9781032405933.

3 Isotope Reactions

INTRODUCTION

In 1913, J.J. Thompson resolved the curious fractional masses of the elements when he discovered that 90% of neon (atomic number 10 and mass 20.2) had mass 20, while about 10% had mass 22, with 12 rather than 10 neutrons. Thompson's student Francis Aston began a systematic measurement of naturally occurring stable isotopes of the elements. Developments in mass spectrometry later allowed their routine measurement in environmental samples (Table 1.2, in Chapter 1), which opened the field of isotope hydrology.

The *environmental* isotopes are naturally occurring isotopes of the major elements that participate in processes at the earth's surface. Isotope measurements complement the geochemical reactions discussed in the previous chapter in understanding the origin and evolution of groundwater. Most elements have isotopes, which are chemically the same but with additional or fewer neutrons and so differ only in atomic mass. Some are stable, making up essentially all the mass of the earth. Others are radioactive and decay to other elements. Most atoms of carbon, for example, have 6 protons and 6 neutrons, making an atomic mass of 12, ${}^{12}_{6}C_6$. One percent of carbon atoms has an additional neutron, making stable carbon-13, ${}^{13}_{6}C_7$ (Figure 3.1). However, the addition of two additional neutrons makes the unstable nucleus of ${}^{14}_{6}C_8$.

The most common stable environmental isotopes are ^{18}O and D used as tracers of the origin and history of water. Others include ^{13}C, ^{15}N, and ^{34}S and are used for tracing the origin and transformations of their various geochemical species. The decay of environmental radioisotopes over time make them useful for dating solutes and groundwater. The most commonly used are tritium, T, with a half-life of 12.32 years and radiocarbon, ^{14}C, with a half-life of 5730 years.

STABLE ENVIRONMENTAL ISOTOPES

Stable isotope concentrations are measured as a ratio of the rare to the abundant isotope and expressed as the normalized difference in this ratio between the sample and a known reference:

$$\delta^{18}O_{\text{sample}} = \frac{\left({}^{18}O/{}^{16}O\right)_{\text{sample}} - \left({}^{18}O/{}^{16}O\right)_{\text{reference}}}{\left({}^{18}O/{}^{16}O\right)_{\text{reference}}}$$

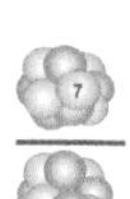

$$= \frac{{}^{\text{amu}}_{z}E_n}{{}^{\text{amu}}_{z}E_n} = \frac{{}^{13}_{6}C_7}{{}^{12}_{6}C_6} = 0.0111$$

FIGURE 3.1 Stable carbon-13 as a ratio with abundant carbon-12, and their atomic mass units, amu, atomic number, z, and neutron number, n.

DOI: 10.1201/9781003353836-3

This normalized difference between the sample and reference is then multiplied by 1000 to express the measurement in permil (‰) units:

$$\delta^{18}O_{\text{sample}} = \left(\frac{\left(^{18}O/^{16}O \right)_{\text{sample}}}{\left(^{18}O/^{16}O \right)_{\text{VSMOW}}} - 1 \right) \times 1000\ \text{‰ VSMOW}$$

where δ, delta, indicates the difference in ratio from the standard, VSMOW (Vienna Standard Mean Ocean Water) is the standard used in this example, and ‰ is the permil notation. A δ-value that is positive, say, $\delta^{18}O = +10‰$, signifies that the sample has 10 permil or 1% more ^{18}O than the reference and so is *enriched* in ^{18}O. Similarly, a sample with $\delta^{18}O = -10‰$ has 10‰ or 1% less ^{18}O than VSMOW and so is *depleted* in ^{18}O.

The δ-‰ notation for stable isotope data is a derived unit of concentration that comes from the requirement to normalize measurements to a reference material for accurate measurements. Isotope measurements could be expressed in concentration units like those used for solutes. For example, the $^{18}O/^{16}O$ abundance ratio in VSMOW is 0.002005 or 2005 ppm (2005 atoms of ^{18}O per million atoms of ^{16}O or 2009 atoms of ^{18}O per million oxygen atoms (^{16}O plus ^{18}O). An example for a water sample is given in Figure 3.2.

The list of isotopes used in hydrology is expanding as new methods and advances in mass spectrometry are developed. Helium isotopes ($^{3}He/^{4}He$) are now often used in studies of groundwater residence time. Others, including ^{9}Li, ^{11}B, ^{37}Cl, and ^{87}Sr, are used to study weathering, sources of salinity, and contaminants. Table 3.1 presents the range of environmental isotopes now routinely used in hydrogeological studies.

Stable Isotope Fractionation

Craig (1961) noted that ^{18}O and D in global freshwaters were unequally distributed on a global basis, with high-latitude samples from Antarctica depleted in ^{18}O by as much as 45‰ compared to tropical samples. Other isotopes, such as ^{13}C and ^{34}S,

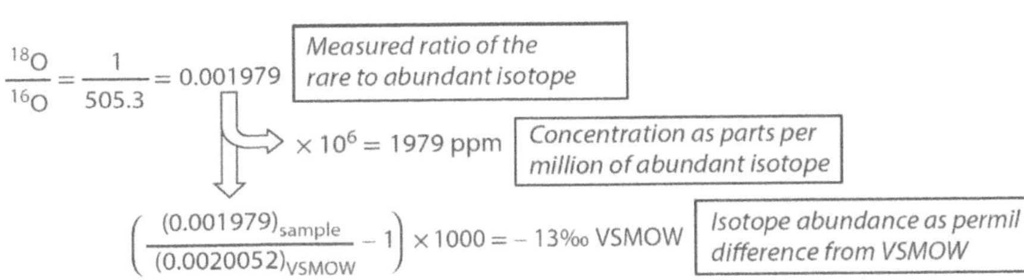

FIGURE 3.2 The concentration of stable isotopes as an atomic ratio, as parts per million of the abundant stable isotope, and in δ-‰ units relative to Vienna Standard Mean Ocean Water (VSMOW) ($^{18}O/^{16}O_{\text{VSMOW}} = 0.0020052$) for a water sample with $\delta^{18}O = -13‰$.

TABLE 3.1
Stable Environmental Isotopes

Isotope	Ratio	Abundance %	Reference (Abundance Ratio)	Common Sample Types
^{2}H or D	D/H	0.015	VSMOW (1.5576×10^{-4})	H_2O, CH_4, clays
^{3}He	$^{3}He/^{4}He$	0.000138	AIR He (1.38×10^{-6})	groundwater, minerals
^{7}Li	$^{7}Li/^{6}Li$	95.15	LSVEC (12.1729)	water, brines, minerals
^{11}B	$^{11}B/^{10}B$	80.1	NBS 951 (4.04363)	water, brines, minerals
^{13}C	$^{13}C/^{12}C$	1.11	VPDB (0.0111105)	DIC, CO_2, $CaCO_3$, CH_4, organic C
^{15}N	$^{15}N/^{14}N$	0.366	AIR N_2 (0.003676)	NO_3^-, NH_4^+, N_2, N_2O, $CaCO_3$
$^{18}O_{water}$	$^{18}O/^{16}O$	0.201	VSMOW (0.0020052)	H_2O, NO_3^-, SO_4^{2-}, O_2
$^{18}O_{carbonate}$	$^{18}O/^{16}O$	0.207	VPDB (0.0020654)see note	Carbonate minerals
^{34}S	$^{34}S/^{32}S$	4.21	VCDT (0.044163)	SO_4^{2-}, H_2S, gypsum, pyrite
^{37}Cl	$^{37}Cl/^{35}Cl$	24.23	SMOC (0.3196)	water, brines, solvents
^{81}Br	$^{81}Br/^{79}Br$	49.314	SMOB (0.9729)	water, brines, contaminants
^{82}Se	$^{82}Se/^{76}Se$	0.9317	NIST SRM3149	water, mine drainage
^{87}Sr	$^{87}Sr/^{86}Sr$	7.0/9.8	Direct measurement	water, brines, minerals

VSMOW, Vienna Standard Mean Ocean Water (Gonfiantini 1976); AIR, N_2, and He from well-mixed open atmosphere, (Rosman and Taylor 1998; Coplen 2002); LSVEC, Lithium carbonate standard (IAEA 2020; Coplen 2011); NBS-951, Boric acid standard, also SRM 951 (Coplen 2002); VPDB, Vienna Pee Dee Belemnite, fossil carbonate (Dunn et al. 2024); VCDT, Vienna Canyon Diablo meteorite Troilite [FeS] (Ding et al. 2001); SMOC, Standard Mean Ocean Chloride (Rosman and Taylor 1998; Xiao et al. 2002); SMOB, Standard Mean Ocean Bromide (Eggenkamp and Coleman 2000). Note: for conversion of ^{18}O data between VSMOW and VPDB scales, $\delta^{18}O_{VSMOW} = 1.03092 \times \delta^{18}O_{VPDB} + 30.92$ (Coplen et al., 1983; Kim et al. 2015).

were also found to have unequal distributions in the environment. This partitioning of stable isotopes within and between different reservoirs allows them to be used as tracers of origin and pathways.

The partitioning of stable isotopes is controlled by the interplay of two processes: the *fractionation* of isotopes during any physical or chemical reaction and the overall *distillation* of isotopes from the reactant reservoir as the reaction proceeds. These two processes work together to partition isotopes into different reservoirs. It is a bit like building a fence from a field of stones, using preferably large stones (like ^{18}O). The selection of stones to build the fence is *fractionating* the large stones from the field into the fence, which is *enriched* in large stones. In a geochemical sense, the fence might be calcite [$CaCO_3$] forming at a spring vent. and the heavy stones are $Ca^{13}CO_3$ molecules.

Although a stable isotope of an element behaves the same chemically, its additional weight increases bond strength. This is *mass-dependent fractionation* (MDF) of stable isotopes. The isotopes of water are a good example. With ^{18}O

(abundance of 0.2%) and deuterium, D or ^2H (abundance of 0.015%), H_2O can have an atomic mass of 18 (H_2O), 19 (HDO), or 20 ($H_2{}^{18}O$). The physical reaction of evaporation favors the light isotopes, enriching water vapor with ^{16}O and ^1H while enriching the water in ^{18}O and D. As a general rule of thumb, heavier isotopes tend to partition into the more condensed phase, gas → liquid → solid, or vapor → water → ice.

Urey (1947) showed that the ratio of isotope vibrational frequencies or partitioning functions, Q, for the reacting molecules are inversely proportional to their mass differences. As an example, the two-times mass difference for D/H is much greater than for $^{18}O/^{16}O$, and so fractionation in a reaction such as condensation is much greater for deuterium than for ^{18}O. This is observed empirically in precipitation data where values for δD are typically 8 times greater than for $\delta^{18}O$. This can also be observed in single elements with more than one isotope. Oxygen, for example, is over 99% ^{16}O, while 0.2% is ^{18}O and 0.04% is trace ^{17}O (Table 1.2). We measure $\delta^{18}O$ as this is the more abundant isotope. Moreover, due to mass-dependent fractionation, $\delta^{17}O$ is generally predictable as $0.5 \times \delta^{18}O$, given that the $^{17}O - {}^{16}O$ mass difference is half that of $^{18}O - {}^{16}O$ and so fractionation for $\delta^{18}O$ is essentially twice that of $\delta^{17}O$.

Although isotope fractionation is largely considered to be mass-dependent, there can be *mass-independent fractionation* (MIF) during photochemical reactions in the stratosphere and in the troposphere leading to enrichments in ^{17}O above those attributable to mass-dependent fractionation (Miller and Pack 2021). Such enrichments are noted as $\Delta^{17}O = \delta^{17}O - 0.5\delta^{18}O$. First observed in chondrite meteorites (Clayton et al. 1973), such enrichments are now found in stratospheric oxygen, which can contribute to tropospheric waters at high latitudes and give new insights into water vapor circulation and precipitation (Aggarwal et al. 2023).

Similar to secondary trace isotopes such as ^{17}O, certain compounds can have the rare combination of *clumped* isotopes. For example, CO_2 can have more than one of the rare isotopes, as $^{13}C^{18}O^{16}O$ with mass 47, $^{12}C^{18}O^{18}O$ (mass 48), and $^{13}C^{18}O^{18}O$ (mass 49) (Eiler 2007). Given the low abundance of ^{13}C and ^{18}O in nature, such multiply substituted isotopologues have abundances less than 100 ppm and so are challenging to measure. Clumped isotopes are expressed as the difference from a randomly distributed isotopologue, in permil units. For $^{13}C\,{}^{18}O^{16}O$, this would be the difference between the measured mass 47/44 ratio ($^{13}C^{18}O^{16}O/^{12}C^{16}O^{16}O$) and the random distribution (^{13}C at 1.1% and ^{18}O at 0.2% for each oxygen site), expressed as Δ_{47}. Similar expressions can be made for Δ_{48} and Δ_{49}. Clumping is a thermodynamic phenomenon and is highly sensitive to temperature, and so in carbonates provides an accurate paleothermometer (Ghosh et al. 2006).

Fractionation Factor, α

In 1947, Harold Urey published the theoretical basis for isotope fractionation as an exchange of isotopes between molecules x and y that are participating in an isotope exchange reaction (e.g., Liu et al. 2010). Isotope fractionation in a physicochemical reaction is a thermodynamic process with a reaction constant or fractionation factor, α.

$$\alpha_{x-y} = \frac{R_x}{R_y}$$

The reaction of freezing water is a good example.

$$H_2O_{water} \leftrightarrow H_2O_{ice}$$

For natural waters, the concentration of ^{18}O in water will be about 2000 ppm but in the forming ice, it would be about 2006 ppm. That makes ^{18}O about 0.3% or 3‰ enriched in ice over water. The fractionation factor, $\alpha_{ice\text{-}water}$, is the ratio $^{18}O/^{16}O$ in the ice (2006 ppm or 2.006×10^{-3}) over the water (2000 ppm or 2.000×10^{-3}) giving a fractionation, $\alpha_{ice-water} = 1.003$

$$\alpha^{18}O_{ice-water} = \frac{R_{ice}}{R_{water}} = \frac{\left(\frac{^{18}O}{^{16}O}\right)_{ice}}{\left(\frac{^{18}O}{^{16}O}\right)_{water}} = \frac{0.002000}{0.002006} = 1.003$$

and written as the reciprocal reaction:

$$\alpha^{18}O_{water-ice} = \frac{1}{\alpha^{18}O_{ice-water}} = \frac{R_{water}}{R_{ice}} = \frac{0.002000}{0.002006} = 0.9970$$

This fractionation factor, α, is the constant for single isotope exchange reactions, as *K* is for geochemical reactions. Isotope fractionation factors are for the most part very close to 1. This is because the difference in the activities of two isotope species, like [$H_2{}^{16}O$] and [$H_2{}^{18}O$] is very minor.

Consider the evaporation of water as an example of a physical reaction and the dissolution of CO_2 to form HCO_3^- as a geochemical reaction. Adding isotopes to the reaction, there will be forward and backward exchange of isotopes until there is an equilibrium distribution with either the reactant or the product having more of the heavy isotope,

$$H_2^{16}O_w + H_2^{18}O_v \leftrightarrow H_2^{16}O_v + H_2^{18}O_w$$

$$^{13}CO_2 + H_2O + H^{12}CO_3^- + H^+ \leftrightarrow H^{13}CO_3^- + H^+ + {}^{12}CO_2 + H_2O$$

Writing the thermodynamic reaction constant, K, with isotope species:

$$K_{Evap} = \frac{[H_2^{16}O_v][H_2^{18}O_w]}{[H_2^{16}O_w][H_2^{18}O_v]} \qquad K_{CO_2} = \frac{[H^{13}CO_3^-][H^+][^{12}CO_2][H_2O]}{[^{13}CO_2][H_2O][H^{12}CO_3^-][H^+]}$$

$$= \frac{[^{18}O_w][^{16}O_v]}{[^{16}O_w][^{18}O_v]} \qquad = \frac{[^{13}C_{HCO_3^-}][^{12}C_{CO_2}]}{[^{13}C_{CO_2}][^{12}C_{HCO_3^-}]}$$

$$= \frac{\left(\frac{^{18}O}{^{16}O}\right)_w}{\left(\frac{^{18}O}{^{16}O}\right)_v} \qquad = \frac{\left(\frac{^{13}C}{^{12}C}\right)_{HCO_3^-}}{\left(\frac{^{13}C}{^{12}C}\right)_{CO_2}}$$

$$= \alpha^{18}O_{w-v} = \frac{R_w}{R_v} \qquad = \alpha^{13}C_{HCO_3^- - CO_2} = \frac{R_{HCO_3^-}}{R_{CO_2}}$$

At 25°C, $\alpha^{18}O_{w-v} = 1.0093$ $\qquad \alpha^{13}C_{HCO_3^- - CO_2} = 1.0097$

From the fractionation factor, the isotope ratio and delta value of the product of an isotope exchange reaction can be predicted from that of the reactant. For example, $\delta^{18}O_{vapor}$ can be calculated from $\delta^{18}O_{water}$ if $\alpha^{18}O_{water-vapor}$ is known. Conversely, the fractionation factor for two compounds involved in an isotope exchange reaction can be calculated using their isotope ratios expressed in δ-permil fashion. This requires substitution of their absolute ratios, R, with measured δ–‰ values, giving the relationship,

$$\alpha^{18}O_{water\text{-}vapor} = \frac{\delta^{18}O_{water} + 1000}{\delta^{18}O_{vapor} + 1000}$$

$$\delta^{18}O_{vapor} = \frac{\delta^{18}O_{water} + 1000}{\alpha^{18}O_{water\text{-}vapor}} - 1000$$

Enrichment Factor, $\varepsilon = 10^3 \ln \alpha$

Experimental and theoretical determination of fractionation factors over a range of temperatures finds the fractionation–temperature relationship best described by a natural logarithmic function, ln α, such as

$$\ln \alpha_{x-y} = aT^{-2} + bT^{-1} + c$$

Multiplying by 10^3 allows the isotope enrichment in isotope exchange reactions to be expressed in permil units, like delta values, δ-‰, which is the enrichment factor, ε,

$$\varepsilon_{x-y} = 10^3 \ln \alpha_{x-y} = \delta_x - \delta_y$$

Using the example of ^{18}O during evaporation,

$$\varepsilon^{18}O_{\text{water-vapor}} = \delta^{18}O_{\text{water}} - \delta^{18}O_{\text{vapor}}$$

In the past, calculations of isotope enrichment used the simplification $\varepsilon = 10^3(\alpha - 1)$, which is only valid for fractionation where α approaches 1 (and so ln α = 0), such as for ^{18}O during evaporation ($\alpha^{18}O_{w-v} = 1.009$ at 25°C), but deviates significantly from $10^3 \ln \alpha$ as α departs from unity in exchange reactions like deuterium during water evaporation ($\alpha D_{w-v} = 1.076$ at 25°C) and reduction of water ($\alpha D_{H_2O-H_2} = 3.762$).

Following from Urey's work, considerable effort has been made to establish the temperature relationship for isotope fractionation in various reactions, such as evaporation, condensation, freezing, aqueous reactions, and mineral crystallization. These relationships allow determination of isotope fractionation at the range of temperatures in natural systems.

At high temperatures, α is very close to 1 but departs from 1 as the temperature of reaction decreases. Fractionation factors are greatest for the low-temperature geochemical reactions in groundwaters and surface waters. This is illustrated for water–vapor fractionation at different temperatures in Table 3.2 and Figure 3.3.

From these data, the trend toward increasing fractionation at lower temperatures is clear. Note the much greater degree of fractionation for D than for ^{18}O between water and vapor. This is a consequence of the mass difference of D over H, which is 100%, whereas ^{18}O is only 12.5% heavier than ^{16}O, and it will not be as strongly fractionated as D. The fractionation–temperature relationships for some isotope exchange reactions useful in groundwater work are shown in Figure 3.4.

TABLE 3.2
Values for Fractionation Factors for ^{18}O and ^{2}H in Water–Vapor–Ice Reactions

T (°C)	$\varepsilon^{18}O_{w-v}$ (‰)	εD_{w-v} (‰)
Water vapor		
0	11.7	106
25	9.3	76
50	7.5	55
75	6.2	39
100	5.1	28
Ice water (0°C)	3.1	19

Based on equations given below (Horita and Wesolowski 1994, for water-vapor; Suzuoki and Kumura 1973, for ice water).

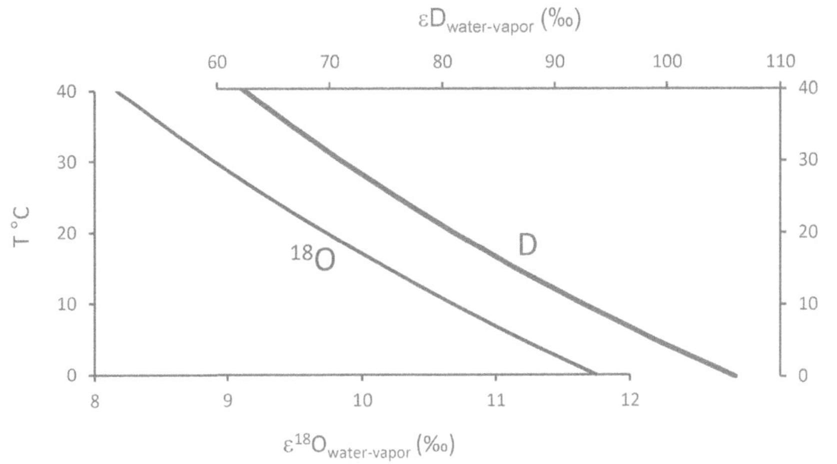

FIGURE 3.3 Temperature effect on fractionation of ^{18}O and D between water and water vapor.

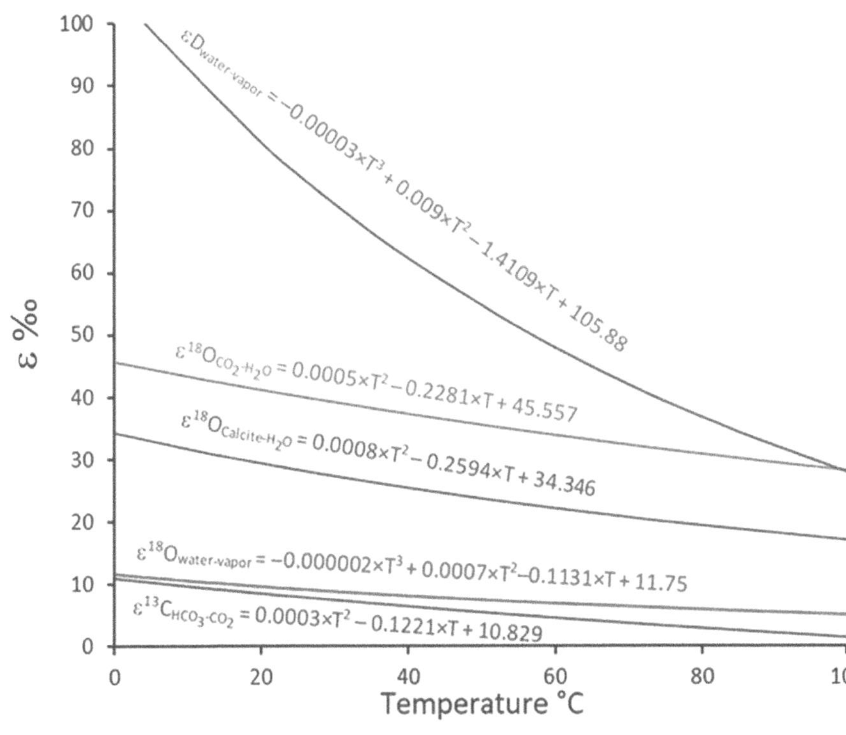

FIGURE 3.4 Variation of $\varepsilon = 10^3 \ln \alpha$ with temperature for some common isotope exchange reactions. Equations regressed for T°C from Horita and Wesolowski (1994), Bottinga (1968), O'Neil et al. (1969), and Mook et al. (1974).

Kinetic and Microbially Mediated Reactions

Physicochemical reactions are accelerated when conditions move away from equilibrium. Such kinetic conditions can reduce the mass discrimination of thermodynamic fractionation when the forward reaction greatly exceeds the backward reaction. An example is the precipitation of calcite from water by rapid freezing. This is observed in permafrost karst, where such cryogenic calcites have $\delta^{13}C$ as high as +17‰. Kinetic reaction favors dissociation of weaker ^{12}C–O bonds over ^{13}C–O bonds, with over 20‰ enrichment above equilibrium values (Clark and Lauriol 1992).

Redox reactions are similar in that there is little to no back reaction. Most redox reactions taking place under near-surface conditions are mediated by bacteria and archaea, which are highly selective in the isotopic bonds they break. Bacteria and archaea can augment the energy they gain from such reactions by preferentially selecting the light isotopes for which bonds are easier to break, for example, ^{32}S–O rather than ^{34}S–O during sulfate reduction. For this reason, biologically mediated redox reactions are highly fractionating. Some of the more common microbially mediated reactions and their associated enrichment factors include the following:

Sulfate reduction	$SO_4^{2-} + 8e^- + 8H^+ \rightarrow H_2S + 4H_2O$	$\varepsilon^{34}S_{SO_4-H_2S} \approx 20‰ - 30‰$
Nitrification	$NH_4^+ + 2O_2 \rightarrow NO_3^- + H_2O + 2H^+$	$\varepsilon^{15}N_{NH_4-NO_3} \approx 30‰$
Denitrification	$NO_3^- + 5e^- + 6H^+ \rightarrow ½N_2 + 3H_2O$	$\varepsilon^{15}N_{NO_3-N_2} \approx 15‰ - 20‰$
Methanogenesis	$CO_2 + 8e^- + 8H^+ \rightarrow CH_4 + 2H_2O$	$\varepsilon^{13}C_{CO_2-CH_4} \approx 60‰ - 80‰$

It is important to consider the case of isotope fractionation during dissolution reactions on the solid surfaces of minerals or organics, where no isotope selection or mixing within the solid reactant reservoir is possible. The reaction is therefore *quantitative* in that the reaction products have the same isotope composition as the reactant surface. Such reactions can be both inorganic and biologically mediated:

Pyrite oxidation	$FeS_2 \rightarrow SO_4^{2-}$	$\delta^{34}S_{SO_4^{2-}} \cong \delta^{34}S_{FeS_2}$
Melting ice	$H_2O_{ice} \rightarrow H_2O_{water}$	$\delta^{18}O_{water} \cong \delta^{18}O_{ice}$
Calcite dissolution	$CaCO_3 \rightarrow HCO_3^-$	$\delta^{13}C_{HCO_3^-} \cong \delta^{13}C_{calcite}$
Aerobic respiration	$CH_2O_{cellulose} \rightarrow CO_2$	$\delta^{13}C_{CO_2} \cong \delta^{13}C_{CH_2O}$

Diffusive Fractionation

Diffusion is an additional process that can fractionate isotopes due to the relative molecular velocity of a heavy isotope species relative to the light isotope species. It is an important process in aquitards where groundwater advection is minimal and transport is dominated by molecular diffusion in the porewaters.

The simplest case is the diffusion of gas in a vacuum, where steady-state fractionation equals the ratio of the velocities of the two isotopes, which resolves to the root of the mass ratio, which is Graham's law:

$$\alpha_{m^*-m} = \frac{v^*}{v} = \frac{\sqrt{\dfrac{kT}{2\pi M^*}}}{\sqrt{\dfrac{kT}{2\pi M}}} = \sqrt{\frac{M}{M^*}}$$

where v is the molecular velocity (cm s^{-1}); k is the Boltzmann constant (1.380658×10^{-23} J/K, gas constant per molecule); M is the molecular mass of the light isotope and M* is the mass of the heavy isotope; and T is the absolute temperature K. Diffusion of CO_2 in air and kinetic uptake was determined to be responsible for strong depletions up to –27.5‰ for $\delta^{13}C$ measured in calcite crusts forming on high pH groundwaters in Oman (Clark et al. 1992).

Derivatives of Graham's law for diffusion through a medium such as air or water include the molecular mass of the medium. However, in groundwaters, diffusive fractionation is limited to aquitards, where low permeability brings solute transport into the domain of diffusion. Here, it is concentration gradients that determine the rate of diffusion together with the diffusion coefficient for the isotope species in water and through the porous network. Fractionation of isotopes by diffusion through aquitards is examined by the rate of diffusion of the individual isotope species (e.g., $H_2^{18}O$, HDO, or $H^{13}CO_3^-$), each with its own coefficient of diffusion. The decoupling of isotopes in this way does not allow us to represent fractionation in terms of α and ε values as shown earlier. It is a mass dependent process like physical–chemical fractionation but can be considered more precisely as diffusive mixing. This is presented in Chapter 4, in the section on groundwater mixing.

Rayleigh Distillation of Stable Isotopes

Isotope fractionation is a physicochemical process that operates at the level of the reaction and favors one isotope over the other. During the reaction, the isotope ratio in the reactant reservoir will evolve as the fractionation reaction *distills* the preferred isotope into the product reservoir. As the reactant reservoir diminishes, this process imparts a significant partitioning of the isotopes between the reactant and product reservoirs. If the reaction goes to completion, then the isotope ratio in the product reservoir will be identical to the starting reactant reservoir.

A good example is a pond of water that is evaporating towards dryness. Evaporation favors the light isotopes, ^{16}O and H, and so ^{18}O and D accumulate in the residual pond water. Over time, this distillation of the light isotopes from the pond leaves the residual fraction of water highly enriched in the heavy isotopes. This process follows a Rayleigh distillation (Lord Rayleigh 1896), represented by the equation,

$$R_f = R_o \times f^{(\alpha-1)}$$

$$\left(\frac{^{18}O}{^{16}O}\right)_f = \left(\frac{^{18}O}{^{16}O}\right)_o \times f^{(\alpha-1)}$$

where R_f is the isotope ratio of the reactant reservoir after reaction to some residual fraction, f. The initial isotopic ratio of the reactant reservoir is R_o; when $f = 1$, α is the fractionation factor for the reaction, written in the form of the product over the reactant, $\alpha_{prod-react}$. Using fractionation in $\delta - ‰$ notation and the enrichment factor, $\varepsilon(‰) = 10^3 \ln \alpha$,

$$\delta_{react} = \delta_{initial\ react} + \varepsilon_{prod\text{-}react} \times \ln f$$

$$\delta^{18}O_{water} = \delta^{18}O_{initial\ water} + \varepsilon^{18}O_{vapor\text{-}water} \times \ln f$$

Plotting this relationship for the residual water in the evaporating pond gives the exponential curve shown in Figure 3.5. Similarly, but the inverse, is the freezing of the pond, where the ice is enriched in ^{18}O.

These two cases represent the isotopic evolution of the reacting reservoir, in this case the water. Also important is the isotopic evolution in the reaction product and in the product reservoir. A good example of this is the Rayleigh distillation of isotopes from vapor masses as they cool and rain out over the continents, which leads to partitioning of heavy isotopes into rain in warm regions with their depletion in the rain and snow of colder regions. In this hypothetical case, the Rayleigh equation is modified to calculate the ideal isotopic evolution in the reaction product, i.e., rain.

Figure 3.6 shows the Rayleigh calculations for (i) depletion of reactant during rainout, (ii) the subsequent evolution of the reaction product, and (iii) the isotopic

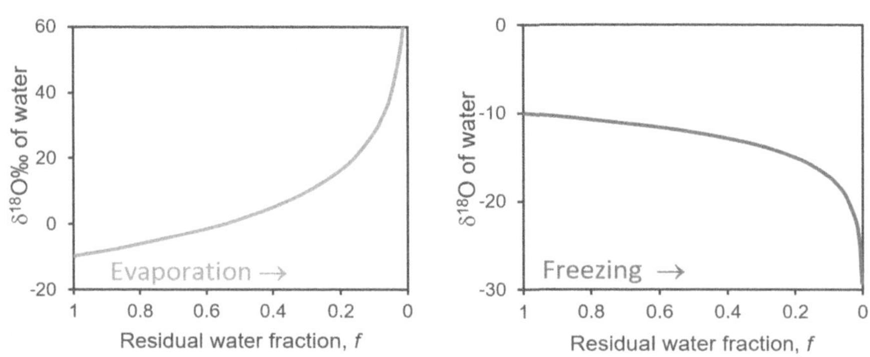

FIGURE 3.5 Rayleigh distillation of ^{18}O in water during kinetic evaporation (left) and in water during freezing (right). The initial $\delta^{18}O$ of the water is set at –10‰ for these examples. For kinetic evaporation, $\varepsilon^{18}O_{v-w} = -16.4‰$ and for freezing, $\varepsilon^{18}O_{v-w} = +3.1‰$.

(i) Evolution of reacting reservoir (vapor)

$$\delta_{react} = \delta_{o\text{-}react} + \varepsilon_{prod\text{-}react} \times \ln f$$

$$\delta^{18}O_{cloud} = \delta^{18}O_{o\text{-}cloud} + \varepsilon^{18}O_{water\text{-}vapor} \times \ln f$$

(ii) Evolution of product (rain)

$$\delta_{prod} = \delta_{o\text{-}react} + \varepsilon_{prod\text{-}react} \times (1 + \ln f)$$

$$\delta^{18}O_{rain} = \delta^{18}O_{o\text{-}cloud} + \varepsilon^{18}O_{water\text{-}vapor} \times (1 + \ln f)$$

(iii) Evolution of product reservoir (surface waters)

$$\delta_{prod\text{-}res} = \delta_{f\text{-}react} - (\varepsilon_{prod\text{-}react} \times \ln f)/(1 - f)$$

$$\delta^{18}O_{surface\ water} = \delta^{18}O_{f\text{-}cloud} - (\varepsilon^{18}O_{water\text{-}vapor} \times \ln f)/(1 - f)$$

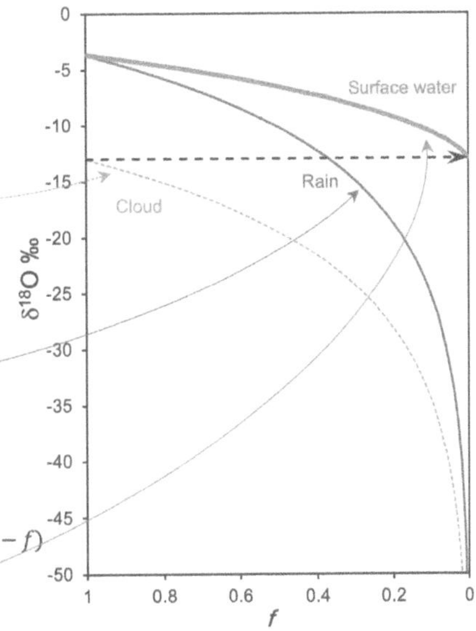

FIGURE 3.6 Isotopic evolution of a reactant reservoir and of the immediate reaction product, in the example of rain produced by a cloud during rainout. Note that the isotopic composition of the product reservoir (surface water) at $f = 0$, when there is no more reactant, is the same as the reactant reservoir (cloud) at $f = 1$, before there is any product.

evolution of the product reservoir, in this case the surface waters accumulating as rainout proceeds, $\delta^{18}O_{surface\text{-}waters}$.

ENVIRONMENTAL RADIOISOTOPES

Radioisotopes are those nuclides having an unstable configuration of neutrons and so decay as the nucleus adjusts towards a more stable configuration. For example, both $^{12}{}_{6}C_{6}$ and $^{13}{}_{6}C_{7}$ are stable, but an eighth neutron produces the radioisotope $^{14}{}_{6}C_{8}$, which decays with a half-life of 5730 years to stable $^{14}{}_{7}N_{7}$. Adding a ninth neutron produces the very unstable nuclide $^{15}{}_{6}C_{9}$ which has a half-life of only 2.5 seconds. Radionuclides with short half-lives, like radiocarbon, are continually regenerated by nuclear reactions in the atmosphere or in the subsurface. Others, with long half-lives like ^{238}U ($T_{½}$ = 4.47 billion years), were created by nucleosynthesis during the formation of our solar system some 5 billion years ago. The environmental radioisotopes are those produced by natural nuclear reactions and are useful for dating and tracing (Table 3.3). These and others can also be produced by anthropogenic activities and can be of concern as radiohazards. However, at natural environmental activities, they are orders of magnitude below levels of concern.

A radioisotope has both a concentration relative to its abundant stable isotope and an activity, measured in becquerels, Bq, as disintegrations per second. For example, radiocarbon is produced in the atmosphere by cosmic rays and has a concentration

TABLE 3.3
The Common Environmental Radioisotopes in Groundwater

Isotope	Half-Life (years)	Decay Mode	Stable Daughter	Atmospheric Ratio	Activity (Bq)	Common Species
^{3}H (T)	12.32	β–	^{3}He	T/H ~ 10^{-17}	~ 1.2 Bq/L	H_2O, CH_2O
^{14}C	5,730	β–	^{14}N	$^{14}C/C = 1.18 \times 10^{-12}$	0.226 Bq/g C	DIC, DOC
^{36}Cl	301,000	EC, β–	^{36}Ar, ^{36}S	$^{36}Cl/Cl \sim 10^{-13}$	10^{-6} Bq/L (10 ppm Cl)	Cl^-
^{39}Ar	269	β–	^{39}K	$^{39}Ar/Ar = 8.2 \times 10^{-16}$	0.016 Bq/m^3 air	groundwater
^{81}Kr	229,000	EC	^{81}Br	$^{81}Kr/Kr = 5.2 \times 10^{-13}$	0.0018 mBq/m^3 air	groundwater
^{85}Kr	10.74	β–	^{85}Rb	$^{85}Kr/Kr = 2.39 \times 10^{-11}$	1.5 Bq/m^3 air	groundwater
^{129}I	15.7×10^6	β–	^{129}Xe	$^{129}I/I \sim 10^{-11}$	10^{-4} Bq/L (1 ppm I)	I^-, IO_4^-, CH_3I

of about 59 billion atoms per gram of carbon in atmospheric CO_2 or an abundance ratio, $^{14}C/C = 1.18 \times 10^{-12}$. Its decay sustains an activity of 0.226 Bq per gram of C in modern carbon.

The most common decay mode is beta decay, β^-, which is the emission of an electron from the nucleus. This converts a neutron to a proton with an *isobaric* transformation to a higher atomic number element in the Periodic Table but with the same atomic mass (Figure 3.7). Decay by electron capture, EC, is also isobaric but with the opposite effect, transforming a proton to a neutron, producing the element of lower atomic number. Isotopes of the actinides, including U and Th, are all radioactive and decay largely by emission of an alpha particle, α, which is a ^{4}He nucleus (2n and 2p). Some of the actinides, such as ^{238}U, can also decay by spontaneous fission into two mid-weight nuclides. Electromagnetic radiation can accompany decay as high-energy gamma, γ, which is a photon from the nucleus.

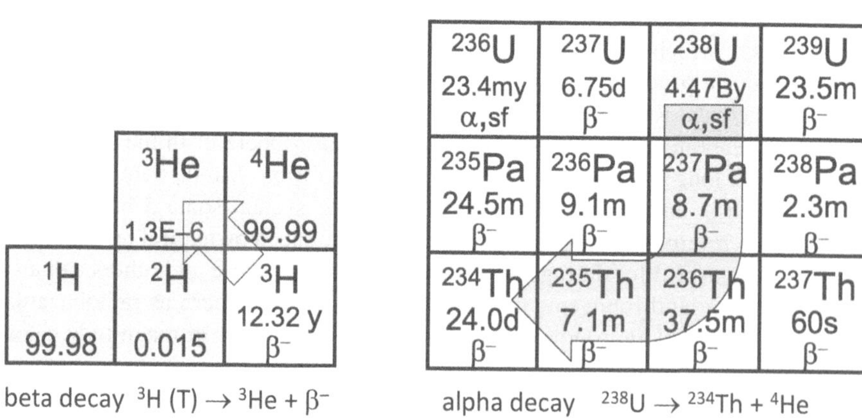

FIGURE 3.7 Principal modes of radioactive decay.

Radioisotope Concentration and Activity

Radioisotopes have both a given concentration, n, relative to the main (stable) isotope at a given time and a given activity, a, at that time, with units of decay events per time (i.e., Bq, or decay per second). The activity of a radioisotope is proportional to its concentration, or the number of atoms, n, of the radioisotope. It is also proportional to its half-life. Tritium, for example, has a short half-life of 12.32 years and so has high activity for a given concentration. Uranium, in contrast, has a half-life close to the age of the earth and has very low activity. The proportionality constant in the activity–concentration relationship is the decay constant, λ, with units of inverse time (y^{-1} or s^{-1}),

$$a = \lambda n$$

Tritium is a good example of the activity–concentration relationship. In studies of tritium emissions from activities related to nuclear power plants and hydrogen fusion weapons, where it is a radiohazard, units of activity are used. Tritium is then expressed in Bq per liter of water, gram of vegetation, or cubic meter of air. In groundwater studies where the exceedingly low levels of natural atmospheric tritium are used for dating, concentration units are used, expressed as tritium units, TU, where one TU is equal to one tritium atom per 10^{18} atoms of hydrogen.

$$1\,\mathrm{TU} = \mathrm{T/H} \times 10^{18}$$

The activity of tritium for this concentration can be derived using the decay constant and the activity–concentration relationship from above:

$$a_\mathrm{T} = \lambda_\mathrm{T} n_\mathrm{T}$$

From above, the decay constant, λ_T, is derived from the half-life (12.32 years) expressed in seconds to be consistent with activity, a, in Bq (s^{-1}). Activity is normalized to a mass of material, in this case, say, 1 L or 1000 g of water:

$$\begin{aligned}\lambda_\mathrm{T} &= \ln 2/T_{½}\\ &= \ln 2/(12.32 \times 365 \times 24 \times 3600)\\ &= 1.78 \times 10^{-9}\,\mathrm{s}^{-1}\end{aligned}$$

$$\begin{aligned}\text{the concentration is: } n_\mathrm{T} &= 1\,\mathrm{TU}\\ &= 1000/18 \times 2 \times 10^{-18} = 111 \times 10^{-18}\,\text{mols T/L}\\ &= 111 \times 10^{-18} \times 6.02 \times 10^{23} = 6.68 \times 10^{7}\,\text{atoms T/L}\end{aligned}$$

$$\begin{aligned}\text{and so activity is: } a_\mathrm{T} &= \lambda_\mathrm{T}\, n_\mathrm{T}\\ &= 1.78 \times 10^{-9} \times 6.68 \times 10^{7}\\ &= 0.119\,\mathrm{Bq/L}\end{aligned}$$

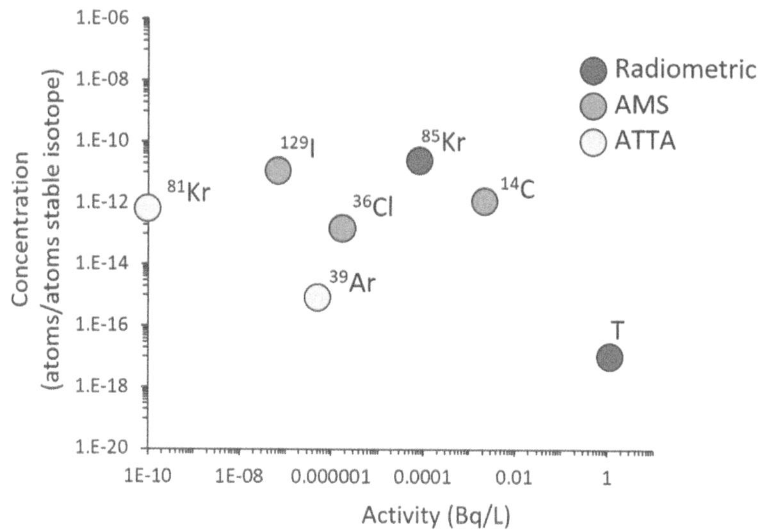

FIGURE 3.8 Concentration–activity relationship for the common radioisotopes present at environmental levels (Ar and Kr as air-saturated water, C as 10 ppm DIC, Cl as 20 ppm, and I as 2 ppb). Tritium and radiocarbon, with high activity, can be measured radiometrically. Radiocarbon is now analyzed by accelerator mass spectrometry (AMS), which accommodates very small (< 1 mg C) samples. For the low activities of ^{36}Cl and ^{129}I, analysis by AMS is required. The noble gas radioisotopes, ^{39}Ar and ^{81}Kr, are only practically measured by the recently developed method of laser atom trap trace analysis (Jiang et al. 2012).

This is a very low level of radioactivity. Natural environmental levels are generally several times higher but nonetheless are many orders of magnitude lower than levels considered to be hazardous.

The activity of a radionuclide then depends on its half-life and its concentration. Over 900 radionuclides with half-lives longer than 1 hour have been identified. However, only a few are readily measured and have half-lives appropriate for dating in groundwater studies (Table 3.3; Figure 3.8).

The environmental concentrations and activities of the radioisotopes have a bearing on how they can be analyzed. Radioisotopes with large decay constants, such as tritium, T ($\lambda = 0.056y^{-1}$), are routinely measured by decay counting, and in the case of tritium, to a limit of detection of 10^{-19} or better. Radioisotopes with small decay constants, such as ^{129}I ($\lambda = 4.39 \times 10^{-8} y^{-1}$), have too low a decay rate for measurement by decay counting and are routinely analyzed by AMS with atom-counting detectors. This activity–concentration relationship is shown in Figure 3.8. Applications of these isotopes in groundwater dating and tracing are discussed in Chapter 7.

Half-Life and Dating

Material with some number of atoms of a given radioisotope will be radioactive, and this radioactivity will decrease through time as those atoms decay. Ernest Rutherford determined that the rate of decay follows a negative exponential function that is equal to the change in the number of atoms of the radioisotope with time, $\partial n/\partial t$. Each radioisotope has a unique and constant rate of decay, represented by the decay constant, λ, which is the constant of this proportionality.

$$-\frac{\partial n}{\partial t} = \lambda n \text{ (decay/time) radioactive decay}$$

or

$$-\frac{\partial n}{n} = \lambda \partial t$$

Integrating gives the general form of the radioactive decay equation that determines the number of atoms after decay from an initial number, n_0, over a given period, t:

$$n_t = n_0 e^{-\lambda t}$$

Taking the natural log of this equation and inverting gives

$$\ln\left(\frac{n_0}{n_t}\right) = \lambda t$$

Solving the decay equation for the period of time required to decay the number of atoms, n, to half ($n_0/n_t = 2$), establishes the half-life, $T_{½}$, for a given radioisotope (Figure 3.9):

$$\ln 2 = \lambda T_{½}$$
$$\lambda = \frac{\ln 2}{T_{½}}\left(\text{time}^{-1}\right)$$

Dating with radioisotopes is based on the loss of the parent from some initial concentration, n_{Po},

$$n_{\mathrm{P}t} = n_{\mathrm{Po}} e^{-\lambda t}$$

or through the gain of the daughter isotope, n_{D},

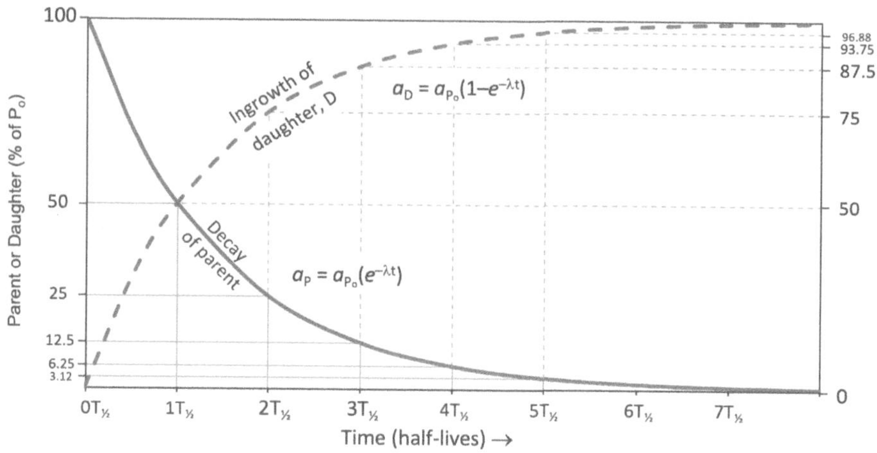

FIGURE 3.9 Radioisotope decay and half-life for loss-of-parent and for gain-of-daughter dating schemes.

$$n_{Dt} = n_{Po}\left(1 - e^{-\lambda t}\right)$$

These functions for the decay of the parent radioisotope and ingrowth of the daughter nuclide are shown in Figure 3.9.

Cosmogenic, Geogenic, and Anthropogenic Radionuclides

Groundwater dating with radioisotopes relies on the decay of atmospheric and surface-derived cosmogenic nuclides that accompany recharge or on the ingrowth of geogenic nuclides produced in the subsurface and gained by groundwaters. There are also anthropogenic radioisotopes that have been emitted to the atmosphere and hydrosphere through nuclear activities that can be used in groundwater dating. Most notable are the large quantities of tritium and ^{14}C produced by atmospheric tests of thermonuclear weapons up until 1963. Releases of ^{85}Kr as well as ^{129}I from nuclear power plants are also important.

The cosmogenic nuclides are produced by the high-energy particles, mainly protons originating from supernovae elsewhere in the galaxy, that constantly bombard our planet. Collisions with the atmosphere generate showers of secondary, lower-energy neutrons that further interact with atmospheric gases to produce a range of cosmogenic radioisotopes that reach the surface as dry fallout or with precipitation as wet fallout (Figure 3.10). The cosmogenic neutron flux can also activate nuclides such as ^{35}Cl on the earth's surface, which is *epigenic* production. These atmospheric or epigenic radioisotopes are then incorporated into groundwaters during recharge, where subsequent decay from their initial value can be used to give age. Tritium and radiocarbon are the dominant atmospheric radioisotopes

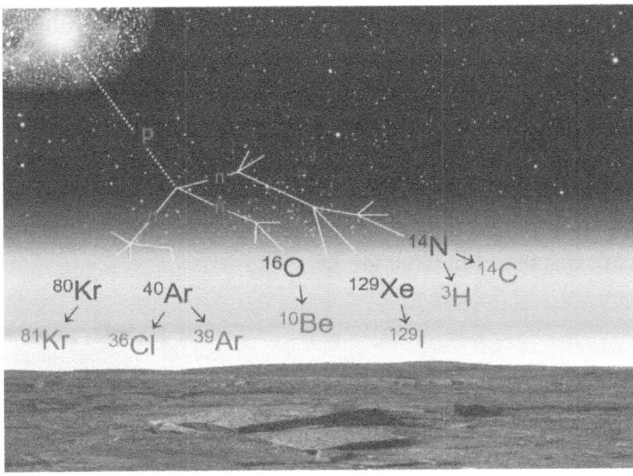

FIGURE 3.10 Production of cosmogenic radionuclides in the atmosphere from the shower of secondary neutrons produced when high-energy particles (mainly protons) originating from supernovae in the galaxy impact gas molecules in the atmosphere.

used in groundwater studies. Both ^{36}Cl and ^{129}I are important in studies of older groundwaters and brines. Noble gas radioisotopes, including ^{39}Ar, ^{85}Kr, and ^{81}Kr, while more challenging to sample, have half-lives appropriate for age ranges not available using T and ^{14}C.

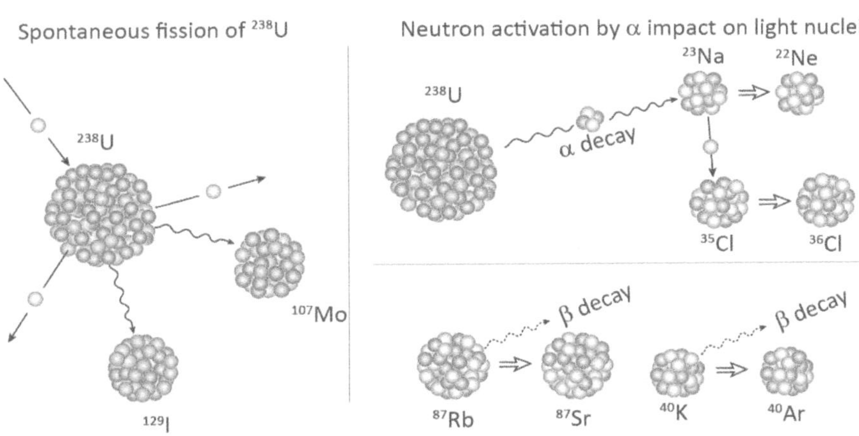

FIGURE 3.11 Radiogenic isotope production in the subsurface through spontaneous fission of ^{238}U, neutron activation following α decay, and radioactive decay of long-lived radioisotopes.

Geogenic nuclides are those produced in the subsurface through the decay and spontaneous fission of uranium and other actinides present in all rocks. The range of radiogenic nuclides produced in the subsurface, which are incorporated by groundwater and can be used for chronology, was established by Andrews et al. (1989). They are generated by three principal mechanisms, which are spontaneous fission of ^{238}U, neutron activation following α-decay, and radioactive decay (Figure 3.11). These geogenic isotope production mechanisms and their application for dating groundwaters are discussed in Chapter 7.

4 Tracing Groundwater

THE GLOBAL WATER CYCLE

Groundwater comprises the largest freshwater reservoir of the global water cycle. This cycle pumps over 300 km^3 of freshwater each day to the continents, the equivalent of 600 Niagara Falls. Of this, over 60% is transpired back to the troposphere during photosynthesis, mainly in the boreal and tropical forests. Most of the balance runs off over the surface or infiltrates into soils and aquifers, ultimately discharging some 115 km^3 of freshwater to the oceans each day (Figure 4.1). Tracing the groundwater component of the water cycle relies on environmental isotopes. The natural isotopes ^{18}O and D are an intrinsic component of the water molecule and so are ideal tracers of groundwater. They fractionate and partition during evaporation and condensation, imparting information on precipitation origin, recharge processes, mixing, and even on extreme situations like geothermal waters, brines, and highly contaminated groundwaters.

In 1961, Harmon Craig published the earliest measurements of $\delta^{18}O$ and δD for freshwaters from around the globe and showed that these two isotopes of water were highly correlated and that they were enriched in warm regions while being depleted in cold regions (Figure 4.2). In 1964, Willi Dansgaard quantified this apparent correlation with mean annual air temperature (Figure 4.3). These observations provide the foundation of isotope hydrology, allowing the tracing of meteorological processes and groundwater recharge.

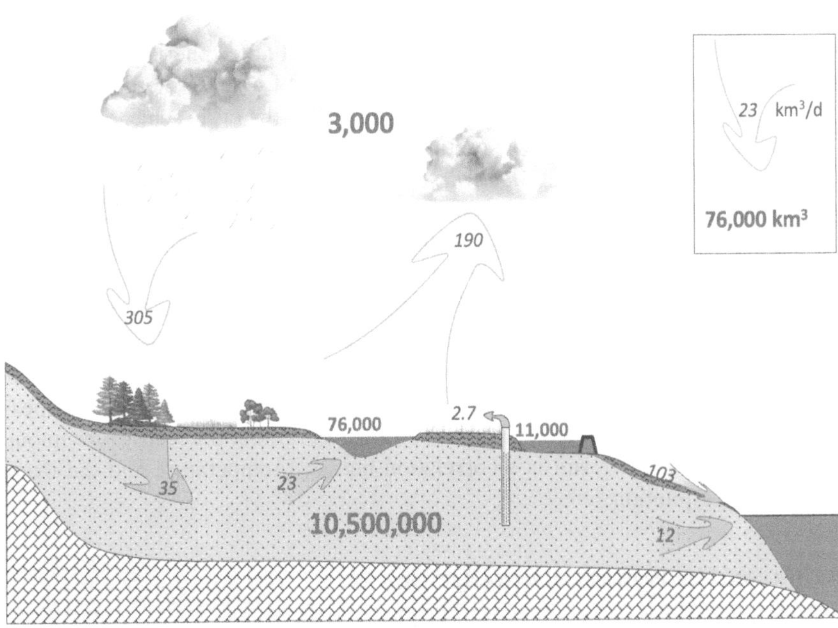

FIGURE 4.1 Components and fluxes in the global water cycle.

DOI: 10.1201/9781003353836-4

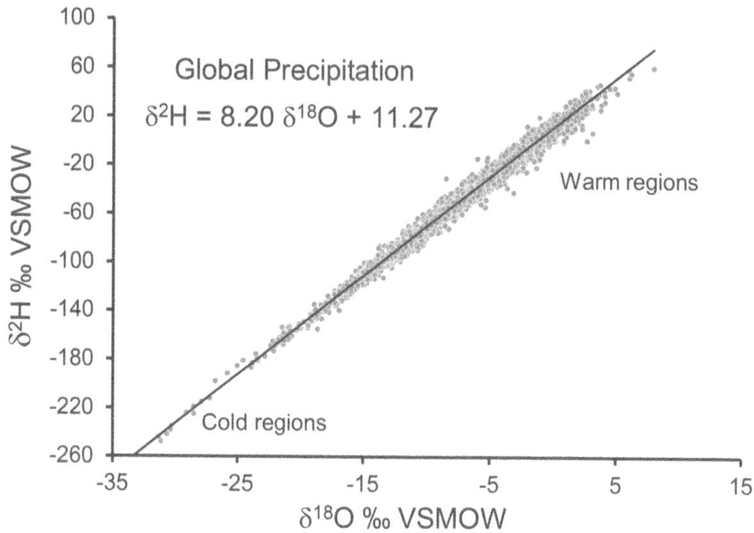

FIGURE 4.2 The $\delta^{18}O$–δD correlation for global precipitation plotted from data on the International Atomic Energy Agency Global Network for Isotopes in Precipitation (GNIP) database (Rozanski et al. 1993; IAEA/WMO 2025). Craig (1961) first observed this correlation in global freshwaters.

ISOTOPES IN PRECIPITATION

The meteorological relationship between the isotopic composition of precipitation and air temperature can be understood by examining the processes that drive condensation and precipitation. It must be understood that atmospheric processes leading to precipitation are highly chaotic and complex, and so any isotope–temperature correlation breaks down over short time periods and at the event scale. However, the δ–T correlation emerges from precipitation data averaged on a monthly to annual basis. This is appropriate for groundwater studies given that dispersion during infiltration attenuates event-scale recharge towards annual averages.

Despite the complexities during the production of vapor in the troposphere and subsequent precipitation, a driving force contributing to the δ–T correlation in precipitation is that of Rayleigh distillation of ^{18}O and D during rainout. From Chapter 3, condensation is accompanied by isotope fractionation, which for $\varepsilon^{18}O_{water-vapor}$ is 9.3‰ and for $\varepsilon D_{water-vapor}$ is 76‰ at 25°C. Accordingly, both ^{18}O and D are preferentially distilled from the vapor mass, with an evolution toward lighter isotopes in the residual vapor, as well as in successive rains, as rainout proceeds. As rainout is driven by cooling, the result is a correlation between temperature and the isotope content of precipitation.

Partitioning of Isotopes during Rainout

Atmospheric water vapor is largely produced by primary evaporation of ocean water, with contributions from evaporation from surface waters and transpiration from vegetation along the air mass trajectory. Precipitation is then generated as the vapor

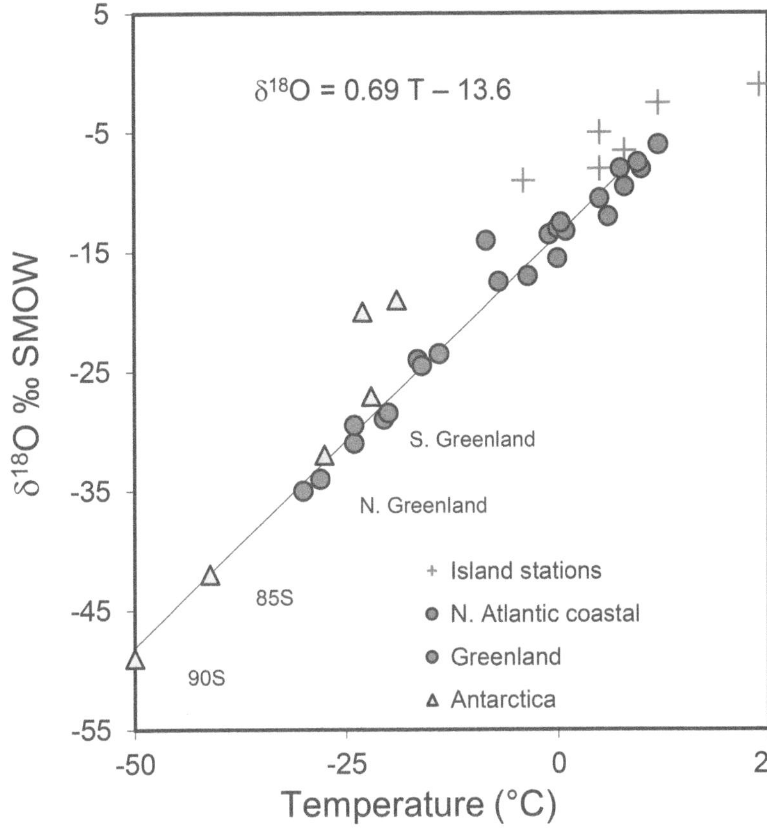

FIGURE 4.3 The global T–$\delta^{18}O$ relationship for precipitation. Temperature is the mean annual surface air temperature (MAAT) at the station. For deuterium, Dansgaard found $\delta D = 5.6\ T_{MAAT} - 100‰$ SMOW (standard mean ocean water). Note that SMOW refers to the laboratory reference water used at that time. It has since run out and been replaced by Vienna Standard Mean Ocean Water (VSMOW). Modified from Dansgaard, W. 1964.

mass cools through adiabatic expansion in convective precipitation systems, during orographic rainfall, or displacement over cold fronts. This is "rainout" and is driven by decreasing temperature because a cooler air mass can hold less water vapor.

Equilibrium fractionation between vapor and water or ice preferentially partitions ^{18}O and D into the rain or snow, and so they are distilled from the vapor phase during rainout. The vapor then becomes progressively depleted in ^{18}O and D during rainout along the vapor mass trajectory (Figure 4.4). As the residual vapor becomes depleted in ^{18}O and D, subsequent rains will also be progressively depleted, and so rainout evolves toward colder, isotopically depleted precipitation. This can be simulated as a Rayleigh distillation process (see Chapter 3), as shown in Figure 4.4.

In reality, weather seldom follows the ideal evolution presented in Figure 4.4. Most systems experience additions to their vapor content through evaporation from lakes and rivers and transpiration from vegetation over the continents. Mixing with other vapor masses in the troposphere along a rainout trajectory also occurs. Aggarwal

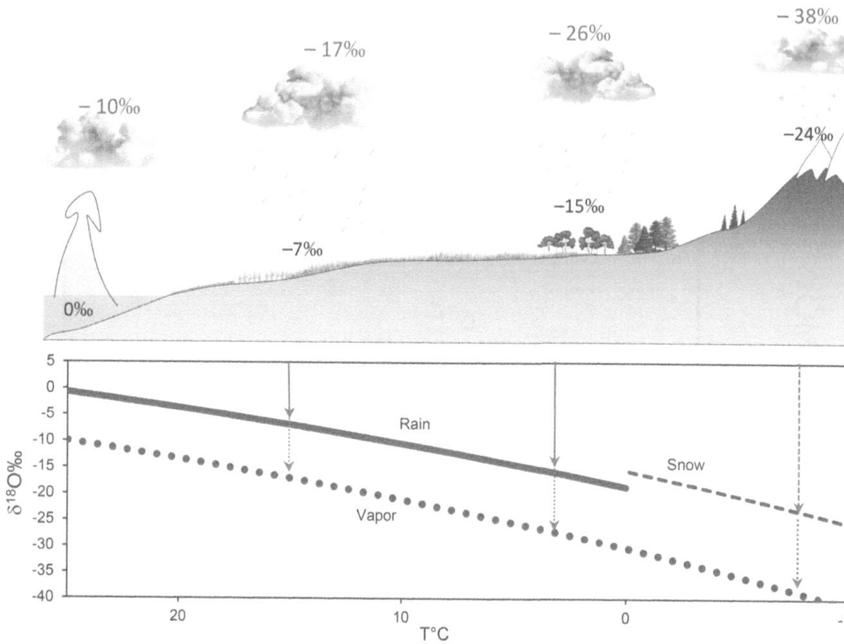

FIGURE 4.4 The evolution in the isotope content of precipitation according to an ideal Rayleigh distillation during rainout. At regional to continental scales, additions of water vapor from freshwater sources and transpiration are gained along a storm trajectory.

et al. (2012) show that the residence time increases mixing with continental sources of water vapor in the upper troposphere. Further, the predominance of stratiform over convective precipitation can play a role in isotope depletion (Aggarwal et al. 2016). For these reasons, the correlation between surface temperature and $\delta^{18}O$ (and δD) breaks down at the scale of individual precipitation events and is only well correlated for mean monthly and mean annual measurements. However, this correlation on longer time scales suits groundwater provenance studies because groundwater comprises mixtures of precipitation over the scale of years to decades and longer due to dispersion, which attenuates any short-term or event-scale signals.

Isotopes in Global and Regional Precipitation

The partitioning of isotopes between cold and warm regions is clearly observed on a global map of $\delta^{18}O$ in precipitation in Figure 4.5. On this global scale, the temperature–$\delta^{18}O$ relationship is clear, with partitioning of ^{18}O into warmer, low-latitude precipitation and depletion in ^{18}O with increasing latitude. Regional features such as the North Atlantic Drift transporting warm ocean waters to high latitudes over the British Isles and Scandinavia or the Tibetan Plateau are also apparent.

The proximity to marine waters exerts a control on the isotope composition of precipitation as vapor masses move inland. Continentality is expressed by Conrad's index, which is a function of the amplitude of the seasonal temperature range. Continentality

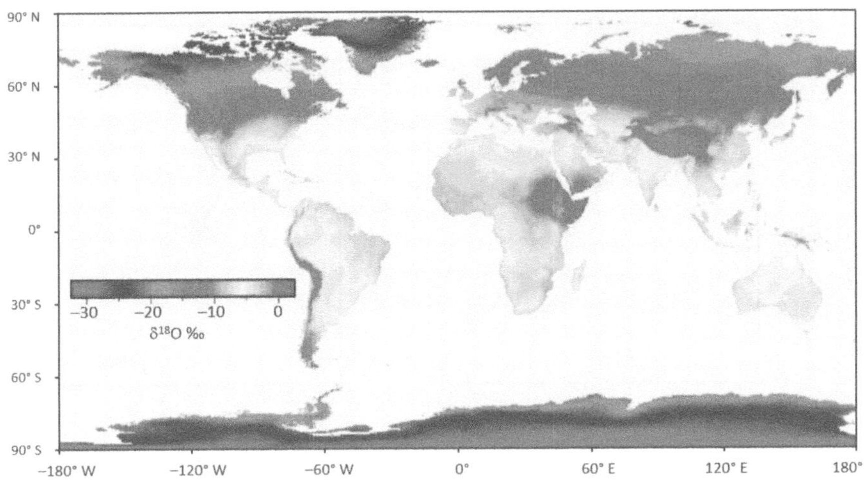

FIGURE 4.5 Global map of $\delta^{18}O$ in precipitation from Bowen and Wilkinson (2002), based on annual averaged data from the Global Network for Isotopes in Precipitation (GNIP) (IAEA/WMO 1998; Rozanski et al. 1993) and interpolated using an altitude and latitude model incorporating vapor-transport effects and model residuals.

increases with distance from the coast and with latitude. Stations with increased continentality have lower $\delta^{18}O$ and δD values for mean annual precipitation.

The continental effect can be observed in Figure 4.5 as inland gradients for $\delta^{18}O$, for example, northwestward from the Gulf of Mexico through the southern United States into northwestern Canada, or eastward through Europe from North Atlantic coastal stations. It is often manifested in the isotope composition of rivers draining from continental-scale basins, which will have a depleted signature relative to local precipitation in the downstream regions.

Altitude Effect in Precipitation

The effect of altitude can also be observed in Figure 4.5, with isotope depletions observed over the Rocky Mountains in western North America and the Andes in South America. This is attributed to the decrease in temperature with elevation, which drives rainout and distillation of isotopes during orographic precipitation. This is also observed at the catchment scale and is an important tool to identify the recharge elevation of groundwater. With this effect, the $\delta^{18}O$ of precipitation typically decreases by 0.15‰ to 0.5‰ per 100-m rise in altitude, and δD decreases by between 1‰ and 4‰.

Precipitation for catchment in the maritime piedmont of the Italian Alps was sampled at elevations from 500 m to over 2000 m, providing the good correlation observed in Figure 4.6. In this case, the altitude effect was –0.31‰ for $\delta^{18}O$ per 100 m rise. For Mont Blanc (Moser and Stichler 1970), the gradient was a much steeper –0.5‰ for $\delta^{18}O$. Warmer regions tend to have lower temperature gradients, and so the altitude effect is generally less than –0.2‰ for $\delta^{18}O$ per 100 m rise.

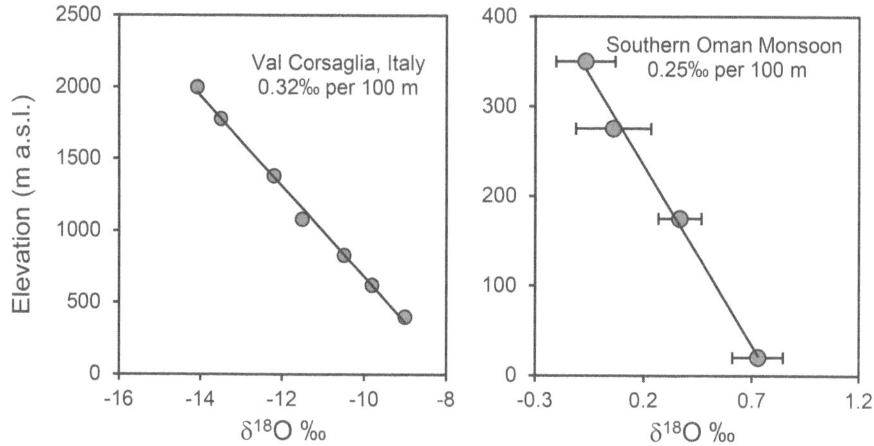

FIGURE 4.6 The altitude effect for $\delta^{18}O$ in precipitation monitoring at three sites: Val Corsaglia, a high-altitude site in the Italian Alps sampled during October 1974 (Bortolami 1978) and the summer monsoon in the Dhofar Mountains of Southern Oman monitored weekly over the 1985 monsoon period (Clark et al. 1987). The Oman site demonstrates that the depletion trend with elevation is apparent even for low-relief catchments.

Lower-elevation catchments with relief contrasts as low as a few hundred meters also demonstrate an altitude effect for ^{18}O and D, based on long-term weighted averages of precipitation. This is worthwhile for monitoring a study site, given that groundwater recharge itself is often a long-term weighted average of precipitation in the recharge area. Accordingly, recharge elevations can be distinguished even in low-relief catchments.

A good example of the altitude effect used to distinguish the origin of groundwater recharge is a case study in northern Chile where alluvial sand and gravel aquifers are bounded inland by the Coastal Mountains up to 2000 m above sea level, rising to over 4000 m in the Andes to the east. Squeo et al. (2006) collected local rainfall and fog precipitation in an isotope study to determine recharge of groundwater in the coastal aquifer system (Figure 4.7).

Seasonal Effects in Precipitation

For temperate and continental regions, a principal control on the isotopic composition of precipitation is the seasonal change in temperature and consequently in the $\delta^{18}O$ and δD of precipitation. These seasonal variations in $\delta^{18}O$ and δD give us an important tool to determine watershed response to precipitation, mean residence time of water in drainage basins, and to monitor groundwater recharge.

Figure 4.8 shows the seasonal variation for $\delta^{18}O$ for stations from a variety of latitudes. The greatest seasonal change is observed for sites with high continentality and high latitude (e.g., Edmonton and Thule). The marine station (St. Helena) shows the least amount of seasonal variation.

FIGURE 4.7 The altitude effect in precipitation recharging coastal groundwaters, showing recharge to the Elqui well from precipitation in the Andes, while the Romeral well is recharged by lower elevation precipitation in the Coastal Mountains, and the Elvira groundwaters are recharged by a mixture of these two. Local rain and fog have little apparent contribution to groundwater recharge.

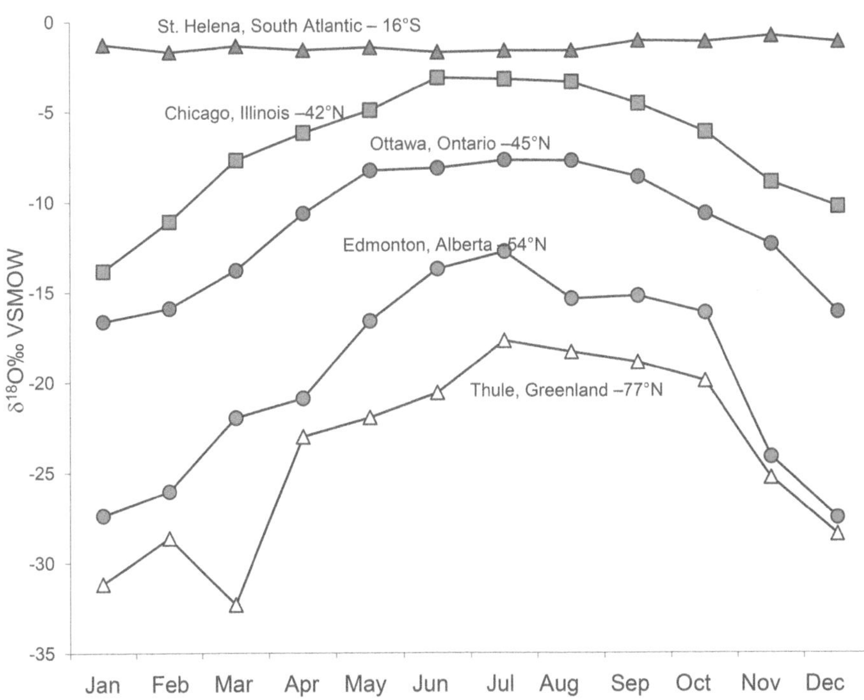

FIGURE 4.8 The seasonal variation in $\delta^{18}O$ in precipitation at stations from low to high latitude in the northern and southern hemispheres. Data are for monthly averages. Note the lack of seasonal effect for the oceanic station—St. Helena, an atoll in the South Atlantic Ocean. Strong seasonality imparts a high amplitude variation for ^{18}O in precipitation at the inland stations—Chicago, Ottawa and Edmonton and the high latitude station—Thule, Greenland.

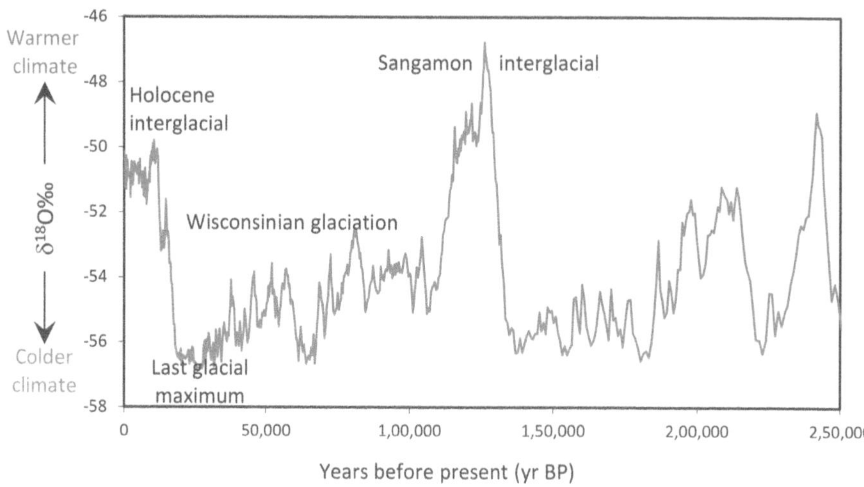

FIGURE 4.9 Change in climate temperature as recorded by the $\delta^{18}O$ in glacier ice from Vostok Station, Antarctica (data from EPICA 2004). Periods characterized by depleted values for $\delta^{18}O$ correspond with glaciations. Peaks of enriched ^{18}O values document the warmer climates of the interglacial periods.

Paleoclimates

Climate change is recorded by the stable isotope contents of precipitation. Measurements in precipitation over the past 30 years show strong T–$\delta^{18}O$ correlations but with amplitudes on the order of only 1‰ to 2‰ and 1°C to 2°C (Rozanski et al. 1993). Longer precipitation records are archived in glacier ice cores. The $\delta^{18}O$ and δD variations in these cores provide a remarkable record of temperature change that now extends back over the past glacial–interglacial cycles, as shown in Figure 4.9.

The Meteoric Water Line

The second observation by Craig (1961) is the strong correlation between ^{18}O and D in global freshwaters, following a regression line with slope $s = 8$ and deuterium intercept of 10. Subsequent analysis of global precipitation samples by Rozanski and others (1993) provides a more precise definition of the *global meteoric water line* (GMWL) (Figure 4.2.):

$$\delta D = 8.2 \times \delta^{18}O + 11.3‰$$

Craig's global line remains in common use as a baseline for isotope hydrology studies, but the Rozanski line is the preferred GMWL given that it is derived from precipitation and so excludes the post-precipitation processes affecting surface waters and groundwater recharge. Further, as a global line, the GMWL is an average of local meteoric water lines (LMWL), which individually incorporate regional and seasonal processes in a given area that affect both the slope and the intercept of the

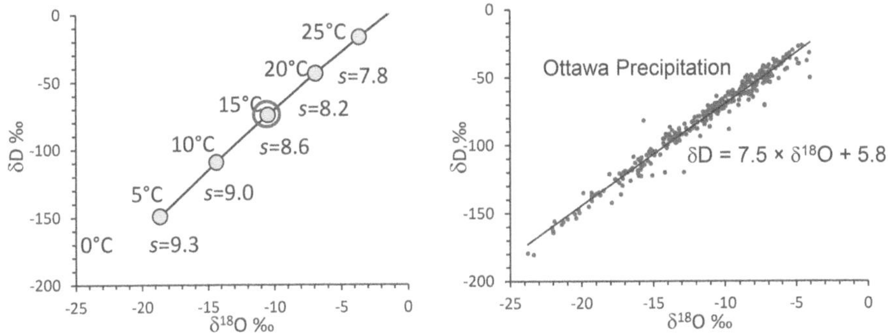

FIGURE 4.10 Left: Theoretical meteoric water line for $\delta^{18}O$ and δD in precipitation during rainout calculated at different temperature steps using an example initial water vapor with $\delta^{18}O = -13‰$ and $\delta D = -100‰$ (open circle) and equilibrium fractionation factors for successive steps as the air mass cools, producing rain starting at 25°C. Right: Actual meteoric water line for Ottawa precipitation.

meteoric correlation for $\delta^{18}O$ and δD. For localized groundwater and surface water studies, it is important to establish LMWLs to observe post-precipitation processes.

Slope of the GMWL

The process of condensation and precipitation is recognized in most cases as an equilibrium reaction and is accompanied by equilibrium fractionation for both isotopes. The slope reflects the greater fractionation for D, which is about 8 times greater than that for ^{18}O at 20°C.

$$s = \frac{\varepsilon D_{w-v}}{\varepsilon^{18}O_{w-v}} \approx 8-9$$

This ratio increases to over 9 at lower temperatures, as shown in Figure 4.10 (left plot), using equilibrium enrichment factors for the range of temperatures through which precipitation forms.

Local Meteoric Water Lines

Contributions to vapor masses from terrestrial evaporation and transpiration along a trajectory, mixing of air masses, and differing mechanisms of precipitation all affect the meteoric relationship of isotopes in precipitation, such that any given region will have a characteristic LMWL. As an example, the LMWL for Ottawa, Canada, has a slope of only 7.5 and an intercept of 5.8 (Figure 4.10, right plot). The GMWL is then an integration of many LMWLs.

The slope of the LMWL can also be affected by evaporation that occurs after condensation. If rain is falling through a dry air column below the cloud base, partial evaporation will impart a selective enrichment in ^{18}O and D. Dansgaard (1964) described such evaporation of rain as the *amount effect* on the isotopic composition of precipitation. For this reason, a LMWL is calculated using isotope measurements for each precipitation event (δ_i) in a month (typically) and the event amount in mm (mm_i),

$$\delta_{\text{mean monthly}} = \frac{\sum_{i=0}^{n} \delta_i \times mm_i}{\sum_{i=0}^{n} mm_i}$$

Amount-weighting as well as excluding the light rain events, which contribute negligibly to groundwater recharge, provides a line with greater slope that is more representative of the precipitation involved in groundwater recharge (Hughes and Crawford 2012).

The second parameter defining an LMWL is the deuterium intercept, which is +11.3‰ for the GMWL. However, this intercept can vary regionally from much less than 10‰ to over 20‰. This excess is due to kinetic evaporation (nonequilibrium) during the formation of the primary vapor mass, and it provides an additional tool to trace the origin of water. Such nonequilibrium or *kinetic* evaporation becomes increasingly kinetic with lower humidity and enhances fractionation of ^{16}O into the vapor, but less so for H. Vapor produced under lower humidity conditions is more depleted in ^{18}O than for equilibrium conditions. As this vapor condenses, it does so with equilibrium fractionation for ^{18}O and D and so produces rain that plots along a LMWL with a slope near 8 but along a line with a greater deuterium intercept. On average, global freshwaters are formed under conditions of about 85% humidity, and so global freshwaters plotting on the GMWL have a deuterium excess of only 11.3‰ (Figure 4.11). An example where primary evaporation takes place with very low humidity is the eastern Mediterranean, where the deuterium intercept is 22‰ (Gat and Carmi 1970; Figure 4.11).

The UN International Atomic Energy Agency (IAEA) in Vienna maintains the GNIP with stations located all around the globe. These data include mean monthly values for $\delta^{18}O$ and δD together with other select isotopes and chemistry, plus precipitation amounts, temperature, and other meteorological data. They are freely available for download from their GNIP Water Isotope System for Data Analysis, Visualization and Electronic Retrieval (WISER) database. This provides local or regional precipitation data to produce LMWLs for groundwater studies. In any study of isotope hydrology, it is useful to use precipitation data that best represent the study area. In the absence of regional or local precipitation data, the GMWL is often substituted. However, this can bring uncertainty, particularly in arid and high-latitude regions.

TRACING GROUNDWATER RECHARGE

The translation of the isotope signal in precipitation into groundwater is an important step in the use of stable isotopes as tracers of groundwater origin. Figure 4.11 shows that the range in stable isotopes of precipitation in given regions varies considerably.

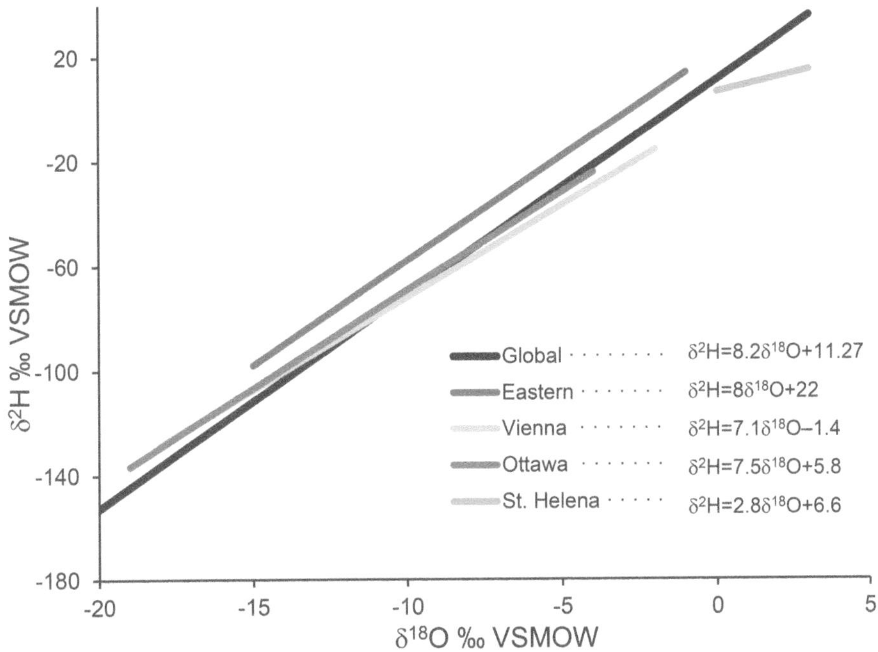

FIGURE 4.11 Local meteoric water lines (LMWL) for selected stations in the International Atomic Energy Agency Global Network for Isotopes in Precipitation (GNIP IAEA/WMO 2013).

How this signal is translated into the isotopic composition of groundwater is central to using these tracers, and several processes are important:

1. Post-precipitation evaporation.
2. Seasonality of recharge.
3. Subsurface mixing.

Although the various temperature-dependent processes of rainout distribute isotopes in precipitation over a wide range, this range is highly attenuated as precipitation enters the subsurface. The seasonality of precipitation plays a role in this, where groundwater may be recharged only at certain times of the year. Hydrodynamic dispersion also plays a role. The effect is such that while the $\delta^{18}O$ content of precipitation may vary over 8‰, in groundwater the variation will usually be less than 1‰.

Evaporation

The $\delta^{18}O$–δD signature of precipitation can be modified in the recharge environment by evaporation prior to infiltration, which imparts a characteristic isotopic enrichment on the residual water. Rivers and lakes also experience evaporative enrichment, which can be transferred to groundwaters if they are a recharge source. Evaporation from the unsaturated zone and ultimately from the water table can also occur. Note, however, that unlike evaporation, the loss of water by transpiration does not fractionate ^{18}O and

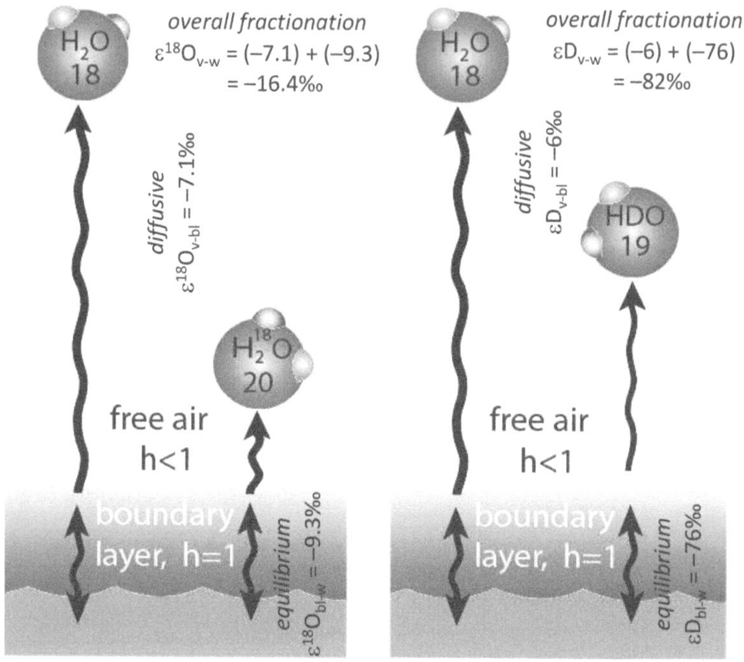

FIGURE 4.12 Kinetic fractionation during evaporation due to enhanced diffusion of H_2O (mass 18) relative to HDO (mass 19) and $H_2{}^{18}O$ (mass 20) from the boundary layer to open air. Kinetic (diffusive) enrichment factors are a function of free air humidity, *h*, determined here for $h = 0.5$ using the equations below from Gonfiantini (1986).

D. The isotope effect of evaporation is the preferential loss of ^{16}O and ^{1}H in the vapor fraction (recall that $\varepsilon^{18}O_{vapor\text{-}water}$ is about −10‰). However, evaporation under conditions of lower humidity, nonequilibrium, imparts an additional *kinetic* fractionation.

The model for evaporative fractionation of water is a two-step process, involving (i) equilibrium fractionation between the water surface and a thin boundary layer, bl, where humidity is 100%, and (ii) diffusive fractionation between the boundary layer and the free air reservoir where humidity is less than 100% (Figure 4.12). The diffusive or *kinetic* fractionation is due to the differences in the velocities of the three isotope species: $H_2{}^{16}O$ (mass 18), $HD^{16}O$ (mass 19), and $H_2{}^{18}O$ (mass 20), as they diffuse through the air column from the boundary layer, bl, of water-saturated air overlying the evaporating water (Figure 4.12). The kinetic enrichment factors are represented in the format of reaction product-reactant, using Δ to designate the kinetic component.

Gonfiantini (1986) determined empirical relationships for the kinetic enrichment factor, $\Delta\varepsilon_{vapor-bl}$, based on humidity:

$$\Delta\varepsilon^{18}O_{vapor\text{-}bl} = -14.2 \times (1 - h)$$

$$\Delta\varepsilon D_{vapor\text{-}bl} = -12.5 \times (1 - h)$$

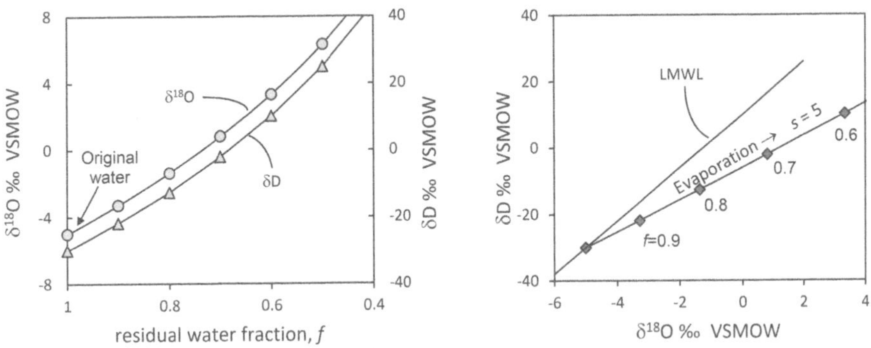

FIGURE 4.13 The isotopic effect of evaporation on water. (a) The diagram shows the Rayleigh enrichment of ^{18}O and D as the residual fraction of water *f* decreases for humidity of 50% and at 25°C. (b) The combined evaporation effect for $\delta^{18}O$ and δ D as a deviation from the local meteoric water line (LMWL).

These kinetic enrichment factors add to the equilibrium enrichment factors to give overall enrichment:

$$\delta^{18}O_{vapor} = \delta^{18}O_{water} + \varepsilon^{18}O_{bl\text{-}water} + \Delta\varepsilon^{18}O_{vapor\text{-}bl}$$

$$\text{and } \delta D_{vapor} = \delta D_{water} + \varepsilon D_{bl\text{-}water} + \Delta\varepsilon D_{vapor\text{-}bl}$$

Significant evaporation of surface waters imparts an isotope enrichment on the residual water, which follows a Rayleigh distillation, as shown in Figure 4.13 (left chart). The ratio of the overall enrichment factors during evaporation, $\varepsilon^{18}O = -16.4‰$ and $\varepsilon D = -82‰$, given in Figure 4.12 for $h = 50\%$, is closer to 5 and much lower than 8 for equilibrium fractionation (left chart, Figure 4.10),

$$s = \frac{\varepsilon D_{bl-w} + \varepsilon D_{v-bl}}{\varepsilon^{18}O_{bl-w} + \varepsilon^{18}O_{v-bl}} = \frac{-82}{-16.4} = 5$$

resulting in a positive deviation from the meteoric water line (MWL) along an evaporation line with this lower slope (Figure 4.13, right chart). Evaporation trends have a slope typically between 4 and 7, depending on the humidity. The influence of humidity h is such that under drier conditions, the slope of the evaporation trend is lower. This can be seen from the ratios of the combined $D/^{18}O$ enrichment factors calculated for different values of humidity from the equations above. The greater the shift away from the MWL along an evaporation trend indicates a greater degree of evaporation (Figure 4.13, right chart). This characteristic evaporative shift in surface waters is preserved during groundwater infiltration, thus providing groundwater with an indication of evaporation prior to their recharge.

Groundwater can also undergo evaporation of soil moisture from the unsaturated zone. In this case, the evaporation effect is stronger due to a longer diffusion distance through the upper dry soil column and results in an isotope shift along a line

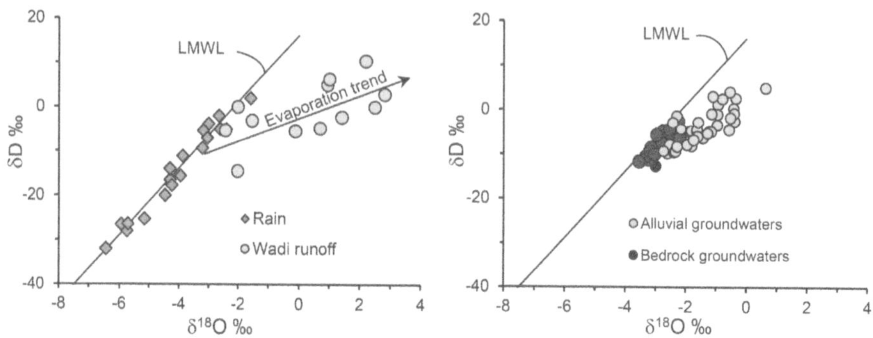

FIGURE 4.14 Evaporation effect observed in surface flow (drainage channels or wadis in arid areas) (a) following precipitation and (b) groundwater for a catchment in the arid country of Oman. Both ^{18}O and D become enriched in water by evaporation during overland flow in the wadi channels. This evaporative signature is retained during infiltration into the alluvial aquifers along the wadi channels. Bedrock aquifers are recharged by direct infiltration during major storm events, with minimal evaporation during recharge.

with a much lower slope than for surface water evaporation, with s often close to 2 (Allison 1982). Evaporation typically occurs during overland flow in arid landscapes, as observed for groundwaters in such regions, in lakes and reservoirs, and from bare soils during infiltration.

Deuterium Excess

The GMWL provides a reference for the precipitation input to freshwaters. Deviations from this reference line by evaporation and other processes are quantified by a second parameter, deuterium excess, or d. Dansgaard (1964) first proposed this value, calculated for any individual water sample as,

$$d = \delta D - 8 \times \delta^{18}O$$

which represents the offset for δD from Craig's global line (Figure 4.15). For any sample plotting on the GMWL this would be, the deuterium intercept 10‰. Surface and groundwaters having experienced evaporation would have values less than 10‰, thus providing a quantification of this deviation. More correctly, this should use the GMWL for precipitation, with $s = 8.2$; however, the Craig line is used consistently in the literature.

The limitation of d is that it normalizes to the global line without considering local meteoric waters in a given study. Landwehr and Coplen (2004) proposed a modification that allows this important parameter to be relevant to local precipitation, using what they call the *line conditioned excess*, d_{Lc}. This uses the slope, s_L, and the deuterium intercept, b_L, of the LMWL,

$$d_{Lc} = \delta D - s_L \times \delta^{18}O - b_L$$

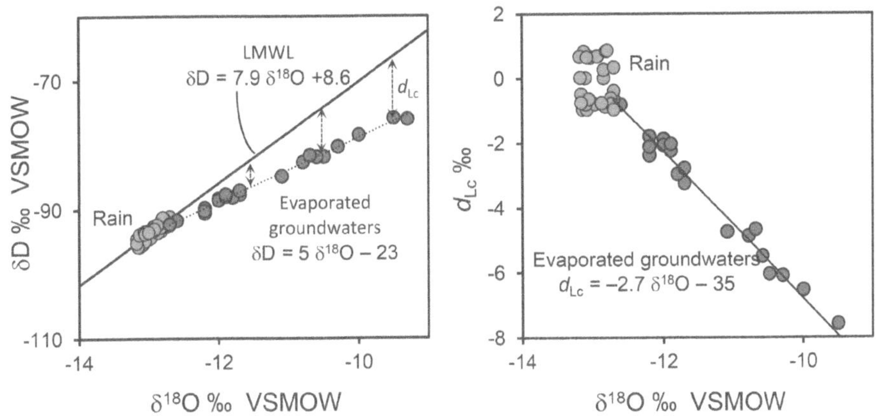

FIGURE 4.15 Deuterium excess for rain (blue) and evaporated waters, normalized to LMWL (d_{Lc}), shown graphically on the left chart and plotted with $\delta^{18}O$ on the right chart.

This allows a more relevant determination of the deuterium offset for a given water sample from local precipitation.

The use of deuterium excess in this manner provides a value that can be used to quantify post-precipitation evaporation that occurs between runoff and recharge. It can also be used to quantify other processes that modify the isotope signature of groundwater, such as exchange with minerals in geothermal settings or with gases in the subsurface, as discussed later.

Seasonality of Recharge

The isotopic composition of groundwater can be traced to that of precipitation in the recharge area. In tropical and arid regions with strong seasonal bias in the amount of precipitation, the weighted mean annual values for $\delta^{18}O$ and δD in precipitation typically represent the isotopic signature of groundwater (Figure 4.16). Using the mean monthly value to represent groundwater would overly bias toward the isotopically enriched dry months, whereas using the weighted mean of monthly precipitation gives a value that is more representative of groundwater.

In temperate climates, the combined effects of winter snow accumulation and transpiration during the summer months restrict groundwater recharge to the spring and fall periods, despite the lack of seasonal bias in amount (Figure 4.16). Although the $\delta^{18}O$ and δD in precipitation during these transition seasons approximate that of mean annual precipitation, there can be deviations of a few permil. In these climates, the weighted mean annual isotopic composition of precipitation is often enriched over local groundwater.

In arid regions, groundwater recharge is largely biased toward the large precipitation events (Figure 4.16). As light rains contribute little to the weighted mean of precipitation, the isotopic composition of groundwater is well represented by the weighted mean annual or long-term weighted value for precipitation.

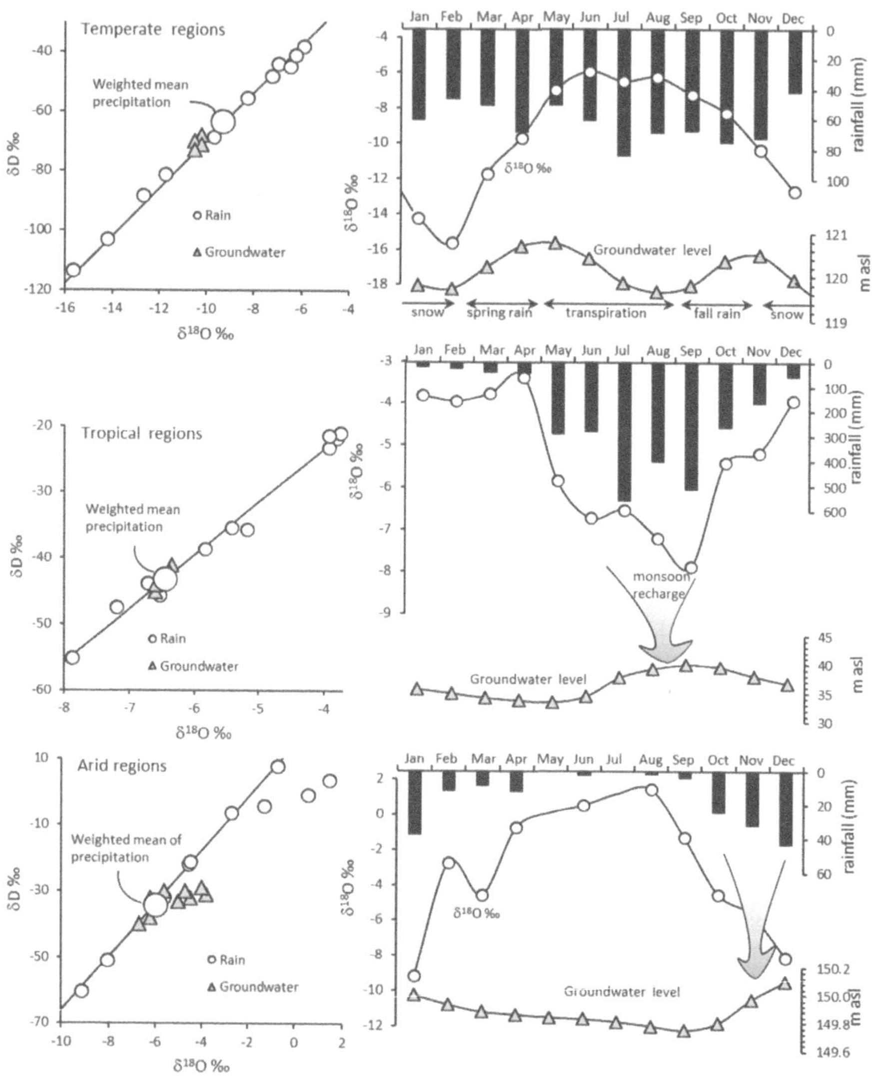

FIGURE 4.16 The seasonal timing of precipitation and transfer of its isotope signal into groundwater, shown for characteristic sites in tropical (Manila), temperate (Ottawa), and arid (Sana'a) regions. Isotopes in precipitation are weighted mean monthly values (IAEA/WMO 2013). Groundwater levels signify periods of seasonal recharge.

GROUNDWATER MIXING

Following recharge to the subsurface, the isotope composition of groundwater is preserved. Only under arid conditions is evaporation significant from the unsaturated zone or water table. Subsequent changes to the $\delta^{18}O$-δD signature can occur only through mixing, or in very exceptional circumstances, by exchange reactions with minerals and gases. Mixing is a dominant process in the recharge environment, where individual recharge events contribute proportionally to the isotope signature of groundwater. This occurs in both the unsaturated zone and below the water table. The discharge area is also a zone of mixing, where convergence of flow paths occurs, such as beneath a river or along lake shorelines. Mixing also occurs in wells, where different aquifers may contribute to discharge during pumping.

Dispersive Mixing during Recharge

Groundwater flow through a porous medium or fracture network is mixed by hydrodynamic dispersion due to the differences in porewater velocities and pathways within connected porosity. This is characterized for a porous medium as its dispersivity, α, in units of length (m). The coefficient of hydrodynamic dispersion, D, is a function of the medium's dispersivity and its average linear groundwater velocity, v, plus molecular or effective diffusion, D_e:

$$D = \alpha v + D_e$$

Molecular diffusion only becomes relevant at very low groundwater velocities and is therefore an important mechanism for groundwater mixing in aquitards, as discussed below.

In recharge environments, hydrodynamic dispersion attenuates seasonal and event-scale variations in isotopes and geochemistry. The shallow groundwater in most phreatic aquifers has an age distribution reflecting contributions from several months to several years of precipitation, increasing to years and decades in confined systems. The effect is attenuation or damping of the event-scale to seasonal-scale variability of isotopes in precipitation. Long-term monitoring of $\delta^{18}O$ or δD in groundwater at increasing depths and distances along the flow system shows the convergence on a long-term mean value. This temporal variation in $\delta^{18}O$ is shown schematically in Figure 4.17 for monitoring in a phreatic aquifer and in a deeper confined bedrock aquifer.

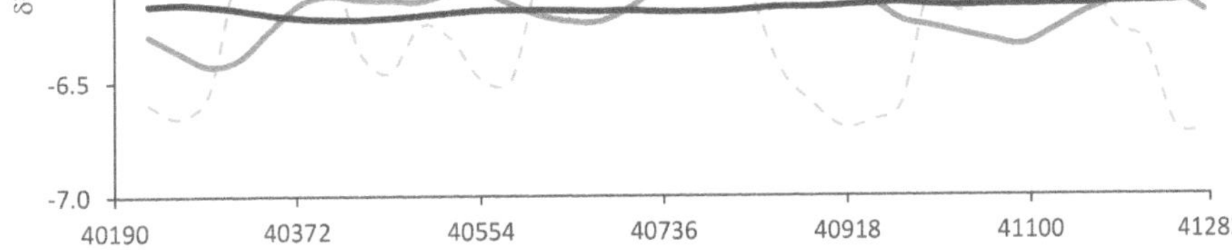

FIGURE 4.17 Time-series monitoring of $\delta^{18}O$ showing attenuation of seasonal trends in precipitation during recharge through overburden into a deeper bedrock aquifer in Eastern Ontario (data from Velderman 1993). Seasonal and event-scale recharge is moderately damped in the phreatic aquifer and fully damped in the confined bedrock aquifer.

Temporal-scale monitoring by repeat sampling at selected sites provides insights on the sensitivity of the groundwater system to recharge frequency and drought. Although considerable damping of event-scale to seasonal isotope variability in precipitation occurs in the unsaturated zone, groundwater near the water table may still fluctuate within 1 to 2 permil for $\delta^{18}O$. This is particularly so where macroporosity from root zones and wormholes can short-circuit recharge to the water table and highlights the uncertainty of characterizing groundwater with one sample on one sampling campaign. It also provides indications of contaminant potential from surface activities. Groundwaters with little to no variability are typically considered to be part of larger, deeper flow systems that are more resilient to temporal variations in recharge than shallow systems showing greater correlation with the temporal signal in precipitation.

Two-component Mixing

Mixing of waters from different origins can occur in wells producing water from multiple zones or where flow from different aquifer zones converges at a discharge point, that is, a spring or lake margin. Resolving mixed waters into their proportions of composite end-members follows simple mass balance equations, where the sum of the end-member contributions, expressed as fractions, f, is equal to 1.

$$\text{Two end-member mixing:} \quad f_A + f_B = 1$$

$$\text{Three end-member mixing:} \quad f_A + f_B + f_C = 1$$

Solving these mass balance equations with two or three unknowns requires an additional equation or two, using either one or two conservative (non-reactive) solutes or isotopes. Consider the case of water from a geothermal well in a hot springs area with springs of different temperatures and local cold groundwater, as shown in Figure 4.18. Two component mixing between cold shallow groundwater and the

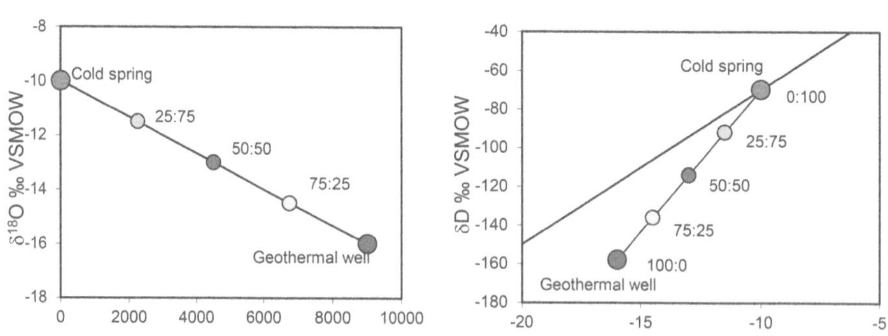

FIGURE 4.18 Two-component mixing trends for conservative species. Cl^-, $\delta^{18}O$ and δD are independent parameters, and mixing follows a linear relationship. Mixing proportions can be determined directly from either axis.

rising geothermal fluid can be calculated from the concentrations of either ^{18}O or Cl^- as two unknown values and using these two mixing equations.

$$\text{hot spring water } f_{hs} = 1 = f_{geo} + f_{cold}$$

$$\delta^{18}O_{hs} = f_{geo}\delta^{18}O_{geo} + f_{cold}\delta^{18}O_{cold}$$

Substituting and rearranging

$$f_{geo} = 1 - f_{cold}$$

$$\delta^{18}O_{hs} = f_{geo}\delta^{18}O_{geo} + f_{cold}\delta^{18}O_{cold}$$

$$\delta^{18}O_{hs} = (1 - f_{cold})\delta^{18}O_{geo} + f_{cold}\delta^{18}O_{cold}$$

and so

$$f_{cold} = \frac{\delta^{18}O_{hs} - \delta^{18}O_{geo}}{\delta^{18}O_{cold} - \delta^{18}O_{geo}}$$

For the hot spring with the greatest geothermal component in Figure 4.18, this gives:

$$f_{cold} = \frac{(-13)-(-14)}{(-10)-(14)} = 0.25$$

and

$$f_{geo} = 1 - f_{cold} = 0.75$$

In this case, hot spring 1 has 25% cold groundwater and 75% geothermal water. The same calculations using Cl^- produce the same results.

Another practical example of groundwater mixing is that of surface water infiltration and mixing with groundwater. The well can be monitored with stable isotopes, providing the surface water source and the groundwater source have well-defined and distinctive isotope signatures. This is shown schematically in Figure 4.19, using the example of binary mixing in wells that induce infiltration from surface waters.

As above, the percent mixing in a given sample is based on the end-member isotope or geochemical tracer values weighted by their mixing fractions, f:

$$\delta_{spl} = \delta_a f_a + \delta_b (1 - f_a)$$

and

$$f_a = \frac{\delta_{spl} - \delta_b}{\delta_a - \delta_b}$$

In this case, the river and reservoir end-members have some variability. The uncertainty of mixing calculations is a function of the uncertainty in the value for the

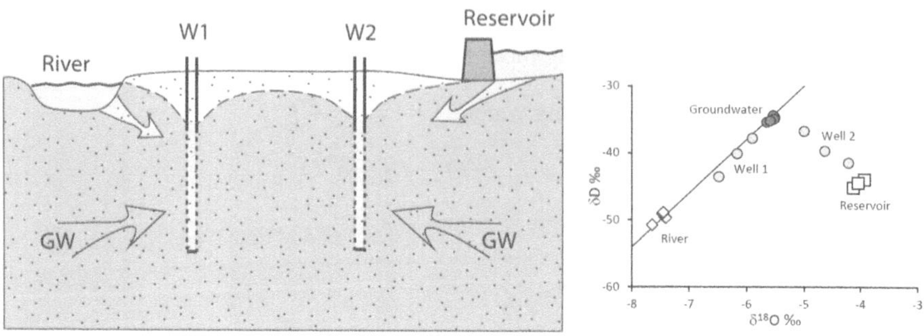

FIGURE 4.19 Infiltration of surface waters to an aquifer supplying groundwater to water supply wells. River waters are typically more depleted in ^{18}O and D than locally derived groundwater because they originate higher in the catchment (elevation effect) and further inland (continental effect). Surface water reservoirs, whether dam-impounded or natural lakes, often have an enriched isotope signal due to evaporation. In this schematic example, the fraction of well water from infiltration of river water and from reservoir water pumped from wells 1 and 2 can be determined from their isotope values, each as a two-component mixing problem.

end-member components and the permil difference between the mixed water and a given end-member. It can be calculated from the isotope values for the end-members a and b, and the mixed sample value (δ_{spl}, δ_a, and δ_b) and the standard deviation, σ, for each end-member value based on the number of replicates made (σ_a and σ_b). The uncertainty of the mixing fraction ($f_a \pm$ uncertainty) is calculated using the following equation:

$$f_a \pm \left[f_a \times \left(\left\{ \frac{1}{(\delta_a - \delta_b)^2} \right\}^2 \times \sigma_{spl}^2 + \left\{ \frac{(\delta_b - \delta_{spl})}{(\delta_a - \delta_b)^2} \right\}^2 \times \sigma_a^2 + \left\{ \frac{(\delta_{spl} - \delta_a)}{(\delta_a - \delta_b)^2} \right\}^2 \times \sigma_b^2 \right)^{\frac{1}{2}} \right]$$

Mixing with Solute Isotope and Concentration

The isotope of a solute, such as $\delta^{34}S_{SO_4^{2-}}$ or $\delta^{13}C_{DIC}$, can also be used to calculate mixing. However, the trend for mixing of two waters follows a curve due to the difference in the concentration of the solute in the two end-members (Figure 4.20). The following solute mass balance and isotope mass balance equations for the hot spring in Figure 4.18 is an example of this trend (where m is the concentration, in this case for ppm SO_4^{2-} and δ is the isotope value of the solute, in this case $\delta^{34}S_{SO_4^{2-}}$).

The mixing fractions of the two end-members are expressed as f_{geo} and f_{cold} in this isotope mass balance equation with substitution and expressed as f_{cold}:

$$\delta_{hs} m_{hs} = f_{geo} \delta_{geo} m_{geo} + f_{cold} \delta_{cold} m_{cold}$$

$$\delta_{hs} m_{hs} = (1 - f_{cold}) \delta_{geo} m_{geo} + f_{cold} \delta_{cold} m_{cold}$$

$$f_{cold} = \frac{\delta_{hs} m_{hs} - \delta_{geo} m_{geo}}{\delta_{cold} m_{cold} - \delta_{geo} m_{geo}}$$

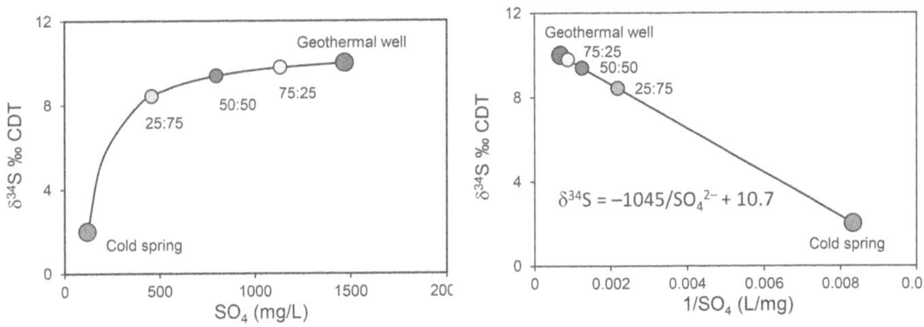

FIGURE 4.20 Two-component mixing trend for a solute and an isotope of the solute. In this case, the mix follows a hyperbolic curve. When plotted as the inverse of the solute concentration, a straight-line relationship is produced, which can be used as a test for binary mixing.

This allows calculation of the end-member mixing components of a given sample, here for a hot spring in Figure 4.18.

Binary mixing of two waters can be readily identified by plotting two conservative tracers, as in Figure 4.18, but is less obvious when plotting a solute concentration against its isotope, as in Figure 4.20, due to the hyperbolic trend. Plotting the isotope value against the inverse solute concentration ($1/SO_4$ in this example) resolves this and produces a straight line (Figure 4.20, right chart).

Three-Component Mixing

It is often the case that a third end-member contributes to mixing. Calculating the mixing fractions will require a second conservative parameter to solve the mixing equations. Near the hot springs in the case above, three additional springs are found near tidewater. In these springs, a third mixing component is suspected—seawater. Plotting this new end-member and the new springs on a two-component graph shows that they fall within the mixing envelope defined by binary mixing between pairs of end-members (Figure 4.21). Samples that fall outside this envelope indicate either non-conservative behavior (reaction) in the subsurface or that the end-member compositions are not correctly defined.

The mixing fractions shown visually in Figure 4.21 can be calculated from solute and isotope mass balance equations. The mixing envelope produced using $\delta^{18}O$–Cl^- provides a good distribution of mixing end-members than for $\delta^{18}O$–δD and is a better choice for these calculations. For three-component mixing, three mass balance equations are needed. The first is a simple mass balance of the fractions of end-member waters (geo, cold, and sea) in the sample. The second and third are mass balance equations for $\delta^{18}O$ and for Cl^- concentration.

$$f_{hs} = 1 = f_{geo} + f_{cold} + f_{sea}$$

$$Cl_{hs} = f_{geo}Cl_{geo} + f_{cold}Cl_{cold} + f_{sea}Cl_{sea}$$

$$\delta^{18}O_{hs} = f_{geo}\delta^{18}O_{geo} + f_{cold}\delta^{18}O_{cold} + f_{sea}\delta^{18}O_{sea}$$

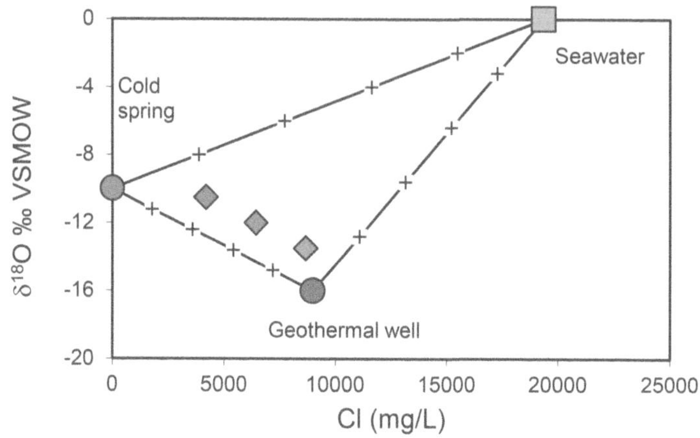

FIGURE 4.21 Three end-member mixing diagrams showing the mixing envelope within which conservatively mixed samples will plot. Note that the use of $\delta^{18}O–Cl^-$ produces a better (broader) mixing envelope than the $\delta^{18}O–\delta D$ due to better separation of the end-member values. The boundary lines are binary mixing lines between two end-members, with crosses plotted at 20% mixing increments. The diamonds are the hot springs with 25%, 50%, and 75% geothermal water; 15%, 40%, and 65% cold groundwater; and each with 10% seawater.

Through rearrangement of the first equation to express it in terms of f_{geo} and then f_{cold}, followed by substitution into the second and third equations, we can determine the fraction contributions of f_{geo} and f_{cold} from the known values for the two parameters (Cl and $\delta^{18}O$) in the three end-members and in the hot spring.

$$f_{geo} = \frac{Cl_{hs} - Cl_{cold} - f_{sea}\left(Cl_{sea} - Cl_{cold}\right)}{Cl_{geo} - Cl_{cold}}$$

$$f_{cold} = \frac{\delta^{18}O_{hs} - \delta^{18}O_{geo} - f_{sea}\left(\delta^{18}O_{sea} - \delta^{18}O_{geo}\right)}{\delta^{18}O_{cold} - \delta^{18}O_{geo}}$$

$$f_{sea} = 1 - f_{geo} - f_{cold}$$

Using two parameters where one is a dependent parameter, such as SO_4^{2-} and $\delta^{34}S_{SO_4}$ again requires an isotope mass balance equation that gives hyperbolic mixing lines (again using m for solute concentration and δ for the isotope of the solute for end-members "geo, cold, and sea") the mixing components of the hot spring sample are determined as follows:

$$1 = f_{geo} + f_{cold} + f_{sea}$$

$$m_{hs} = f_{geo}m_{geo} + f_{cold}m_{cold} + f_{sea}m_{sea}$$

$$\delta_{hs}m_{hs} = f_{geo}\delta_{geo}m_{geo} + f_{cold}\delta_{cold}m_{cold} + f_{sea}\delta_{sea}m_{sea}$$

Then substitute $f_{cold} = 1 - f_{geo} - f_{sea}$ and $f_{geo} = 1 - f_{cold} - f_{sea}$ in the solute and isotope mass balance equations and rearrange to give three equations to solve for the mixing end-members:

$$f_{geo} = \frac{m_{hs} - m_{cold} - f_{sea}\left(m_{sea} - m_{cold}\right)}{m_{geo} - m_{cold}}$$

$$f_{cold} = \frac{\delta_{hs}m_{hs} - \delta_{geo}m_{geo} - f_{sea}\left(\delta_{sea}m_{sea} - \delta_{geo}m_{geo}\right)}{\delta_{cold}m_{cold} - \delta_{geo}m_{geo}}$$

$$f_{sea} = 1 - f_{geo} - f_{cold}$$

The mixing envelope using these tracers for the hot springs above is shown in Figure 4.22. Providing all parameters are conservative in the subsurface, they will provide the same values for the f_{geo} and f_{cold} mixing fractions. The seawater end-member, f_{sea}, is calculated by iteratively solving the three equations. On a spreadsheet program, a goal-seek function is useful for this exercise.

Isotopes are commonly used in mass balance equations because they can be measured with high precision. Using $\delta^{18}O_{H_2O}$ or δD_{H_2O} provides a parameter that, unlike a solute, conservatively tracks the water. Using a dependent parameter such as $\delta^{34}S_{SO_4}$, $\delta^{13}C_{DIC}$, or even $\delta^{15}N_{NH_3}$, for example, allows calculations using one species only. This helps identify non-conservative behavior because both parameters will change with reaction and move samples outside the mixing envelope.

These mixing calculations are the basis of hydrograph separation (Figure 4.23), where contributions to basin discharge can be resolved as a two- or three-component mixing problem. Hydrograph separations can be done by solving these mixing equations using the measured parameters in stream discharge at discrete times. In most cases this can be done with some confidence, given that groundwater baseflow is generally geochemically and isotopically distinct from the storm or snowmelt events

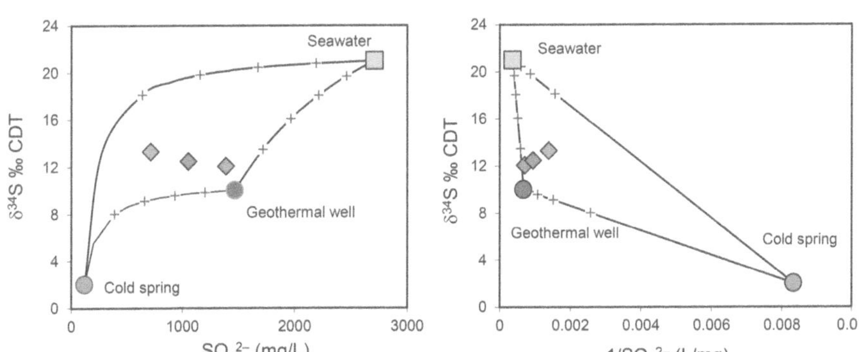

FIGURE 4.22 Mixing envelope for the hot spring data using a dependent parameter, $\delta^{34}S$, with SO_4^{2-} concentration. Expressing SO_4^{2-} as inverse concentration produces straight mixing lines, which are diagnostic for conservative mixing.

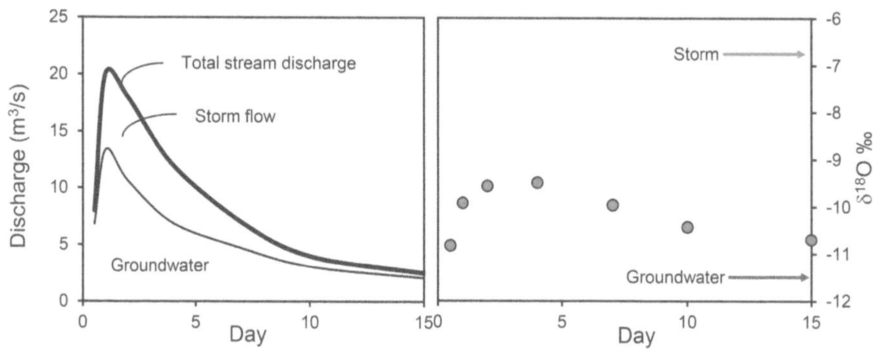

FIGURE 4.23 Storm hydrograph separation resolved as a two-component mixing system using $\delta^{18}O$.

that contribute to surface flow. Greater uncertainty arises when there is seasonal or spatial variability in the mixing end-members.

Diffusive Mixing in Aquitards

Understanding the movement of groundwater through aquitards is critical where such units are used as barriers to contaminant migration and for the protection of water resources from waste storage and subsurface disposal. This is the case for surface waste management facilities such as sanitary landfills, for fracking activities, and for the siting of geological repositories for nuclear waste. Isotopes are important to identify aquitards where diffusive transport dominates over advective transport.

Diffusive transport is represented by Fick's Law, which characterizes the diffusive flux of a solute through a low-permeability material:

$$J = -D_e \frac{\partial C}{\partial x}$$

In the case of isotope diffusion, J is the flux of the isotope (i.e., mol/m^2), C is the isotope concentration (mol/m^3), x is the diffusion distance (m), and D_e is the effective diffusion coefficient for the isotopic species in that medium (m^2/s). For diffusion in aquitards, D_e is equal to the free-water diffusion coefficient, D, for the species, multiplied by dimensionless parameters that characterize the porosity and tortuosity of the porous medium. Although D is known for a wide range of solutes and isotopes in water, D_e must be determined for the particular aquitard materials under investigation.

Hendry and Wassenaar (1999) present a 180-m isotope profile for porewaters in clay-rich sediments in southern Saskatchewan. The aquitard comprises 80 m of clay till deposited by the Pleistocene Laurentide Ice Sheet overlying 76 m of Cretaceous marine clay. Looking here at the upper 50 m of their profile, δD in porewaters extracted from the clay till shows a smooth profile from modern values of –130‰ (15.04 mol/m^3 H_2O) at the surface to a glacial meltwater value of –180‰ (14.18 mol/m^3 H_2O) at 40-m depth in the till (Figure 4.24).

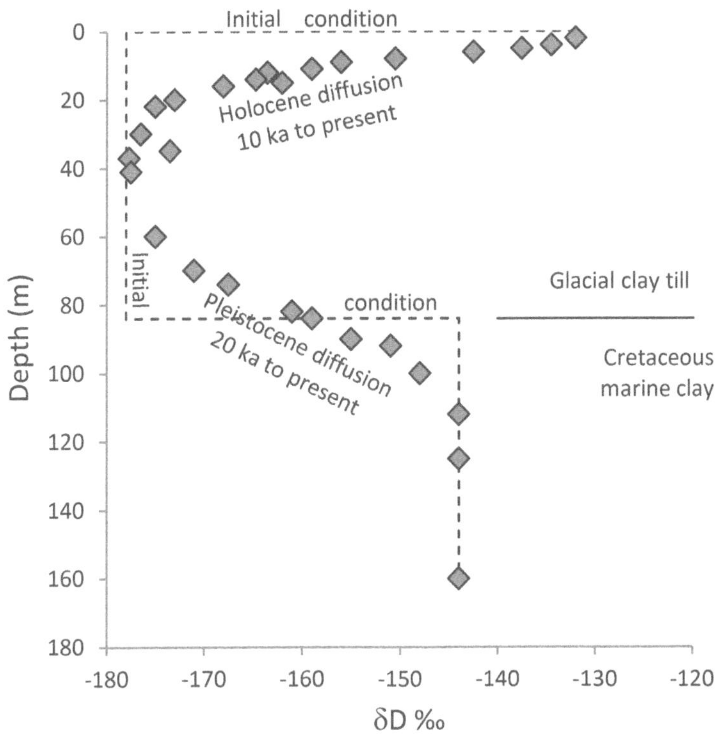

FIGURE 4.24 Diffusion profile for deuterium in porewaters from clay till in southern Saskatchewan. Data from Hendry and Wassenaar 1999.

The use of a one-dimensional advective-diffusion model shows that the upper profile (0–40 m) developed since the glacier retreated 10,000 years ago when modern HDO began diffusing downward into the D-depleted glacial meltwaters in the till (Hendry and Wassenaar 1999). Shown in Figure 4.24 are the stages of the porewater isotope profile as it evolved by diffusion from an initial condition with glacial meltwater in the till to the profile observed today. A minor advective flux is apparent from the downward flexure in the measured profile near the surface.

ISOTOPE EXCHANGE WITH MINERALS AND GASES

The use of stable isotopes to trace groundwater from their recharge origin as precipitation is based on a preservation of the meteoric signal during its subsurface flow. In this way, recharge at different elevations, different seasons, or different climates can be recognized. Under rare circumstances, the isotope composition of groundwater can be modified by exchange reactions in the subsurface (Figure 4.25). Such exchange can only take place through a geochemical reaction, such as mineral alteration reactions or gas production reactions. However, because the "concentration" of water is 55.6 mol/kg, to have a measurable impact requires reaction between a small volume of water and a large volume of solids or gas. This is the case for groundwater in settings with very

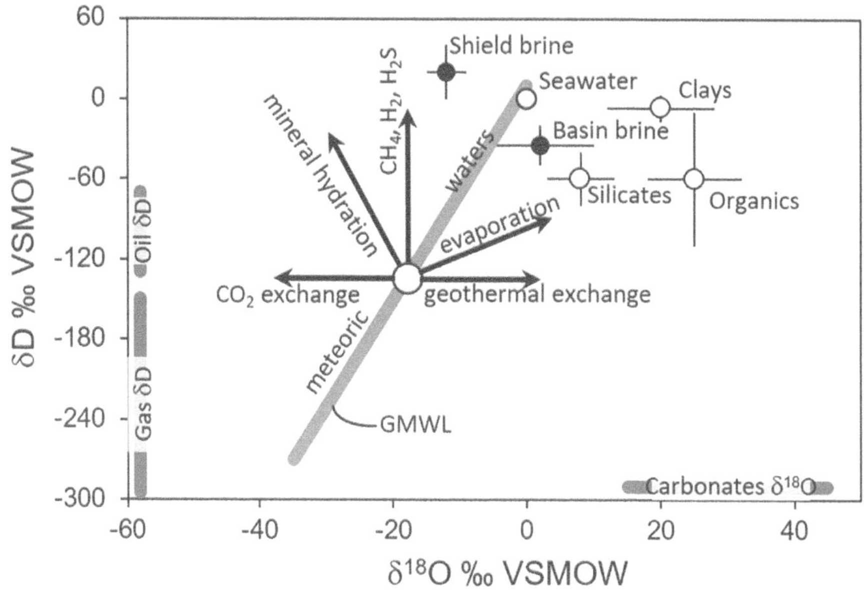

FIGURE 4.25 $\delta^{18}O$ and δD ranges for major crustal reservoirs and principal subsurface exchange mechanisms that modify the isotopic composition of water. (Sources of data: Savin and Epstein 1970; Craig 1963; Clayton et al. 1966; Feurdean et al. 1997; Clark and Fritz 1997; Pearson 1987; Edwards et al. 1985; Gori et al. 2013.)

low water-to-rock ratios, such as deep crystalline rocks and geothermal settings, or in settings with very high gas production, such as landfills or volcanic settings.

The remarkable deuterium enrichment in shield brines, discussed in Chapter 6 on sources of salinity, is due to hydration of primary silicates in deep crystalline settings with very high rock-to-water ratios. The strong ^{18}O enrichment noted by Craig (1963) for geothermal water is due to high-temperature isotope exchange between water and minerals, where the degree of isotope fractionation between the water and mineral is reduced. For example, the enrichment factor for ^{18}O between water and calcite decreases from $\varepsilon^{18}O = 28‰$ at 25°C to 19‰ at 150°C and for silica, from 35‰ at 25°C to 23‰ at 150°C. With a small water reservoir reacting with a large mineral reservoir, the isotope value of the water will shift towards that of the mineral, rather than the reverse case for systems with high water-to-rock ratios.

Under circumstances where substantial production of gas takes place in the subsurface, isotope partitioning with the gas can impart a measurable enrichment or depletion on the residual water phase. This can occur for ^{18}O during the migration of CO_2 from deep injection wells for carbon storage and sequestration or for deuterium with high production of methane in landfills (see Chapter 8).

5 CO_2 and Weathering

THE CARBON CYCLE

Carbon has a crustal abundance of only 200 ppm, yet its redox cycling through photosynthesis and respiration leaves its geochemical imprint throughout the hydrosphere. In groundwater, both dissolved organic carbon (DOC) and CO_2 as dissolved inorganic carbon (DIC) play fundamental roles in weathering reactions and the geochemical evolution of groundwater. Most crustal carbon resides in carbonate rocks and graphite, but active cycling takes place through the atmosphere, hydrosphere, and biosphere (Figure 5.1).

Global Carbon Reservoirs and Fluxes

The atmosphere holds approximately 875 billion tons (Gt) of carbon as CO_2, with a much smaller 4 GtC present as methane and other carbon gases (IPCC; Ciais et al. 2013, Canadell 2021). The CO_2 partial pressure in 2025, $P_{CO_2(atm)}$, is 0.00043 atm ($10^{-3.37}$ atm) or about 430 ppmv in 2025, increasing by about 2 ppm annually (science.nasa.gov). Living terrestrial biomass amounts to over 800 GtC. Another

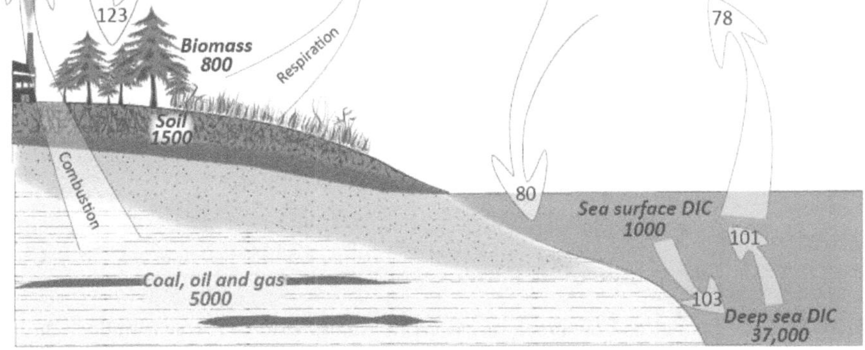

FIGURE 5.1 The carbon cycle with major carbon reservoirs and fluxes of carbon. Values in GtC (gigatons or billion tons of carbon) for reservoirs and GtC/y for fluxes. Sources of data: U.S. Department of Energy; IPCC – Ciais 2013; Canadell 2021.

DOI: 10.1201/9781003353836-5

1500 GtC exists as labile soil organic carbon (SOC), most of which is returned to the atmosphere by bacterial respiration. Respiration in soils maintains $P_{CO_2(soil)}$ at two or more orders of magnitude above that in the atmosphere, and so greatly increases weathering potential.

Terrestrial forests and grasslands sequester about 123 GtC or 15% of atmospheric CO_2 annually through gross primary production (GPP). Both GPP and terrestrial biomass are increasing in response to increased CO_2 fertilization, particularly in the boreal forests (Chen et al. 2022; Hyvönen et al. 2007) and agricultural areas (Erda et al. 2005). Changes in soil temperature and humidity are responsible for much of the seasonal and year-to-year variability in photosynthesis and respiration (Humlum 2013; Sarmiento and Gruber 2002; Jones et al. 2003), which affect atmospheric CO_2 and give life to the idea of Gaia—the breathing Earth (Figure 5.2).

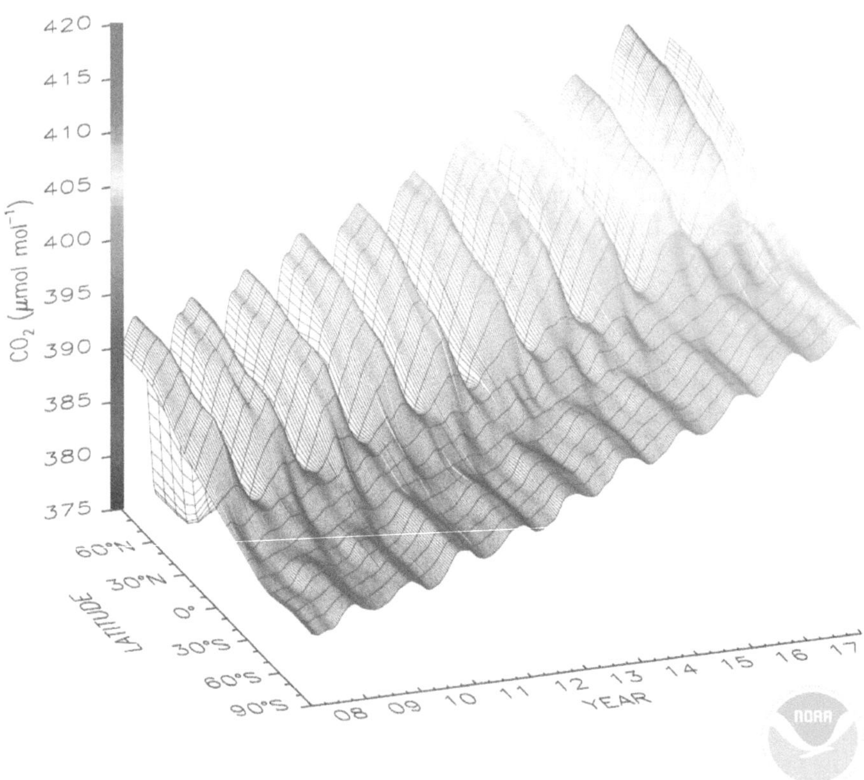

FIGURE 5.2 Spatial and temporal variations in atmospheric CO_2 concentrations due to annual variations in photosynthesis and respiration. The strong latitudinal effect reflects the greater participation of the northern boreal forests and disproportionate continental land mass in the northern hemisphere. The overall increase reflects contributions from fossil fuels, land use change, and respiration. From NOAA 2025.

Terrestrial and the marine fluxes of CO_2 exchange about one quarter of atmospheric CO_2 annually. Emissions from combustion of fossil fuels and production of cement add about 10 GtC per year to the active carbon cycle. Volcanic degassing contributes about 0.2 GtC of mantle CO_2 to the atmosphere each year, with a similar amount removed by continental weathering to the oceans as DIC.

$\delta^{13}C$ and ^{14}C Tracers in the Carbon Cycle

Both stable ^{13}C and radiocarbon, ^{14}C, are important tracers of carbon sources and pathways among various carbon reservoirs (Figure 5.3). Photosynthesis is a highly fractionating process that discriminates against ^{13}C during transformation of CO_2 to carbohydrate, CH_2O. Most plants follow the Calvin or C_3 photosynthetic pathway, which involves an inefficient step of CO_2 respiration from the plant. As a result, the $\delta^{13}C$ of C_3 plants is much lower than that of atmospheric CO_2, with a kinetic fractionation of close to 20‰ ($\varepsilon^{13}C_{CH_2-CO_2} \approx -20‰$). The C_3 pathway is used by almost 95% of plants, including most terrestrial trees and shrubs as well as marine plants and algae. Carbon fixed by terrestrial C_3 plants has $\delta^{13}C$ values of about −27‰.

In response to declining atmospheric CO_2 concentrations during the past 15 million years, the more efficient Hatch–Slack or C_4 photosynthetic pathway evolved, which accelerates the fixation of CO_2 by reducing respiration. This results in lower fractionation ($\varepsilon^{13}C_{CH_2O-CO_2} \approx -6‰$) and $\delta^{13}C$ values between about −10‰ and −14‰. C_4 plants include various tropical grasses. For geochemists, the contrast in $\delta^{13}C$ between

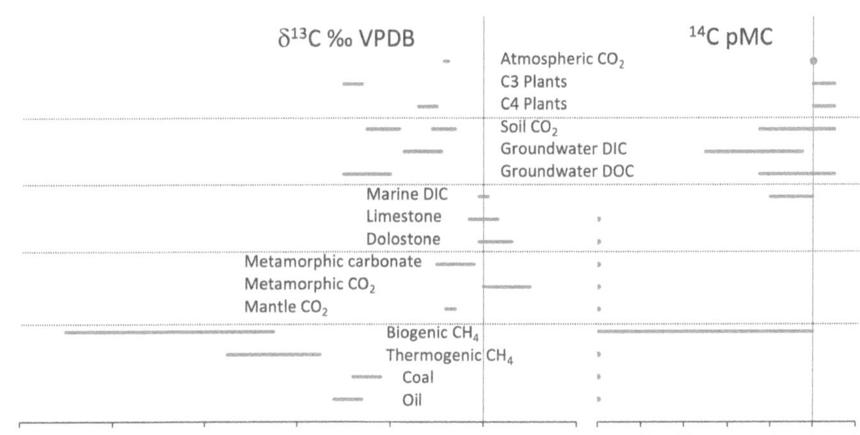

FIGURE 5.3 Typical ranges of $\delta^{13}C$ and ^{14}C for major terrestrial reservoirs of organic and inorganic carbon. The Vienna Pee Dee Belemnite (VPDB) standard for ^{13}C is a synthesized calcite powder prepared by the International Atomic Energy Agency in Vienna and closely represents the original PDB standard from a fossil belemnite from the Pee Dee Formation in South Carolina. pMC is percent Modern Carbon, where MC = 0.226 Bq/g C.

C_3 and C_4 vegetation provides a useful tracer for DIC, particularly in agricultural areas cropped with C_4 plants:

C_3 crops: Wheat, rice, barley, oats, rye, tobacco, sugar beets, dry beans, soybeans, sunflowers, groundnuts, bluegrass, and most lawn grasses and fescue
C_4 crops: Corn (maize), sugarcane, sorghum (millet), and prairie and dryland grasses

Marine organic carbon is accompanied by a greater fractionation during photosynthesis and so has $\delta^{13}C$ values closer to –30‰. The degradation of marine organic carbon during diagenesis and hydrocarbon formation produces enriched CO_2 as a by-product, and so oil and gas have additional depletion in ^{13}C.

SOIL CO_2 AND WEATHERING

The chemical weathering of carbonate and silicate minerals is driven by acidity. The solubility of most minerals in pure water is very low. Carbonate minerals, such as calcite or dolomite, have low dissociation constants ($K_{cal} = 10^{-8.48}$ and $K_{dol} = 10^{-17.09}$) and simple dissociation reactions, producing Ca^{2+} and Mg^{2+}, and CO_3^{2-} allows dissolution of only a few ppm. However, the addition of acid greatly enhances carbonate dissolution. Weathering of aluminosilicate minerals, such as feldspars, follows hydrolysis reactions that consume hydrogen ions and raise pH. The addition of acid therefore increases the solubility of such minerals and so greatly accelerates weathering.

In natural soils, the acidity required for weathering is principally generated by oxidation of organics to generate carbonic acid (H_2CO_3) produced from hydration of CO_2. The atmosphere contains about 400 ppmv of CO_2, whereas CO_2 from respiration of SOC reaches 3,000–10,000 ppmv. In Precambrian and early Paleozoic times, the atmosphere hosted CO_2 at concentrations over an order of magnitude higher than today, which supported geochemical weathering before the arrival of terrestrial plants and soils. With the development of the terrestrial biosphere and reduction of atmospheric CO_2 concentrations, continental weathering became constrained to the soil horizon overlying bedrock.

P_{CO_2} AND CARBONIC ACID IN THE RECHARGE ENVIRONMENT

Respiration of dead biomass in soils by aerobic bacteria produces high P_{CO_2} typical of soils. The process results in the humification of soil, with the accumulation of less labile and more refractory organics as humic substances. These large 10,000-plus Da (Dalton, unified atomic mass units) compounds include solid humin and more soluble humic acids and lighter-weight fulvic acids, composed of aromatic carbon structures with OH and COOH functional groups. Humic acids dissociate near pH 2–4, making them weak acids. Fulvic acids are soluble over a wide pH range. In oxic groundwater these organic acids are largely consumed by aerobic bacteria, such that below the water table, DOC concentrations are generally less than 1 ppm.

The aerobic decay of vegetation can generate P_{CO_2} levels generally between $10^{-2.5}$ and $10^{-1.5}$ atm (0.0032–0.032 atm), whereas in the open atmosphere P_{CO_2} is much lower ($10^{-3.40}$ or 0.0004 atm). These high partial pressures contribute significant carbonic acid to recharging groundwaters. Recalling that the activity of carbonic acid is established by the Henry's Law constant:

$$CO_2 + H_2O \Leftrightarrow H_2CO_3$$

$$K_{CO_2} = \frac{a_{H_2CO_3}}{P_{CO_2}} = 10^{-1.47} = 0.034$$

$$P_{CO_2} = 0.032$$

$$a_{H_2CO_3} = 0.034 \times 0.032 = 0.00107 = 10^{-2.97}$$

Carbonic acid then dissociates, and so we can calculate from this the pH and bicarbonate activity during recharge:

$$H_2CO_3 \Leftrightarrow H^+ + HCO_3^-$$

$$K_1 = \frac{a_{HCO_3^-} \times a_{H^+}}{a_{H_2CO_3}} = 10^{-6.35}$$

$$a_{HCO_3^-} \times a_{H^+} = 10^{-6.35} \times 10^{-2.97} = 10^{-9.32}$$

$$a_{HCO_3^-} = a_{H^+} = \sqrt{10^{-9.32}}$$

$$= 10^{-4.66}$$

$$pH = 4.66$$

So, for a partial pressure of, say, 0.032 atm CO_2 ($10^{-1.5}$), recharging groundwaters would be quite acidic, and the carbonic acid concentration supported by the high P_{CO_2} would be $10^{-2.97}$ molal or $0.00107 \times 62 \times 1000 = 66.4$ ppm as H_2CO_3 or 12.9 ppm as DIC. Bicarbonate, at this low pH, is not significant (Figure 2.14). The range of pH and carbonic acid concentration as initial conditions for recharging groundwaters before weathering is shown in Figure 5.4 for a range of soil P_{CO_2} conditions.

The oxidation of sulfides [FeS_2] and other ferrous iron minerals in the recharge environment can be a source of acidity for weathering. As soils develop, these minerals become exposed and oxidized by bacteria. Precipitation of ferric iron hydroxide ($Fe[OH]_3$) in soils and on outcrops is evidence for such sources of acidity.

$$FeS_2 + 3\tfrac{1}{2}O_2 + H_2O \rightarrow Fe^{2+} + 2SO_4^{2-} + 2H^+ \quad \text{sulfuric acid from pyrite oxidation}$$

$$Fe^{2+} + \tfrac{1}{4}O_2 + 2\tfrac{1}{2}H_2O \rightarrow Fe(OH)_3^{\circ} + 2H^+ \quad \text{ferrous iron oxidation and hydration}$$

CARBONATE WEATHERING

Carbonate minerals are ubiquitous, not just as limestone and dolomite formations, but as secondary phases filling fractures in granitic and metamorphic rocks. Carbonic acid from soils is the dominant process for carbonate dissolution, generating high karstic permeability in carbonate formations. Despite a low relative abundance of

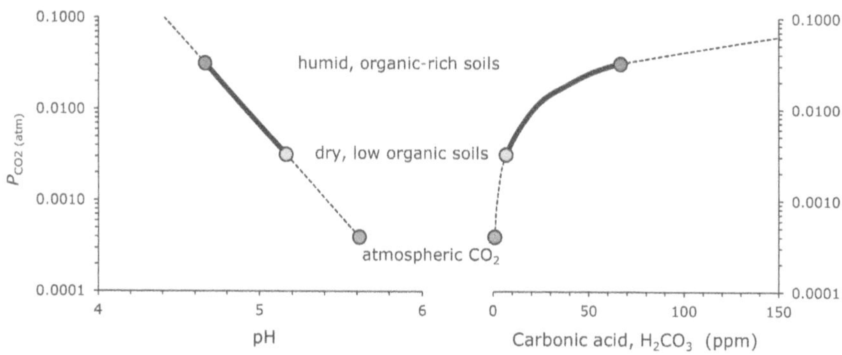

FIGURE 5.4 pH and carbonic acid for groundwater in equilibrium with varying soil P_{CO_2}. Weathering potential increases with P_{CO_2} and carbonic acid concentration.

less than 0.1% of the lithosphere, limestone and dolomite formations are important aquifers, and their weathering has a major effect on watershed geochemistry, producing hardness in water quality. Even in crystalline terrains, weathering of carbonate minerals lining fractures can dominate over silicate weathering (Leybourne and Goodfellow, 2010). Two principal reactions dominate–calcite and dolomite dissolution, each driven by consumption of carbonic acid:

$$CaCO_3 + H_2CO_3 \rightarrow Ca^{2+} + 2HCO_3^-$$

$$CaMg(CO_3)_2 + 2H_2CO_3 \rightarrow Ca^{2+} + Mg^{2+} + 4HCO_3^-$$

The extent of carbonate weathering is a function of the P_{CO_2} of soils in the recharge area and the degree of openness of the system during infiltration.

Open and Closed System Carbonate Weathering

The action of weathering by carbonic acid is greatly enhanced when the carbonic acid can be replenished as it is consumed. This is the case for unsaturated soils and fractures, where mineral weathering takes place with open contact with CO_2-rich soil air. As carbonic acid in the soil water is consumed by weathering, it is replenished by CO_2 from the soil air. Contact with soil CO_2 maintains a constant P_{CO_2} and so a constant concentration of carbonic acid. This is weathering under *open system conditions*. As carbonate weathering proceeds under these conditions, the system reaches calcite saturation with no further net dissolution (Figure 6.6).

In the saturated zone below the water table, weathering takes place under *closed system conditions*, as groundwater is no longer in direct contact with the soil atmosphere. Here, carbonic acid consumed by weathering is not replenished. The degree of mineral weathering is then limited to the amount of CO_2 dissolved during recharge through the unsaturated soil zone. As this carbonic acid is consumed, P_{CO_2} in the water decreases, pH increases, and weathering is greatly diminished. Closed system conditions exist in recharge areas characterized by high water tables and rapid infiltration, where percolation to the water table occurs before minerals are encountered

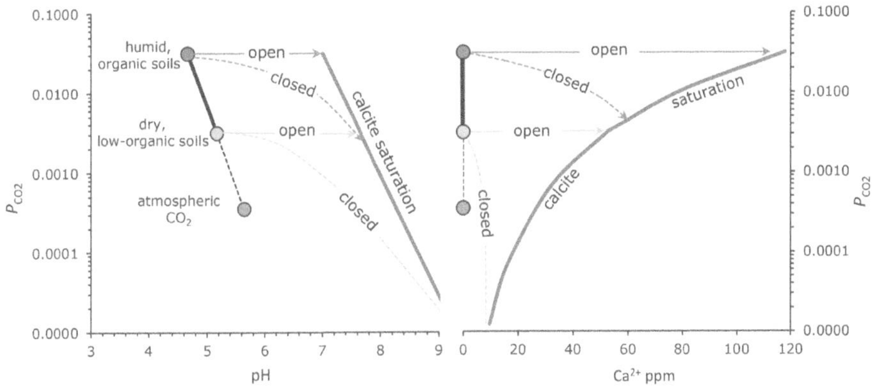

FIGURE 5.5 Evolution of groundwater P_{CO_2} with pH (left chart) and Ca^{2+} (right chart) during carbonate weathering, starting at the initial point of equilibration with soil CO_2 and proceeding to the point of calcite saturation. An open system has constant P_{CO_2}, while pH and Ca^{2+} increase to the point of calcite saturation. In a closed system, P_{CO_2} of the groundwater decreases during weathering, limiting carbonate dissolution.

and weathering can begin. This contrasts with open system weathering, which is more aggressive and results in higher concentrations of solutes.

P_{CO_2} and Carbonate Weathering

Much of carbonate weathering begins under open system conditions where the reservoir of soil CO_2 and organics replenishes the carbonic acid consumed by mineral dissolution and may continue to a limited extent below the water table. The amount of carbonate dissolved under open system conditions is ultimately controlled by the partial pressure of the soil CO_2 (Figure 5.5). This contrasts with carbonate dissolution under closed system conditions, where the limited CO_2 gained during infiltration through the soil precludes extensive weathering. Here, the P_{CO_2} will evolve to very low values with correspondingly high pH before reaching calcite saturation, as shown in Figure 5.5. Less calcite is dissolved, as seen with lower Ca^{2+}, before reaching calcite saturation at these higher pH values. In contrast, continual access to soil CO_2 under open system conditions allows considerably more calcite to be dissolved for a given initial soil P_{CO_2}. The concentration of Ca^{2+} as well as the DIC and pH of the groundwater can be calculated using the basic equations that control DIC speciation and shows the importance of open system weathering.

SILICATE WEATHERING

Crystalline rocks, including metamorphic shield terrains, plutonic intrusions, and volcanics, make up most of the earth's crust. These are largely composed of aluminosilicate feldspar minerals, including anorthite [$CaAl_2Si_2O_8$], alkali feldspars, albite [$NaAlSi_3O_8$] and orthoclase [$KAlSi_3O_8$], and quartz [SiO_2]. Silicate terrains also include varying amounts of high-temperature mafic (Mg and Fe) minerals, including olivine [$(Fe,Mg)_2SiO_4$], pyroxene [$Ca(Mg,Fe)Si_2O_6$], amphibole [$KCa_2(Mg,Fe,Al)_5[Al,Si)_4O_{11}]_2(OH)_2$], and biotite [$K(Mg,Fe)_3AlSi_3O_{10}(OH)_2$], plus muscovite [$KAl_2(AlSi_3O_{10})(OH)_2$] and quartz [$SiO_2$] (see Chapter 1, Silicates). Unlike carbonate rocks, silicate terrains are more resilient to weathering, with lower reaction rates and lower mineralization in groundwater.

The acidity available for weathering in silicate terrains is largely determined by the P_{CO_2} conditions that exist in the recharge area. The slow rates of silicate weathering are such that much may take place under closed system conditions below the water table. If no subsurface sources of CO_2 (e.g., oxidation of methane, metamorphic CO_2) or other acids (e.g., H_2SO_4 from oxidation of pyrite) are available, then P_{CO_2} will drop to very low levels and pH will increase to values of 10 or higher. An example of such a case comes from the Ahouset geothermal spring on the Pacific Coast of southern Canada with a pH of 10.05 (see example in the Groundwater Geochemistry and Isotopes Workbook). More extreme are the pH 12 groundwaters from the alkaline Lovozeo Massif in the Kola Peninsula, where dissolved silica reaches up to 13 g/L (Kraynov et al. 1969).

Silicate Weathering Reactions

Aluminosilicate weathering releases cations into solution, according to incongruent dissolution reactions, where the primary minerals, such as feldspars, are altered to more stable clay minerals:

$$NaAlSi_3O_{8\ \mathrm{albite}} + 4½H_2O + H_2CO_3 \rightarrow Na^+ + HCO_3^- + ½Al_2Si_2O_5(OH)_{4\ \mathrm{kaolinite}} + 2H_4SiO_4^\circ$$

$$CaAl_2Si_2O_{8\ \mathrm{anorthite}} + H_2O + 2H_2CO_3 \rightarrow Ca^{2+} + 2HCO_3^- + Al_2Si_2O_5(OH)_{4\ \mathrm{kaolinite}}$$

The alteration of these high-temperature feldspars and formation of low-temperature clay minerals (incongruent dissolution) is such that thermodynamic equilibrium is seldom reached. In the near-surface environment where water circulation is relatively rapid, weathering will continue as long as there is a source of acidity to drive the reactions. For the alkali feldspars, the concentrations of Na^+ and K^+ can be used as an indication of the extent of weathering, as there are few mineralogical controls on their concentrations. For anorthite and some mafic silicates, Ca^{2+} concentrations will become limited by carbonate mineral saturation.

$$CaAl_2Si_2O_{8\ \mathrm{anorthite}} + H_2O + H_2CO_3 \rightarrow Al_2Si_2O_5(OH)_{4\ \mathrm{kaolinite}} + CaCO_{3\ \mathrm{calcite}}$$

$$CaMgSi_2O_{6\ \mathrm{pyroxene}} + 2H_2CO_3 + 2H_2O \rightarrow 2H_4SiO_4^\circ + CaMg(CO_3)_{2\ \mathrm{high\ Mg-calcite}}$$

Weathering of the ferromagnesian silicates, such as pyroxene and olivine, can proceed congruently, with release of silica and either Fe^{2+} or Mg^{2+} to solution:

$$CaMgSi_2O_{6\ \mathrm{pyroxene}} + 4H^+ + 2H_2O \rightarrow Ca^{2+} + Mg^{2+} + 2H_4SiO_4^\circ$$

$$MgFeSiO_{4\ \mathrm{olivine}} + 4H^+ \rightarrow Mg^{2+} + Fe^{2+} + H_4SiO_4^\circ$$

As these minerals contain no aluminum, clay minerals, such as kaolinite, do not form, although under conditions of low water/rock ratios, serpentinization proceeds [$(Mg,Fe)_3Si_2O_5(OH)_4$].

These reactions are all driven by the consumption of acidity, shown above as carbonic acid. The slow rate of silicate weathering is such that weathering continues below the water table under closed system conditions. If no subsurface sources of acidity are available, then P_{CO_2} will drop to very low levels and pH will increase to values as high as 10 or higher.

Differentiating between silicate and carbonate weathering can be approached using ion ratios. From the reactions presented here, the production of alkali elements, in particular Na^+, can be an indicator of feldspar weathering. However, it must be corrected for chloride salinity by subtracting the moles of Na^+ that are supported by Cl^- derived from marine and halite sources. Gaillardet et al. (1999) show that the weathering in carbonate watersheds produces runoff that is enriched in the alkaline earths over the alkali cations, with Ca/Na molar ratios greater than 10. In contrast, feldspar weathering yields higher alkali cations, and so the Ca/Na ratio is generally less than 1 (Figure 5.6, left chart). Bicarbonate provides an additional constraint when normalized to Na^+, with strong enrichment of the HCO_3/Na^+ ratio in drainage from carbonate terrains but with ratios closer to 1 in silicate terrains. Silica provides an additional constraint. Plotting the molar ratio of bicarbonate to silica provides further distinction of carbonate and silicate weathering fields (Figure 5.6).

Silica Solubility

Silica is released into solution by the weathering of the alkali feldspars (albite and orthoclase), although not necessarily by the weathering of quartz and anorthite, in which an extra Al^{3+} replaces one Si^{4+} to maintain the charge balance with divalent

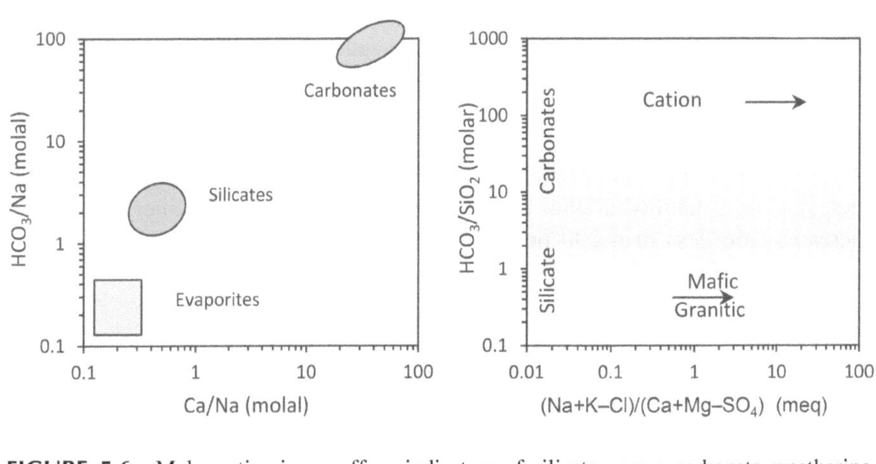

FIGURE 5.6 Molar ratios in runoff as indicators of silicate versus carbonate weathering. (Modified from Gaillardet 1999.) Corrections are made to molal concentrations of cations to account for chloride (Na+K–Cl) and gypsum (Ca+Mg–SO_4) contributions in the right-hand chart.

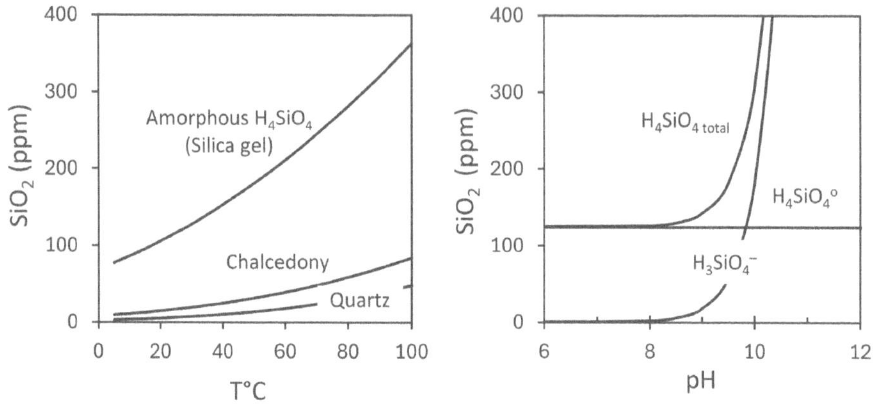

FIGURE 5.7 Solubility of quartz (SiO_2) and its polymorphs with temperature (left). Right chart shows amorphous silica solubility increasing at high pH due to silicic acid dissociation above pH 9.83 ($K_{H_4SiO_4} = 10^{-9.83}$).

Ca^{2+}. It is also released by weathering of ferromagnesian silicates. As the concentration of silicic acid [$H_4SiO_4°$] increases through silicate weathering, quartz solubility can be exceeded. Although quartz saturation at 25°C is less than 10 ppm (Figure 5.7), the solubility of silicic acid is controlled by lower-temperature polymorphs, including chalcedony and ultimately amorphous silica (H_4SiO_{4am}), which can precipitate at low temperatures. Over time, this phase can recrystallize into more stable opal ($SiO_2{\cdot}n[H_2O]$), a hydrated form ($n < 1$).

Opal can also be directly precipitated from solution, for example, in plant cells as phytoliths, which can be analyzed for δ^{18} O as a climate indicator (Webb and Longstaffe 2000). With increased temperature and time, amorphous to poorly crystalline silica forms transform to chalcedony (SiO_2) – a microcrystalline form of quartz. However, high concentrations of silicic acid are generally only achieved by silicate weathering in geothermal systems, with precipitation of amorphous and poorly crystallized phases occurring during cooling after discharge.

In low-temperature systems, silicic acid concentrations are low, typically in the range of a few ppm. This is due to the low rate of silicate mineral weathering. However, silicic acid dissociates above pH 9 (Figure 5.7, right), which greatly increases its solubility at low temperature. The solubility of amorphous silica increases from less than 200 ppm in neutral pH waters to very high levels near pH 10, where $H_3SiO_4^-$ dominates.

^{13}C AND ^{14}C TRACERS OF WEATHERING

As recharge evolves through open and closed system weathering, pH rises, and so DIC evolves from dominantly carbonic acid to bicarbonate (Figure 2.14). The high P_{CO_2} in soils is generated by degradation of biomass, and so its $\delta^{13}C$ is in the range for C_3 and C_4 vegetation (Figure 5.3) but with a slight enrichment due to the preferential outward diffusion of $^{12}CO_2$ from the high P_{CO_2} in soils. As ^{12}C is more diffusive, this

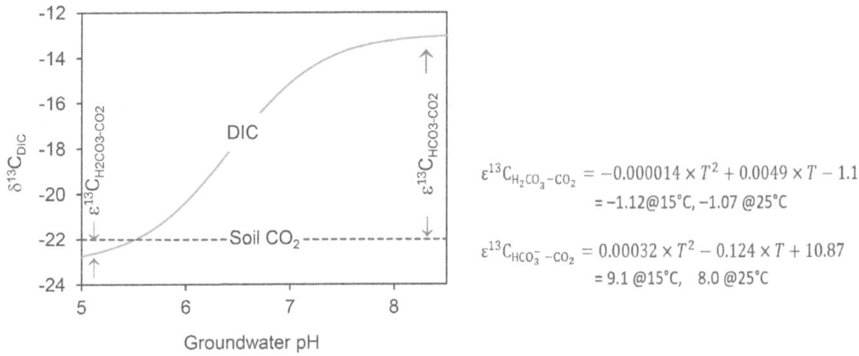

FIGURE 5.8 Effect of recharge pH on the $\delta^{13}C$ of dissolved inorganic carbon (DIC) in equilibrium with soil CO_2 (open system conditions). Calculated for a case where the $\delta^{13}C$ of soil CO_2 is –22‰ at 15°C.

imparts a slight enrichment on the remaining soil CO_2 (Cerling et al. 1991). In most C_3 landscapes, the soil CO_2 has $\delta^{13}C$ in the range of –20 to –24‰.

The $\delta^{13}C$ of DIC is a composite of the enrichment factors between CO_2 in the soil and the DIC species, weighted by their relative abundance. The ^{13}C content of carbonic acid is about 1‰ lower than that of the soil CO_2 ($\varepsilon^{13}C_{H_2CO_3-CO_2} = -1‰$) while $\delta^{13}C_{HCO_3^-}$ is about 9‰ greater than $CO_{2(g)}$ ($\varepsilon^{13}C_{HCO_3-CO_2} = 9‰$). As the relative abundance of these DIC species is controlled by pH, the value for $\delta^{13}C_{DIC}$ is then a function of pH, as long as the recharge waters are in contact with soil CO_2 (open system conditions). This is illustrated in Figure 5.8, which shows the enrichment in ^{13}C of DIC as the pH rises and the DIC becomes increasingly dominated by bicarbonate, according to the isotope mass balance equation:

$$\varepsilon^{13}C_{DIC-CO_2} = f_{H_2CO_3} \times \varepsilon^{13}C_{H_2CO_3-CO_2} + f_{HCO_3^-} \times \varepsilon^{13}C_{HCO_3^--CO_2}$$

$$\delta^{13}C_{DIC} = \delta^{13}C_{CO_2} + \varepsilon^{13}C_{DIC-CO_2}$$

and the relative fractions of carbonic acid and bicarbonate in DIC can be determined from pH and the dissociation constant for carbonic acid, $K_{H_2CO_3}$, as outlined in Chapter 2:

$$f_{H_2CO_3} = 1 - f_{HCO_3^-}$$

$$f_{HCO_3^-} = \frac{K_{H_2CO_3}}{K_{H_2CO_3} + 10^{-pH}}$$

$$K_{H_2CO_3} = 10^{-6.35} \text{ at } 25°C$$

In addition to ^{13}C, radiocarbon can provide an additional constraint on the origin and evolution of DIC. The high activity of ^{14}C in the living biosphere is also present in soil CO_2. However, any material of geologic age (limestone, dolostone, metamorphic CO_2, fossil fuel, bitumen, coal, etc.) will have long ago lost ^{14}C through radioactive decay (Figure 5.3). Vegetation formed within the past 4 decades will have values greater than 100 pMC due to contributions from atmospheric testing of thermonuclear weapons in the 1960s. Today, the ^{14}C activity of atmospheric CO_2 has returned to near pre-nuclear levels of 100 pMC due to dilution in the biosphere and oceans and by additions from combustion of fossil fuel.

Considering the recharge scenarios for groundwater in carbonate terrains presented earlier, ^{14}C provides a useful constraint on the origin of the DIC. In the case of open system recharge, the exchange of CO_2 between the DIC and soil atmosphere during carbonate weathering will impart the ^{14}C activity of the soil CO_2 on the DIC, maintaining $a^{14}C_{DIC}$ at 100 pMC or greater.

An important consideration is that, like ^{13}C, ^{14}C undergoes fractionation during the dissolution of soil CO_2. For an enrichment in $\varepsilon^{13}C_{DIC-CO_2}$ of 7‰, as an example, the enrichment in the pMC value for ^{14}C would be 2× greater due to the two times mass effect. So this would amount to 14‰ or 1.4%. For $^{14}C_{CO_2}$ of 100 pMC, $a^{14}C_{DIC}$ would be 101.4 pMC.

By contrast, groundwaters that dissolve carbonate under closed system conditions will derive up to half of DIC from ^{14}C-free geogenic sources. Radiocarbon is important for tracing fugitive gases, such as CO_2 and CH_4, from CO_2 sequestration and shale fracking activities.

$\delta^{13}C_{DIC}$ AND $a^{14}C_{DIC}$ IN CARBONATE WEATHERING

Carbonate weathering under both open and closed conditions contributes DIC from both the soil CO_2 and the dissolved carbonate minerals,

$$CaCO_3 + H_2CO_3 \rightarrow Ca^{2+} + 2HCO_3^-$$

Closed system: In a closed system, this would result in half the DIC coming from soil CO_2 and half from the carbonate source, usually limestone or dolostone. The resulting value for $\delta^{13}C_{DIC}$ is then a 50:50 mixture of the two sources. This is the same for $a^{14}C_{DIC}$, which is a mixture of modern radiocarbon and ^{14}C-free carbonate (Figure 5.9).

$$\begin{aligned}\delta^{13}C_{DIC} &= \tfrac{1}{2}\delta^{13}C_{CO_2} + \tfrac{1}{2}\delta^{13}C_{CaCO_3}\\ &= \tfrac{1}{2}(-22‰ + 0‰)\\ &= -11‰\end{aligned}$$

and

$$\begin{aligned}a^{14}C_{DIC} &= \tfrac{1}{2}a^{14}C_{CO_2} + \tfrac{1}{2}a^{14}C_{CaCO_3}\\ &= \tfrac{1}{2}(100 + 0)\\ &= 50\,\text{pMC}\end{aligned}$$

Open system: However, both $\delta^{13}C_{DIC}$ and $^{14}C_{DIC}$ evolve differently in an open system, and considerably more DIC is produced (Figure 5.5). Under full open system conditions, with the continued uptake of soil CO_2 and active exchange between soil

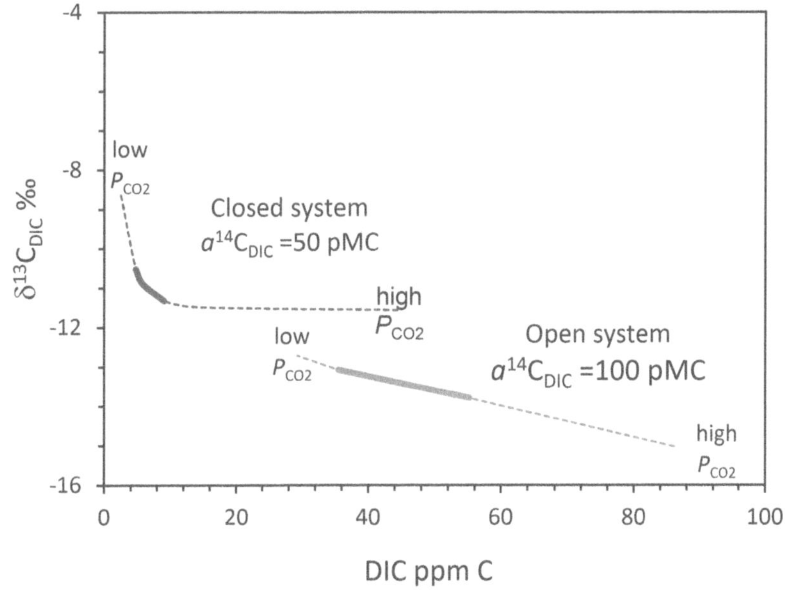

FIGURE 5.9 $\delta^{13}C_{DIC}$ after carbonate weathering under open and closed system conditions, calculated for a range of initial soil P_{CO_2} from 0.001 to 0.03 atm. In both scenarios, $\delta^{13}C_{CO_2} = -22‰$.

CO_2 and DIC, $\delta^{13}C_{DIC}$ will be in equilibrium with the soil CO_2 at the recharge pH with $\varepsilon^{13}C_{DIC\text{-}CO_2}$ close to 7‰, (Figure 5.8) and $^{14}C_{DIC}$ will be near that of soil CO_2 (Figure 5.9),

$$\begin{aligned}\delta^{13}C_{DIC} &= \delta^{13}C_{CO_2} + \varepsilon^{13}C_{DIC-CO_2}\\ &= -22 + 7‰\\ &= -15‰\end{aligned}$$

and

$$\begin{aligned}a^{14}C_{DIC} &= a^{14}C_{CO_2} + \varepsilon a^{14}C_{DIC-CO_2}\\ &= 100 + (0.7\% \times 2)\\ &= 101.4\,\text{pMC}\end{aligned}$$

(Note: Fractionation for $a^{14}C$ is 2× that of ^{13}C due to the mass effect, with conversion from permil to percent.)

The difference in $\delta^{13}C_{DIC}$ for fully open and fully closed conditions during recharge is only a few permil. However, together with the DIC concentration and ^{14}C activity, it can be diagnostic.

$\delta^{13}C_{DIC}$ and $^{14}C_{DIC}$ in Silicate Weathering

The change in $\delta^{13}C$ during weathering in silicate terrain is less complicated than in carbonate aquifers, as the only source of carbon is the soil CO_2, providing no carbonate fracture minerals are involved. As a consequence, the only controlling factor is the pH

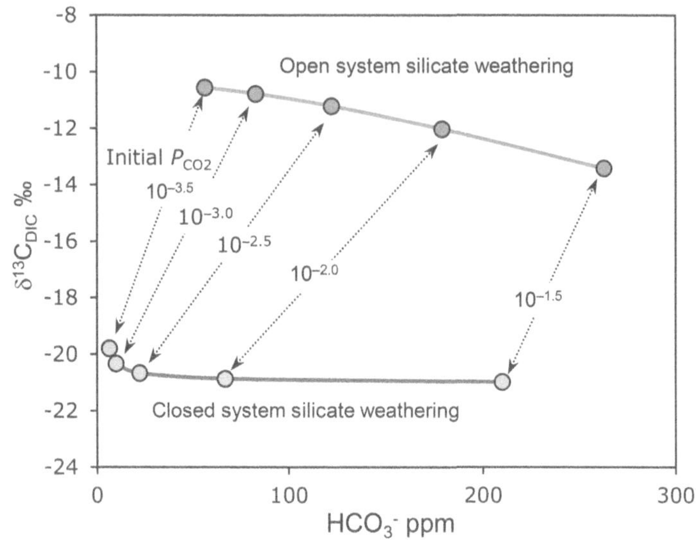

FIGURE 5.10 $\delta^{13}C$ of DIC during open and closed system silicate weathering, for $\delta^{13}C$ of soil CO_2 of −20‰ and at 10°C. Note that in both cases, the $\delta^{13}C$ value is controlled by equilibrium exchange with the soil CO_2 according to the enrichment factors, $\varepsilon^{13}C_{H_2CO_3-CO_2}$ and $\varepsilon^{13}C_{HCO_3^--CO_2}$.

at the point that the recharge waters enter closed system conditions. After this point, the $\delta^{13}C_{DIC}$ will not change, even as the pH continues to increase, providing no subsurface sources of carbonate are encountered. Figure 5.10 shows the evolution of $\delta^{13}C_{DIC}$ for both open and closed system weathering in silicate bedrock.

The pH under open system weathering of silicate minerals is controlled by the P_{CO_2} in the unsaturated zone and so controls the distribution of HCO_3^- and H_2CO_3, which in turn controls $\delta^{13}C_{DIC}$ in isotopic equilibrium with soil CO_2 (Figure 5.10). In closed system silicate weathering, the initial pH from dissolution of CO_2 and carbonate acid dissociation, as shown in Figure 5.5 (left chart), will establish the value for $\delta^{13}C_{DIC}$.

Similarly, radiocarbon in DIC during pure silicate mineral weathering is unaffected by dilution with matrix carbonate. Only minor fractionation during exchange with soil $^{14}CO_2$ will modify the activity of radiocarbon in DIC. The enrichment factors for exchange with soil CO_2 in Figure 5.10 are multiplied by 2 for ^{14}C exchange. For neutral pH water, $\varepsilon^{13}C_{DIC-CO_2} = 9.1‰$; therefore, $\varepsilon^{14}C_{DIC-CO_2} = 18.2‰$ or 1.8%. For soil CO_2 with $a^{14}C_{CO_2}$ equal to 100 pMC, the DIC would have $a^{14}C_{DIC} = 101.8$ pMC.

^{87}Sr as a Tracer of Weathering

The weathering of carbonate rocks and silicate rocks can be distinguished not only on the basis of DIC and its isotopes, but also with the abundance of stable ^{87}Sr as normalized to another strontium isotope, ^{86}Sr. As a divalent alkaline earth element, Sr^{2+} substitutes for Ca^{2+} in carbonates, sulfates, feldspars, and other rock-forming minerals. However, its $^{87}Sr/^{86}Sr$ ratio varies greatly based on the potassium content of the parent rock. Potassium-bearing rocks contain the alkali element rubidium, Rb^+, which can substitute for K^+. The radioisotope ^{87}Rb decays to radiogenic ^{87}Sr and over time enriches the $^{87}Sr/^{86}Sr$ ratio in K-bearing rocks.

$$^{87}Rb \rightarrow {}^{87}Sr + \beta^- \qquad T_{1/2} = 47.5 \text{ billion years}$$

The primordial $^{87}Sr/^{86}Sr$ ratio that existed at the time of nucleosynthesis some 5 billion years ago was 0.699, as derived from meteorites. Contributions from the decay of primordial ^{87}Rb have increased this ratio over the life of the earth. Alkaline rocks such as granites with abundant K-feldspars have abundant Rb^+ and so have enriched $^{87}Sr/^{86}Sr$ ratios of 0.71 to 0.74 and higher. Potassium-poor terrains such as basalts, which are dominated by Ca-feldspars, have low $^{87}Sr/^{86}Sr$ ratios below 0.709. Submarine alteration of K-poor basalts along mid-oceanic ridges is the dominant source of Sr in seawater, which today has a low $^{87}Sr/^{86}Sr$ ratio of 0.709. This has varied between 0.707 and 0.709 over geological time due to tectonic controls on rates of continental weathering, which contribute ^{87}Sr from K-rich shield terrains. The seawater $^{87}Sr/^{86}Sr$ curve through Phanerozoic time (Veizer 1989) is an important baseline for tracing ^{87}Sr in groundwaters. Like seawater, marine carbonates have a low $^{87}Sr/^{86}Sr$ ratio between about 0.707 and the modern seawater value of 0.709, in contrast with greater $^{87}Sr/^{86}Sr$ in drainage from crystalline (granitic) terrains. Accordingly, the $^{87}Sr/^{86}Sr$ of Sr^{2+} is a good indicator of the dominant terrain in which groundwater has recharged and weathering took place (Figure 5.11).

In a study to identify groundwater flow paths, Frost and Toner (2004) found well-defined ranges for $^{87}Sr/^{86}Sr$ and Sr^{2+} concentrations in wells completed within carbonate-cemented Casper sandstone and overlying carbonate and evaporite Redbeds of the Mesozoic Laramie Basin and underlying Precambrian crystalline basement (Figure 5.12). They show groundwaters are recharged in their respective aquifer outcrop regions, where they gain their strontium isotopic signature from weathering. The springs in this basin are shown to be flowing from both the Casper sandstones and the Precambrian basement aquifer.

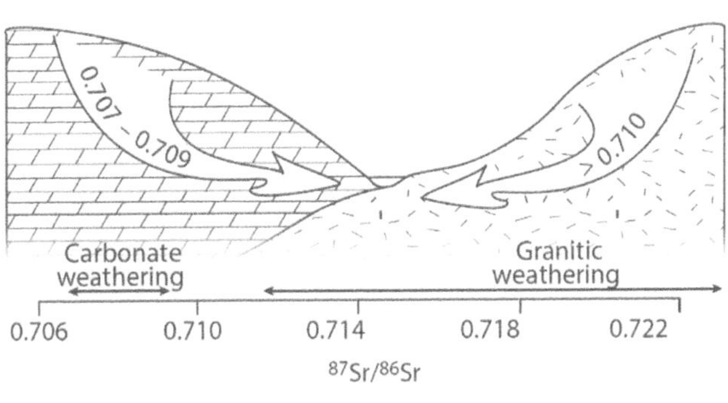

FIGURE 5.11 Strontium isotope ratios in groundwaters from weathering of carbonate terrain and crystalline (granitic) terrain.

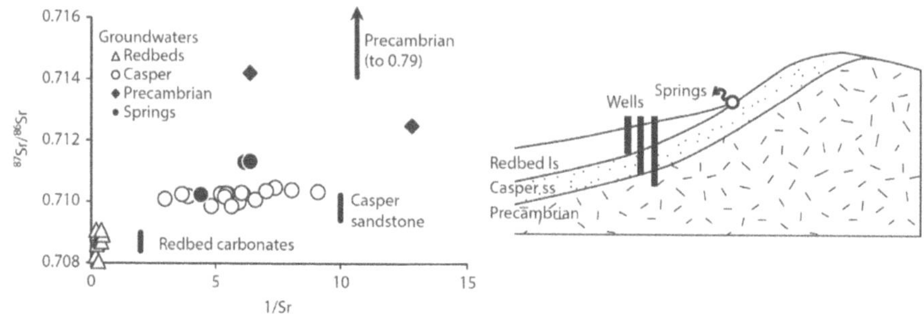

FIGURE 5.12 Groundwaters from wells completed in the three main aquifers have $^{87}Sr/^{86}Sr$ ratios that closely associate with those of the host aquifer materials in the Laramie Basin, Wyoming. Modified from Frost and Toner 2004.

WEATHERING AND ALKALINITY

The consumption of acidity during mineral weathering results in the production of alkalinity (alk)—a measure of the anions in water that can buffer pH during the addition of acid. As carbonic acid is the dominant weathering agent in most natural groundwaters, carbonate alkalinity provides the principal acid buffering capacity in water. The major anions that provide this acid-buffering capacity are HCO_3^- and CO_3^{2-}, with minor contributions in some groundwaters from dissociated weak acids, including hydrogen sulphide (HS^-) at neutral pH plus borate ($H_2BO_3^-$), silicate ($H_3SiO_4^-$), and even hydroxide (OH^-) at high pH. Their acid buffering reaction is simply the inverse of the dissociation of these weak acids, here with their dissociation pH (see also Table 2.3).

$$HCO_3^- + H^+ \Leftrightarrow H_2CO_3 \qquad \text{pH } 6.35$$

$$HS^- + H^+ \Leftrightarrow H_2S \qquad \text{pH } 6.99$$

$$H_2BO_3^- + H^+ \Leftrightarrow H_3BO_3 \qquad \text{pH } 9.24$$

$$H_3SiO_4^- + H^+ \Leftrightarrow H_4SiO_4 \qquad \text{pH } 9.83$$

$$CO_3^{2-} + H^+ \Leftrightarrow HCO_3^- \qquad \text{pH } 10.3$$

$$OH^- + H^+ \Leftrightarrow H_2O \qquad \text{pH } 14.0$$

Measurements of alkalinity have units of equivalents per liter, and so the molal concentration of all contributing species is multiplied by their valence. Most alkaline anions are monovalent. As CO_3^{2-} is bivalent, its molality is doubled to give equivalents.

DIC, P_{CO_2} AND ALKALINITY

As H_2CO_3 is consumed during weathering, the pH increases and the groundwater gains alkalinity, which are the anions of weak acids that can buffer acidity. Alkalinity is the sum of the concentrations of all acid-consuming anions and is expressed in units of equivalents. Weathering of carbonates contributes carbonate alkalinity with increasing concentrations of HCO_3^- and CO_3^{2-}, which in most systems are the dominant contributors to alkalinity. At high pH, the anions of dissolved silica ($H_3SiO_4^-$) and boron ($H_2BO_3^-$) can contribute if present in significant concentrations. Similarly, if present, HS^- will contribute at neutral pH. For most waters, total alkalinity (alk_T) is represented by the following:

$$\text{Total alkalinity, alk}_T = m_{HCO_3^-} + 2m_{CO_3^{2-}} + m_{H_3SiO_4^-} + m_{HS^-} + m_{H_2BO_3^-} + m_{OH^-} - m_{H^+}$$

Carbonate alkalinity is the sum of the carbonate species alone:

$$\text{Carbonate alkalinity, alk}_C = m_{HCO_3^-} + 2m_{CO_3^{2-}}$$
$$= \text{alk}_T - m_{H_3SiO_4^-} - m_{HS^-} - m_{H_2BO_3^-} - m_{OH^-}$$

In circum-neutral pH groundwater where bicarbonate dominates, this is the species that contributes the most to alkalinity and DIC. In lower pH groundwater, carbonic acid contributes to DIC but not to alkalinity:

$$\text{pH} \approx 7 \text{ to } 9 \qquad \text{alk}_C = \text{DIC} = m_{HCO_3^-}$$
$$\text{pH} <\sim 7 \qquad \text{alk}_C = m_{HCO_3^-} \qquad \text{DIC} = m_{HCO_3^-} + m_{H_2O_3}$$

Alkalinity can be measured in the field by titration with acid (see section in *Groundwater Geochemistry and Isotopes* Workbook), and is routinely taken as a measurement of bicarbonate, which is the dominant anion in non-saline groundwaters. Take, for example, the following water quality analysis for a karst spring, with field measurements: $T = 10°C$, pH 7.95 and $alk_T = 1.9$ meq/kg.

	SO_4^{2-}	Cl^-	Ca^{2+}	Mg^{2+}	Na^+	K^+
ppm	6.2	1.1	39	2.8	4.3	0.18
meq/kg	0.13	0.03	0.19	0.23	0.03	0.005

The charge balance of the analysis without bicarbonate is highly skewed ($2m_{Ca^{2+}} + 2m_{Mg^{2+}} + m_{Na^+} + m_{K^+} = 2.06\,\text{meq/kg}$, $2m_{SO_4^{2-}} + m_{Cl^-} = 0.16\,\text{meq/kg}$). At this neutral pH and with no evidence of hydrogen sulfide, then,

$$\text{alk}_T = \text{alk}_C = m_{HCO_3^-} = 1.92 \text{ meq/kg.}$$
$$m_{HCO_3^-} + 2m_{SO_4^{2-}} + m_{Cl^-} = 2.06 \text{ meq/kg} \qquad \therefore \text{charge neutrality}$$

The relationship between P_{CO_2} and DIC, discussed in Chapter 2, relates to carbonate alkalinity through the distribution of DIC species. As groundwater below the water table (i.e., with no additional CO_2) consumes carbonic acid through mineral weathering, P_{CO_2} drops and carbonate alkalinity increases. Alkalinity measurements are often made instead of DIC measurements to determine DIC species, and can be used in calculations of P_{CO_2} and calcite saturation.

Extreme Alkalinity: High pH versus High P_{CO_2}: A Tale of Two Travertines

The following two case studies are remarkably different examples of extremely high alkalinity in natural groundwaters. In Jordan, the Maqarin springs discharge with pH 12.5 from a contact metamorphic zone, producing remarkable stalactites and stalagmites (Figure 5.13). The Edziza springs discharge with pH 6.59 from basalts on the flank of a recently active volcano producing extensive fields of travertine (Figure 5.14). The spring waters in these two different settings have almost identical alkalinities of 40 meq/kg. They are also both rapidly depositing calcite travertine.

	Maqarin	Edziza
T	25	45.9
pH	12.5	6.59
alk_T meq/kg	39.5	40.2
OH^- mmol/kg	39.5	0.000185
CO_3^{2-} mmol/kg	0.02	<0.01
HCO_3^- mmol/kg	<0.01	40.2
Ca^{2+}+Mg^{2+} mmol/kg	27.9	9.87

FIGURE 5.13 The hyperalkaline groundwaters of Maqarin, Jordan (Khoury et al. 1992; Clark et al. 1993) discharge with a pH of 12.5. They flow from a contact metamorphic zone with cement-like minerals such as portlandite ($Ca[OH]_2$) ($K = 10^{-5.19}$); upon discharge, uptake of atmospheric CO_2 by these Ca-OH waters causes $CaCO_3$ supersaturation and formation of travertine. Inset shows meter-tall stalagmites growing from drips of hyperalkaline groundwater from the roof of an adit constructed through the metamorphic zone.

FIGURE 5.14 Travertine deposits from hot springs on the Mount Edziza volcanic complex (Clark et al. 1989). The hot spring waters have a very high alkalinity of 40.2 meq/kg but with the low pH of only 6.59, their alkalinity is essentially from bicarbonate. They discharge on the flank of a recently active (Holocene) volcano, where the hot magma is decomposing limestone to wollastonite and producing metamorphic CO_2 that is dissolved by circulating groundwater. CO_2 degassing at spring vents (inset photo) produces extensive fields of travertine draping the hill slopes.

However, the origin of these travertines could not be more different. The key feature differentiating the origin of their travertine is P_{CO_2}.

Calculating P_{CO_2} for Maqarin:

$$m_{CO_3^{2-}} = 2\times 10^{-5}\,\text{mol/kg}$$

$$\text{alk}_C = 2\times m_{CO_3^{2-}} = 0.04\text{ meq/kg or } 4\times 10^{-5}\text{ eq/kg}$$

$$m_{HCO_3^-} = \frac{m_{CO_3^{2-}}\times 10^{-pH}}{K_2} = \frac{2\times 10^{-5}\times 10^{-12.5}}{10^{-10.32}} = 1.32\times 10^{-7}$$

$$m_{H_2CO_3} = \frac{m_{HCO_3^-}\times 10^{-pH}}{K_1} = \frac{1.32\times 10^{-7}\times 10^{-12.5}}{10^{-6.34}} = 9.14\times 10^{-14}$$

$$P_{CO_2} = \frac{a_{H_2CO_3}}{K_{CO_2}} = \frac{9.14\times 10^{-14}}{10^{-1.48}}$$

$$= 2.76\times 10^{-12} = 10^{-11.56}$$

$$\log P_{CO_2} = -11.56$$

Calculating P_{CO_2} for Edziza:

$$m_{HCO_3^-} = \text{alk}_C = 40.2 \text{ meq/kg}$$

$$K_1 = \frac{a_{HCO_3^-} \times a_{H^+}}{a_{H_2CO_3}} \approx \frac{m_{HCO_3^-} \times m_{H^+}}{m_{H_2CO_3}}$$

$$m_{H_2CO_3} = \frac{m_{HCO_3} \times m_{H^+}}{K_1} = \frac{0.0402 \times 10^{-6.59}}{10^{-6.26}}$$

$$= 0.0188 \text{ mol/kg} = 18.8 \text{ meq/kg}$$

$$P_{CO_2} = \frac{a_{H_2CO_3}}{K_{CO_2}} = \frac{0.0188}{10^{-1.69}}$$

$$= 0.94 = 10^{-0.03}$$

$$\log P_{CO_2} = -0.03$$

Comparing these P_{CO_2} values with air ($\log = P_{CO_2} - 3.4$) we see that the Maqarin waters are *undersaturated* with CO_2 by over 8 orders of magnitude. Upon discharge to the open air, these Ca-OH waters take up atmospheric CO_2 and precipitate $CaCO_3$,

$$Ca^{2+} + 2OH^- + CO_{2\,\text{atm}} \rightarrow CaCO_3 + H_2O$$

In contrast, the Edziza spring waters are 2500 times *oversaturated* with CO_2 compared with the atmosphere, and so rapidly degas CO_2 as seen in the photo inset, precipitating calcite travertine:

$$Ca^{2+} + 2HCO_3^- \rightarrow Ca^{2+} + CO_2 + CO_3^{2-} + H_2O \rightarrow CO_2 + H_2O + CaCO_3$$

The resulting travertines for both are spectacular and extensive, forming very rapidly. The carbon isotopes in these are also diagnostic of their origin. The $\delta^{13}C$ of the DIC from the Edziza hot spring is –1.2‰, in the range for marine limestone. Its ^{14}C activity is 0 pMC.

In contrast, the $\delta^{13}C$ of the Maqarin travertines is –25‰, far lower than atmospheric CO_2 ($\delta^{13}C_{CO_2\,\text{atm}} = -8‰$) due to rarely observed kinetic effects during the reaction of $CO_{2(aq)}$ with OH^- in hyperalkaline Ca–OH^- waters (Clark et al. 1992). The Maqarin travertines also have an interesting feature in that their ^{14}C is modern, with 100 pMC.

ADVANCED WEATHERING: SAPROLITES AND LATERITES

Terrestrial landscapes are dominated by silicate and aluminosilicate minerals, from the metamorphic shield terrains at the core of the continents to the extensive flood basalts and plutonic rocks in igneous terrains. Continental-scale weathering of feldspar in silicate landscapes is the primary sink for atmospheric CO_2 on geological time scales, as observed for the alteration of anorthite and alkali feldspars to clay:

$$CaAl_2Si_2O_{8\,\text{anorthite}} + H_2O + 2H_2CO_3 \rightarrow Ca^{2+} + 2HCO_3^- + Al_2Si_2O_5(OH)_{4\,\text{kaolinite}}$$

However, weathering does not necessarily stop at this point, as weathering under humid tropical climates leads to further leaching of silica. The result is the formation of highly insoluble aluminum hydroxide, gibbsite.

$$Al_2Si_2O_5(OH)_{4\ \text{kaolinite}} + 7H_2O \rightarrow 2Al(OH)_{3\ \text{gibbsite}} + H^+ + 2H_4SiO_4^\circ$$

Extensive weathering of ferromagnesian minerals, such as pyroxene, leads to the formation of iron hydroxide:

$$Fe_2Si_2O_{6\ \text{pyroxene}} + 2H_2O + 4H_2CO_3 \rightarrow 2Fe^{2+} + 4HCO_3^- + 2H_4SiO_4^\circ$$

$$2Fe^{2+} + \tfrac{1}{2}O_2 + 5H_2O \rightarrow 2Fe(OH)_{3\ \text{ferrihydrite}} + 4H^+$$

Such alteration of feldspars and mineral dissolution weathers silicate bedrock to a residual mantle of saprolite – claystones with iron and aluminum hydroxide weathering products. Saprolites are typically found in lower-latitude regions with substantial seasonal precipitation, which favors growth of vegetation and high soil P_{CO_2}. Continued weathering and leaching of silica from these sections generates extensive laterites, which are essentially iron and aluminum hydroxide soils.

The weathering of the basalts in western and northern India is a good example. Extensive weathering over millions of years in the monsoon climate of western India has produced deep accumulations of laterites overlying the Deccan Trap flood basalts. The major oxide geochemistry of the basalts is presented in the column chart in Figure 5.15, showing the basalts to be dominated by the calcic anorthite [$CaAl_2Si_2O_8$] with significant pyroxene [$Ca(Mg,Fe)Si_2O_6$] and other ferromagnesian

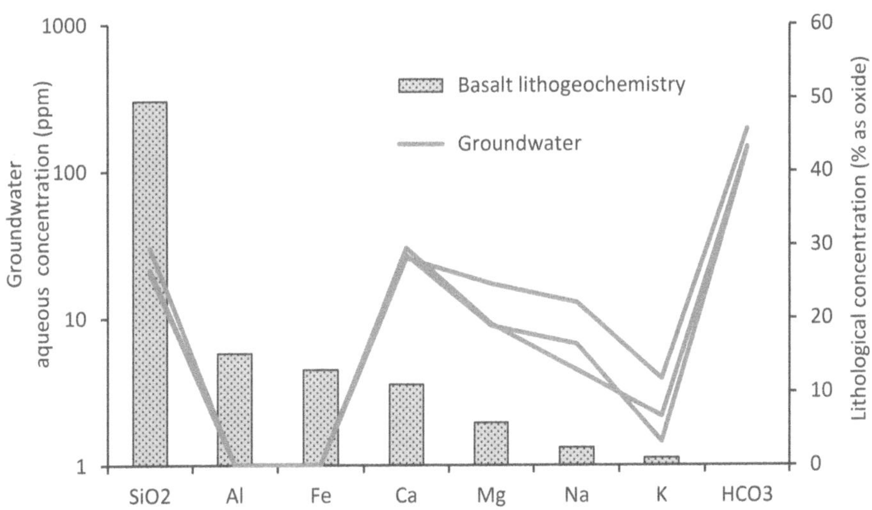

FIGURE 5.15 Geochemistry of drainage from laterite development on basalt, from the Deccan Traps volcanics in the Western Ghats of India. Lithogeochemistry of the basalts is from Cox and Hawkesworth (1985), and aqueous geochemistry of discharge waters from the region is from Das et al. (2005).

FIGURE 5.16 Solubility of the three most abundant elements of silicate rocks (after oxygen). The higher solubility of silica from feldspar and clay weathering removes this from the accumulating saprolite, leaving behind highly insoluble ferric iron hydroxides (laterite) and aluminum hydroxides (bauxite) capping weathered igneous and metamorphic parent rocks.

minerals. The drainage from weathering of these formations has a Ca-HCO_3 geochemical facies, with high dissolved Si and the soluble base cations (Ca^{2+}, Mg^{2+}, Na^+, and K^+). The high bicarbonate concentration is generated by the open system weathering of the basalts with carbonic acid incorporated from these tropical soils with high P_{CO_2}. Neither Fe nor Al is significant in drainage from these rocks, due to the very low solubility of Fe and Al hydroxides compared with silica, as shown in Figure 5.16.

Under the seasonal monsoon climate with an extensive dry season, these hydroxides stabilize to oxides and oxi-hydroxides including ferrihydrite (Fe[OH]$_3$), goethite (FeOOH), and hematite (Fe_2O_3), with bauxite ores of gibbsite (Al[OH]$_3$) and diaspore (AlOOH). Over millions of years of weathering, laterites become increasingly dominated by Fe and Al oxy-hydroxide minerals. Given that the solubility of gibbsite is even lower than ferric iron hydroxide (Figure 5.16), over time the aluminum-to-iron ratio in laterites increases, generating bauxite ore suitable for aluminum smelting.

6 Geochemical Evolution

INTRODUCTION

The major ion geochemistry of groundwater is established in the recharge environment, where oxygen and carbonic acid in soil drive mineral weathering reactions. Shallow groundwater recharged through most soils creates an initial Ca^{2+}–HCO_3^- geochemical facies. Below the water table, access to O_2 and CO_2 diminishes and natural groundwater geochemistry evolves through a range of processes, including cation exchange, redox reactions, and increased salinity (Figure 6.1).

CATION EXCHANGE

Aquifer minerals are bristling with mostly negative surface charges generated by oxygen ions at the mineral surface. Cations in solution compete for these negative sites where they bind electrostatically. The predominance of aluminosilicate clay minerals in aquifers makes them perhaps the most available substrate for cation exchange. Others include ferric oxyhydroxides, humics, and in rare natural environments, zeolite.

Clay minerals are comprised of alumino-silicate sheets in tetrahedral layers—sheets of Si^{4+} coordinated with O^{2-}, and octahedral layers—layers of Al^{3+} coordinated with O^{2-} (the gibbsite [$Al(OH)_3$] layer) (Figure 6.2). During the weathering of ferromagnesian minerals (olivine, pyroxene, and amphibole), Al^{3+} is replaced by Mg^{2+} (the brucite [$Mg(OH)_2$] layer). The sheets are joined by weak van der Waals' bonds into a single crystal.

There is a great range of clay minerals, but three are common in low-temperature systems.

- Kaolinite [$Al_2Si_2O_5(OH)_4$] is a two-layer clay that forms in humid climates from acid weathering of granites. Note that kaolinite has no interlayer positions, which limits its ion exchange capacity to surface charge sites. Other two-layer clays include serpentine, in which a brucite layer replaces the gibbsite layer.

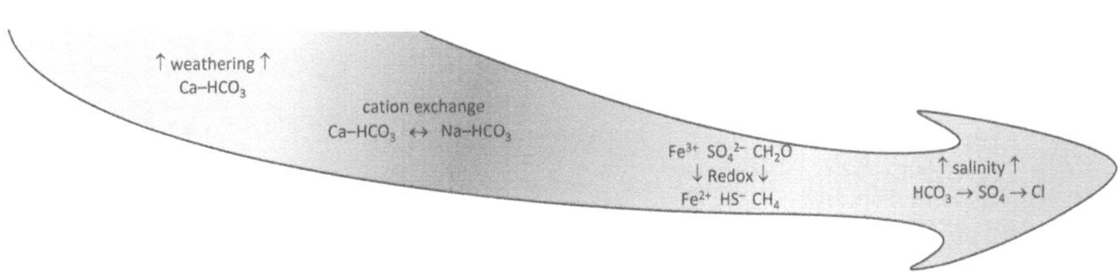

FIGURE 6.1 Generalized trends in geochemical evolution for groundwater. Each process depends on the mineralogy encountered along the flow path and mixing with other flow systems.

DOI: 10.1201/9781003353836-6

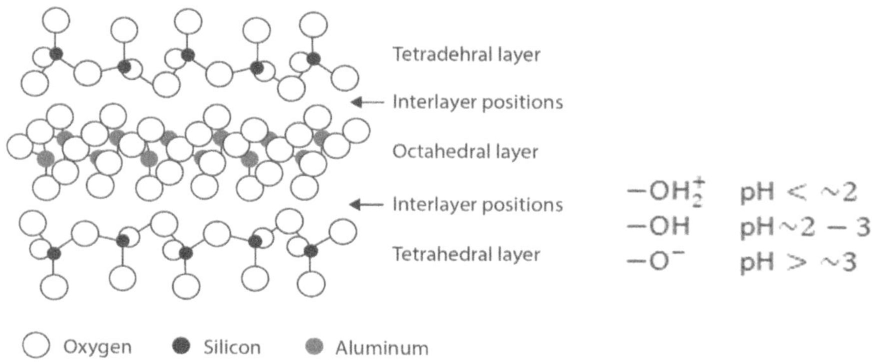

FIGURE 6.2 Tetrahedral and octahedral layers of clay minerals. Oxygen atoms on surfaces provide a negative surface charge at most pHs.

- Smectite $[(Na,Ca_{0.5})(Mg,Fe,Al)_2Si_4O_{10}(OH)_2]$ is a three-layer clay that forms from weathering of mafic (Fe-, Mg-, and Ca-rich, K-poor rocks). These clays, and particularly the Na-montmorillonite variety, swell from the uptake of water and cations to satisfy excess negative interlayer charges. This gives the smectite clays a high cation exchange capacity (CEC).
- Illite $[K(Fe,Mg,Al)_2(AlSi_3)O_{10}(OH)_2]$ a three-layer clay that forms from weathering of potassic rocks. Unlike smectites, the interlayer ion, K^+, is firmly positioned and stabilizes the structure with no swelling and so reduces its CEC.

Other important exchange surfaces in soils and aquifers include organic humic substances. These are the principal organic components of soils and develop through the biodegradation of vegetation. They are heterogeneous, high-molecular-weight C–O–H–(N, S) compounds (Figure 6.3). Humic acid is soluble only at higher pH, whereas humin is the insoluble, refractory component in soil that retains K^+ and other nutrients for plants. Structurally, humic substances are three-dimensional networks of aromatic rings that are fringed with various functional groups, including

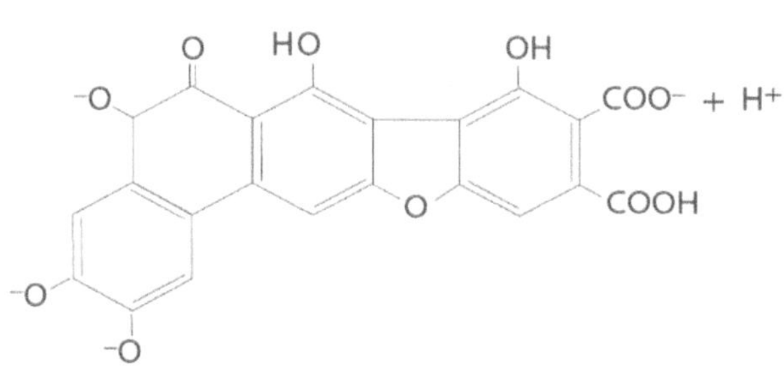

FIGURE 6.3 Schematic of a component of humic matter showing negatively charged exchange sites, which can be occupied by H^+ in low-pH waters by major cations from solution or trace metals.

carboxyl (–COOH) and phenol (–OH). These groups can shed their H^+, providing negative sites for other cations in the pore waters (Figure 6.3).

Iron and manganese released by weathering readily oxidize to form very insoluble, amorphous oxyhydroxides like ferrihydrite [$Fe(OH)_3$] and [MnO_2OH] as coatings on mineral grains in soils and aquifers. Such amorphous solids represent excellent substrates for cation exchange and adsorption of trace metals.

Substrates with the highest CEC are zeolites, a low-temperature class of highly hydrated tectosilicates with an open porous structure formed by silica tetrahedrons linked corner to corner. They occur as a product of weathering of volcanic rocks and feldspar-rich sediments under conditions of high pH and high salinity:

$$NaAlSi_3O_{8\,\text{albite}} + 3H_2O \rightarrow NaAlSi_3O_8 \cdot H_2O_{\text{analcime}} + H_4SiO_4$$

$$CaAl_2Si_2O_{8\,\text{anorthite}} + 2H_2SiO_4 \rightarrow CaAl_2Si_4O_{12} \cdot 4H_2O_{\text{laumontite}}$$

Both unbalanced charge deficiencies and the high surface area of these minerals make them excellent substrates for cation exchange. Combined with their open water-filled lattice, zeolites are important industrial minerals used as molecular sieves to purify waters and gases.

Cation Exchange Capacity

The most versatile model for cation exchange involves a double layer of cations electrostatically bound to the clay surface (Figure 6.4). The electrostatic potential is greatest along the negatively charged surface and maintains a fixed layer of cations. Outward from the fixed layer, the electrostatic potential decreases, creating a diffuse layer characterized by a cation excess over anions that diminishes with distance. Cations in the diffuse layer can be expected to exchange more readily than those in the fixed layer.

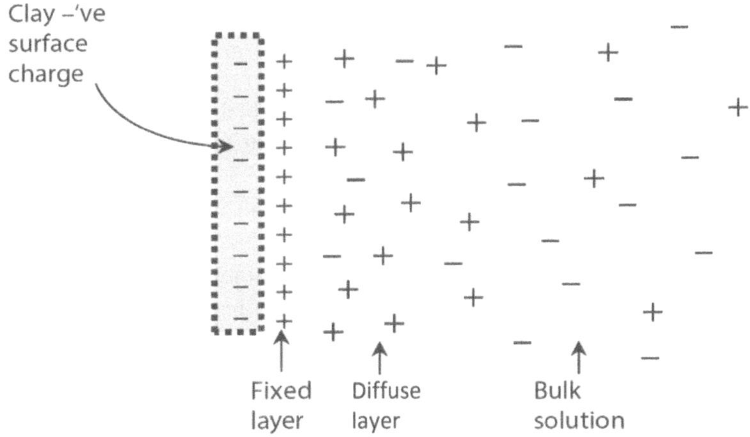

FIGURE 6.4 Double-layer model of the negative charge distribution on a clay mineral and cations attached at exchange sites.

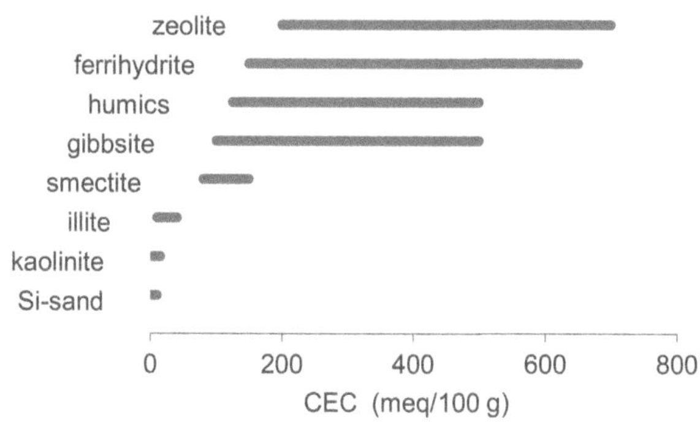

FIGURE 6.5 Ranges of cation exchange capacity for different aquifer materials at neutral pH (clay data from Grim 1968).

The density of exchange sites on this surface is its CEC determined as the number of exchange sites per unit dry weight of sample. It is measured by saturating the sample with a 1 molar ammonium acetate solution at pH 7 and analyzing the release of NH_4^+ following equilibration with a (usually) Na^+ solution. CEC is expressed as meq/100 g dry sample.

Of all natural materials, zeolite has the highest range for CEC (Figure 6.5), making it a good material for treating contaminated groundwater with in-situ barriers (see contaminant sorption in Chapter 8). Of the principal clay groups identified above, smectites have the highest CEC. By contrast, K^+ is retained in interlayer positions of illite clays, reducing their CEC to less than 50 meq/100 g. Kaolinites have tightly spaced layers with no interlayer cations and so have low CEC, generally less than 10 meq/100 g (Figure 6.5). The CEC of ferric iron oxyhydroxides (ferrihydrite) is even greater than that of smectite clays at neutral to high pH. Below approximately pH 4, ferrihydrite becomes increasingly soluble and retains no capacity for cation exchange.

Exchange Selectivity Coefficient

The exchange of a cation in solution with another adsorbed to a clay surface can be represented by the following reaction, here using the example of Na^+ exchange with Ca^{2+}:

$$\text{Ca-clay} + 2\text{Na}^+ \Leftrightarrow 2\text{Na-clay} + \text{Ca}^{2+}$$

Under steady-state conditions, an equilibrium is established between the solution and the surface that governs the concentration of adsorbed cations according to their concentration in solution. The constant for exchange on a given surface and solution is the selectivity coefficient:

$$K_{\text{Na-Ca}} = \frac{\left[\text{Clay}_{\text{Na}}\right]^2 \times \left[\text{Ca}^{2+}\right]}{\left[\text{Clay}_{\text{Ca}}\right] \times \left[\text{Na}^+\right]^2}$$

The activities of the dissolved cations are determined from ionic strength and activity coefficients. The adsorbed cation activities are most easily considered by their molar fraction of the total of exchange sites on the clay surface:

$$\left[\mathrm{Clay_{Na}}\right] = \frac{\text{moles } \mathrm{Na^+} \text{on clay}}{\text{moles } \mathrm{Na^+} + \mathrm{Ca^{2+}} \text{on clay}}$$

Each cation has its own selectivity for sorption relating to the surface charge of the cation and the stability of its hydration sheath, which reduces its electrostatic interaction. Divalent cations have a greater tendency for exchange onto surfaces than monovalent ions. Cations with smaller hydrated radii (Table 2.4) are also more readily adsorbed:

$$\textit{more soluble} \rightarrow$$

$$\mathrm{Ba^{2+} > Sr^{2+} > Ca^{2+} > Mg^{2+}}$$

$$\mathrm{Cs^{+} > Rb^{+} > K^{+} > Na^{+} > Li^{+}}$$

Groundwater evolution by cation exchange is a geochemical response to a disequilibrium between the groundwater geochemistry and the exchange medium. Road salt contamination in groundwater is common in Nordic countries. Typically, salt-impacted groundwater recharge takes place through a shallow glacial till with Ca-clays produced through dissolution of carbonates and feldspars (Figure 6.6). As the plume of Na–Cl water infiltrates the till, Na^+ begins to displace Ca^{2+} from the exchange sites.

The salt-impacted groundwater moving through the till begins to evolve from a Na–Cl water to a Ca–Cl water. If Na–Cl water continues to infiltrate through the till, it will eventually reestablish an equilibrium with the clays, converting them from Ca-clays to Na-clays, and the discharging flow will return to an Na–Cl facies. In such cases, cation exchange is a transient geochemical condition. Nonequilibrium conditions are also generated by a change in groundwater flow paths through new material. Examples include erosion and uplift, allowing groundwater circulation through deeper bedrock horizons, and the reestablishment of groundwater flow through previously inundated (marine) sediments.

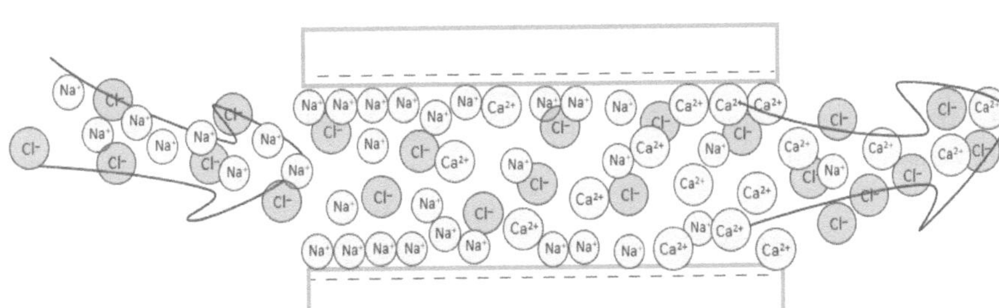

FIGURE 6.6 Sketch of road-salt-impacted groundwater moving through a pore throat in clay till that previously hosted Ca^{2+}–HCO_3^- groundwater. The cation population in inflowing Na–Cl water is dominated by Na^+, which exchanges with and releases Ca^{2+} to the outflowing Ca–Cl groundwater.

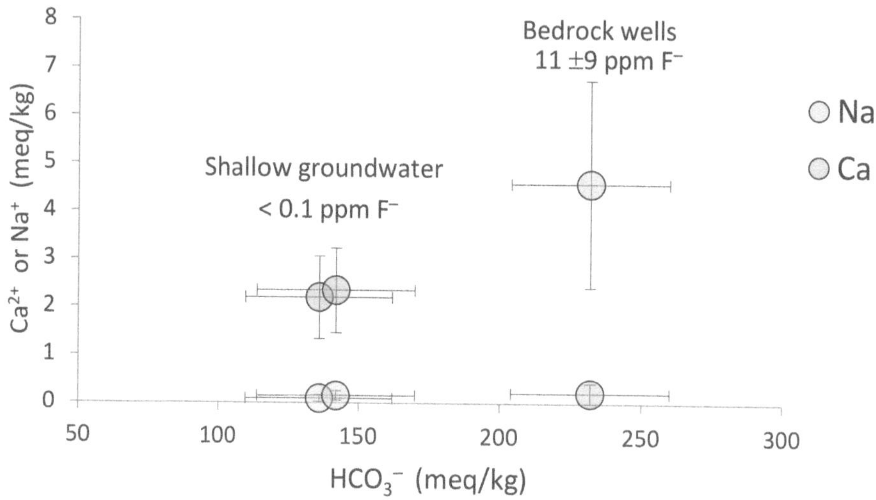

FIGURE 6.7 Natural water softening in deep groundwaters in Gaspé, Quebec (data from Boyle and Chagnon 1995). Decreased Ca^{2+} activity in the deep groundwaters has allowed dissolution of fluorite, CaF_2. Consequently, the high F^- in these groundwaters has caused cases of skeletal fluorosis in the local population.

Natural Groundwater Softening

The process of natural water softening by cation exchange is observed in the Tertiary shales of northern New Brunswick (Boyle and Chagnon 1995). Here, the gradual erosion and exposure of deeper marine shales has allowed Ca–HCO_3 groundwaters to circulate more deeply and exchange with the Na-clays of these sediments, transforming these waters to a Na–HCO_3 facies (Figure 6.7). The reduction of Ca^{2+} activity has also reduced the saturation index for calcite, allowing dissolution of this mineral and an increase in HCO_3^-.

Although softened water is generally considered an attribute for domestic water supplies, in this case it has caused a serious water quality issue. Fluorite [CaF_2] is common in sediments and has low solubility in most groundwaters maintained by elevated Ca^{2+}. As Ca^{2+} is removed from solution by exchange onto the clays, fluorite can dissolve, adding F^- to the groundwaters. The drinking water limit for F^- is set at 2 ppm. Dental fluorosis can occur above this level, with risk of skeletal fluorosis above about 5 to 10 ppm.

REDOX EVOLUTION

The change in redox conditions during flow from the oxidizing conditions in the recharge environment to reducing conditions represents a second evolution pathway for groundwater geochemistry. In natural and contaminated settings, a range of redox reactions can occur, depending on the availability of both electron donors, such as organic carbon or sulfides, and electron acceptors, including elemental oxygen, nitrate, ferrihydrite, and sulfate.

In most cases, redox evolution is driven by an excess of organic carbon, whether transported with groundwater from the recharge environment or gained in the subsurface from buried organics, such as peat in Quaternary sediments, hydrocarbons from marine sediments, or from contaminated settings such as landfills. The sequence of reactions during redox evolution takes place within the range of electromotive potentials that define the pe-pH stability field for water (Figure 6.8; see Chapter 2, Redox).

O_2/H_2O: Aerobic Oxidation

The first step along the sequence of redox reactions is the aerobic oxidation of organic carbon, although other aerobic oxidation reactions can occur with electron donors from lower on the electromotive potential scale in Figure 6.8. The reaction is highly energetic, and so aerobic bacteria outcompete those operating with electron acceptors at lower electromotive potential.

Half-cell reaction: $\frac{1}{2}O_2 + 2e^- + 2H^+ \rightarrow H_2O$ $\quad \Delta G^\circ_r = -237.14\,\text{kJ/mol}$

Respiration: $CH_2O + O_2 \rightarrow CO_2 + H_2O$ $\quad \Delta G^\circ_r = -501.84\,\text{kJ/mol}$

Nitrification: $NH_4^+ + 2O_2 \rightarrow NO_3^- + H_2O + 2H^+$ $\quad \Delta G^\circ_r = -266.5\,\text{kJ/mol}$

Iron oxidation: $Fe^{2+} + \frac{1}{4}O_2 + H^+ \rightarrow Fe^{3+} + \frac{1}{2}H_2O$ $\quad \Delta G^\circ_r = -192.89\,\text{kJ/mol}$

Pyrite oxidation:
$FeS_2 + 3\frac{1}{2}O_2 + H_2O \rightarrow Fe^{2+} + 2SO_4^{2-} + 2H^+$ $\quad \Delta G^\circ_r = -1177.8\,\text{kJ/mol}$

The O_2/H_2O redox half-cell reaction defines the upper stability boundary for water. Respiration proceeds through the unsaturated zone and below the water table, to the point where dissolved oxygen is depleted or no more electron donors are

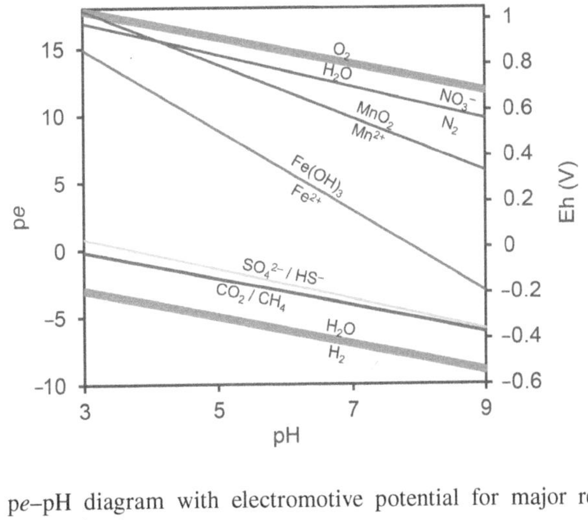

FIGURE 6.8 pe–pH diagram with electromotive potential for major redox pairs (see Chapter 2, Redox Reactions).

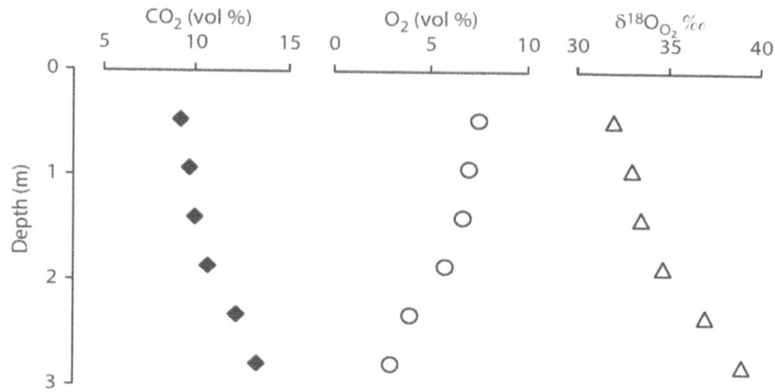

FIGURE 6.9 Profiles for CO_2, O_2, and $\delta^{18}O_{O_2}$ in unsaturated sand, showing the downward increase in CO_2 and decrease in O_2 consistent with aerobic degradation of organic carbon (modified from Birkham et al. 2007). The increase in $\delta^{18}O_{O_2}$ demonstrates reactive loss of O_2 where the aerobic bacteria discriminate against ^{18}O during respiration, leaving the residual O_2 enriched in ^{18}O.

available. This O_2 depletion is observed in the vadose zone of fine sand in northern Alberta, where vertical profiles of CO_2 and O_2 concentrations in the soil atmosphere were measured (Figure 6.9; Birkham et al. 2007). The profiles of these gases show a downward trend in reactant and product concentrations, which is confirmed by the enrichment trend in $\delta^{18}O$ of the O_2 due to fractionation by microbial activity.

NO_3/N_2: Denitrification

Nitrate is one of the most soluble anions, and its removal from surface and groundwaters is mainly achieved by bacterial reduction to elemental nitrogen, N_2. In this reaction, an electron donor, such as organic carbon, is oxidized. The resulting release of energy is significant and so is mediated by bacteria. Organic carbon, as either dissolved organic carbon (DOC) or solid organic substrate in the aquifer, is typically required, although other electron donors with lower electromotive potential, like pyrite, have been observed (e.g., Tesoriero and Puckett 2011).

$$\text{Half-cell reaction: } NO_3^- + 5e^- + 6H^+ \rightarrow \tfrac{1}{2}N_2 + 3H_2O \qquad \Delta G^\circ_r = -602.68\ \text{kJ/mol}$$

Denitrification:

$$CH_2O + \frac{4}{5}NO_3^- + \frac{4}{5}H^+ \rightarrow \frac{2}{5}N_2 + CO_2 + \frac{7}{5}H_2O \qquad \Delta G^\circ_r = -509.7\ \text{kJ/mol}$$

Denitrification is evident from the enrichment in both $\delta^{15}N$ and $\delta^{18}O$ in the residual nitrate (Böttcher et al. 1990). Denitrifying bacteria produce ^{15}N-depleted N_2, leaving the residual nitrate enriched in ^{15}N but also in ^{18}O. As the nitrate is consumed, the isotope enrichment for both ^{15}N and ^{18}O follows a Rayleigh distillation (Figure 6.10). The organic carbon substrate is transformed to inorganic carbon (or sulfate if pyrite

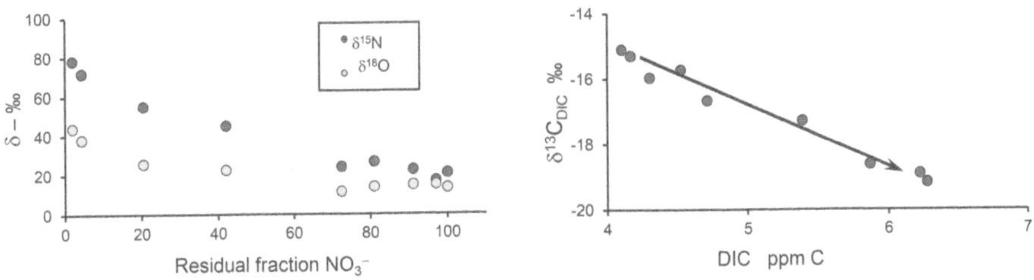

FIGURE 6.10 Left: Enrichment in $\delta^{15}N$ and $\delta^{18}O$ for residual nitrate during denitrification, following a Rayleigh distillation trend toward very enriched residual nitrate. Right: Depletion in $\delta^{13}C$ of DIC during denitrification due to contributions to total DIC from oxidation of organics. Modified from Suchy 2008.

is providing electrons), contributing ^{13}C-depleted carbon to the dissolved inorganic carbon (DIC) reservoir.

MnO_2/Mn^{2+}: Manganese Reduction

Like iron, the oxidized form of this transition metal is highly insoluble manganese oxide pyrolusite [MnO_2], often found with iron oxyhydroxides as a secondary precipitate on grains in clastic aquifers or on fracture surfaces in bedrock. The manganese redox boundary occurs at a considerably higher *pe* than iron, following nitrate on the electromotive ladder of redox couples.

$$\text{Half-cell reaction: } MnO_2 + 2e^- + 4H^+ \Leftrightarrow Mn^{2+} + 2H_2O \qquad \Delta G^\circ_r = -237.45\,\text{kJ/mol}$$

Manganese reduction:

$$CH_2O + 2MnO_2 + 4H^+ \rightarrow CO_2 + 2Mn^{2+} + 3H_2O \qquad \Delta G^\circ_r = -502.05\text{ kJ/mol}$$

Manganese is very soluble in groundwater in its reduced form as Mn^{2+}. This is the case for a clastic aquifer near Beijing, China (Liu et al. 2023), where Mn^{2+} is found at concentrations up to 522 µg/L, exceeding the guideline limit for drinking water of 120 µg/L based on aesthetic criteria and even the World Health Organization (WHO) health limit of 400 µg/L. Groundwater in crystalline bedrock wells in Finland (Kousa et al. 2020) with Mn^{2+} concentrations up to 6560 µg/L and an inverse correlation with dissolved O_2. The landscape hosts extensive peat, which is a likely electron donor providing DOC during recharge.

The mixing of Mn^{2+} groundwater with shallow O_2 bearing groundwater in discharge areas or in lake beds results in the oxidation and mineralization of manganese and iron oxyhydroxides, often in remarkably colorful patterns and structures (Figure 6.11).

$Fe(OH)_3/Fe^{2+}$: Iron Reduction

Iron is the crust's fourth most abundant element yet is rarely found in surface or oxic groundwaters at concentrations over 1 ppm due to its low solubility in its oxidized

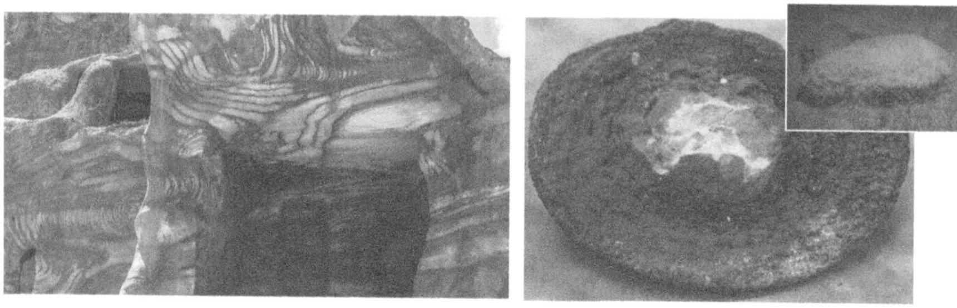

FIGURE 6.11 Left: Manganese and ferric iron oxyhydroxide banding in the late Cambrian Umm Ishrin quartz sandstone of the ca. 300 BCE Nabatean city of Petra, Jordan, where local Mn ore deposits exist (Khoury 2019). Right: Manganese lacustrine nodule (15 cm dia) formed by diffusion of high Mn^{2+} and Fe^{2+} groundwater into an oxic lake bed environment, inset showing growth position (Hayles et al. 2021).

form. Nonetheless, ferric iron mineralization is common in soils following oxidation of Fe^{2+} from weathering of common mafic silicates, such as amphiboles and pyroxenes, as well as iron sulfides (pyrite) and oxides (magnetite).

Amorphous and poorly structured forms have the general formula of ferrihydrite [$Fe(OH)_3$] or limonite [$FeO(OH) \cdot nH_2O$]. Through burial and heating, water is lost and crystalline forms develop, including goethite [FeOOH] and hematite [Fe_2O_3].

The energy released per mole of CH_2O oxidized by iron reduction is substantial:

$$\text{Half-cell reaction: } Fe(OH)_3 + e^- + 3H^+ \Leftrightarrow Fe^{2+} + 3H_2O \qquad \Delta G^\circ_r = -102.23\,\text{kJ/mol}$$

Iron reduction:

$$CH_2O + 4Fe(OH)_3 + 8H^+ \rightarrow CO_2 + 4Fe^{2+} + 11H_2O \qquad \Delta G^\circ_r = -436.48\,\text{kJ/mol}$$

The solubility of ferrous iron in freshwater is limited by highly soluble ferrous iron hydroxide [$Fe(OH)_2$] at 0.01 molal or over a gram per L, and so concentrations in reducing groundwaters are generally limited by the availability of iron hydroxide coatings in the flow system. A case study at the Canadian Nuclear Laboratories research campus in eastern Ontario studied groundwater discharging from a late Pleistocene complex of sand dunes. In the outcrop area, these sands show abundant buried biomass and iron-oxide coatings on the sand. The groundwaters discharging into a wetland at the base of the dune complex are depositing extensive mats of bacterial iron oxides (BIOS) composed dominantly of ferrihydrite (Figure 6.12).

Chapelle and Lovley (1992) present an example of iron reduction based on the availability of ferrihydrite and organic carbon in a coastal aquifer in South Carolina. This confined sandy aquifer contains sedimentary organic carbon with ferrihydrite coatings on the sand grains over a 45-km stretch along the groundwater flow path. Dissolved iron concentrations up to 3 ppm have degraded the quality of these groundwaters that supply a number of local communities. The authors show the redox potential of groundwaters to evolve from aerobic conditions in the recharge area to iron reduction through the ferrihydrite zone and into sulfate reduction in the ferrihydrite-free zone (Figure 6.13). Beyond the

Field parameters			
T°C	9.2	Eh (mV)	212
pH	5.85	DO (ppm)	0.05
Major ions, ppm			
Na	2.55	Fe(II)	3.16
K	0.76	Fe(III)	0.22
Ca	2.93	Mn	0.06
Mg	1.11	SO_4	4.36
Si	5.14	Cl	1.20
DIC	9.83	DOC	4.70
$\delta^{13}C_{DIC}$	–20.3	$\delta^{13}C_{DOC}$	–27.3

FIGURE 6.12 Biogenic ferrihydrite deposits from a dune complex where groundwaters evolve to low p*e* due to bacterial reduction of iron hydroxide coatings on the sand, using buried organics as the electron donor. Note the high ferrous Fe(II) concentration but also elevated ferric iron. Fe(III) 0.22 mg/L is above equilibrium concentration with ferrihydrite at that pH (Figure 6.14) of 0.013 mg/L, leading to precipitation of the extensive ferrihydrite seen in the photo (Grant Ferris, photo credit). The high DOC (4.7 ppm C) drives iron reduction in the aquifer, and high DIC indicates the extent of iron reduction. $\delta^{13}C_{DOC}$ indicates typical C3 biomass. Low $\delta^{13}C_{DIC}$ from respiration of DOC.

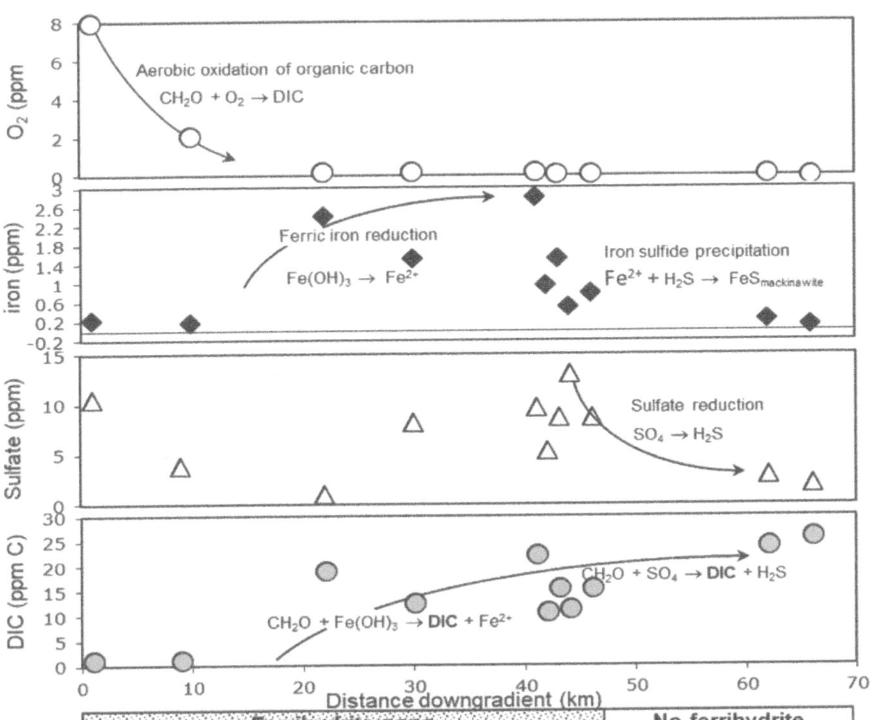

FIGURE 6.13 Redox evolution along groundwater flow path in sand aquifer with 2% to 5% organic carbon and ferric iron oxyhydroxide coatings (Chapelle and Lovley 1992).

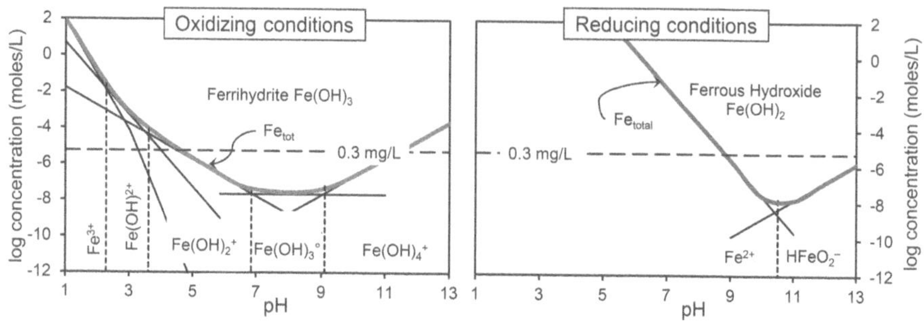

FIGURE 6.14 Solubility of ferric iron (left diagram) and ferrous iron (right diagram) according to pH. The WHO water quality objective of 0.3 ppm for total dissolved iron is shown.

ferrihydrite zone, sulfate that has diffused into the groundwater from confining beds is reduced to hydrogen sulfide (discussed in the following section), which sequesters the ferrous iron produced up-gradient and precipitates it as iron sulfide, mackinawite [FeS].

This study illustrates the enhanced solubility of iron under reducing conditions. However, both ferrous and ferric iron have constraints on their solubility due to the control of pH on the stability of iron hydroxide species and hydroxide minerals like ferrihydrite. Ferrihydrite precipitates are common in aquifers, as almost all soils will have some ferrihydrite coatings on mineral grains from the weathering of iron-bearing minerals under oxidizing conditions. Figure 6.14 shows the concentration of the main dissolved ferric and ferrous iron hydroxide species in equilibrium with their precipitated hydroxides.

Figure 6.14 shows that the solubility of Fe^{2+} hydroxide is much greater than Fe^{3+} hydroxide (ferrihydrite) over the pH range of most natural waters. At pH 7, the concentration of Fe^{2+} can exceed 1000 ppm. In contrast, ferrihydrite [$Fe(OH)_3$] maintains iron solubility below the drinking water standard of 0.3 ppm over a wide pH range from below pH 5 to above pH 11. Only under acidic conditions or highly alkaline conditions, well beyond the range for most groundwaters, will ferric iron become soluble.

This is important for the attenuation of heavy metals and other contaminants, which are strongly sorbed onto the highly charged surface of ferrihydrite. However, under anoxic conditions, reduction of ferrihydrite to ferrous iron not only solubilizes iron but also releases all the sorbed metals and other compounds into solution. An example is the high arsenic groundwater of Bangladesh, which is released by the reduction of ferric iron compounds by high DOC groundwaters in the deltaic sediment aquifers (see case study in Chapter 8).

Both sulfide and carbonate place important limits on Fe^{2+} solubility. In the presence of high DIC at high pH, ferrous iron will precipitate as siderite [$FeCO_3$] (Figure 6.15). Hydrogen sulfide in groundwater wells is associated with the familiar black amorphous FeS precipitate that encrusts the pump and well casing. In the absence of O_2, steel will oxidize to Fe^{2+} by reducing H^+ to H_2. In the presence of H_2S and HS^- from sulfate reduction (below), the Fe^{2+} will precipitate as amorphous

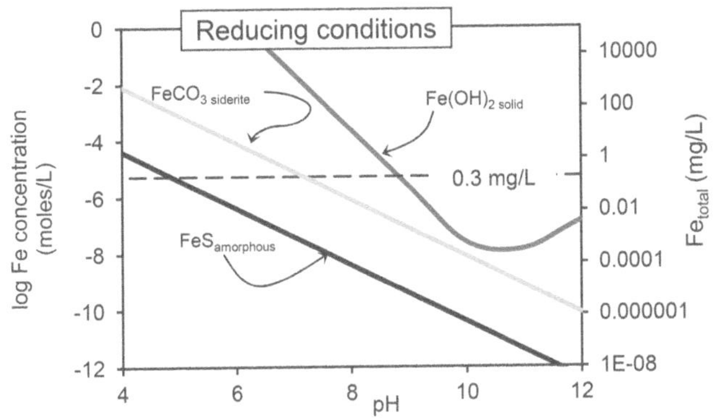

FIGURE 6.15 Solubility of iron under reducing conditions (Fe^{2+}) in the presence of HCO_3^- (200 ppm) and in the presence of hydrogen sulfide (10 ppm).

ferrous iron sulfide or mackinawite [FeS]. Crystalline forms such as pyrrhotite [FeS] and pyrite [FeS_2] are thermodynamically more stable but require time and a higher temperature to form, and so are not controlling phases in near-surface geochemical environments.

SO_4/H_2S: Sulfate Reduction

The reduction of sulfate to sulfide is one of the most common redox buffers, owing to the high solubility and variety of sources of SO_4^{2-} in natural waters. In marine sediments, the high sulfate content in seawater provides a large reservoir of this electron acceptor, giving it an important role in diagenetic reactions involving sedimentary organic carbon. Sulfate can be found as gypsum deposits in fractures or in evaporitic horizons. It can be derived from oxidation of pyrite if present within the oxidizing zone of aquifers. Sulfate also is supplied by high-salinity brines found in deep crystalline settings and deep sedimentary basins.

Sulfate reduction is a common reaction that is mediated by a group of sulfate-reducing bacteria (SRBs). They operate on very different organic substrates ranging from fixed organic carbon sources such as peat and soil organics to petroleum, bituminous solids, and methane. Sulfate reduction involves an 8 electron transfer, from S^{6+} to S^{2-}.

$$\text{Half-cell reaction: } SO_4^{2-} + 8e^- + 10H^+ \rightarrow H_2S + 4H_2O \qquad \Delta G^\circ_r = -232.26\,\text{kJ/mol}$$

$$\text{Sulfate reduction: } CH_2O + SO_4^{2-} \rightarrow H_2S + 2HCO_3^- \qquad \Delta G^\circ_r = -197.9\,\text{kJ/mol}$$

Hydrogen sulfide is a weak acid in water ($K_{H_2S} = 10^{-6.99}$), and so ionized HS^- dominates above pH 7, and H_2S dominates at low pH. H_2S is toxic to most organisms at very low concentrations, and as a heavy gas, can accumulate in wells and confined spaces. Hydrogen sulfide can be detected by odor at less than 0.01 ppmv in air and is an irritant at 20 ppmv. Above 100 ppmv, it neutralizes the olfactory nerve so it is

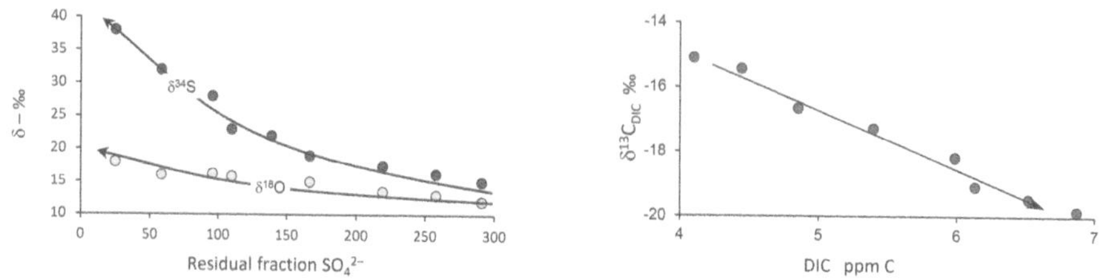

FIGURE 6.16 Sulfate reduction trend observed for $\delta^{34}S$ and $\delta^{18}O$ in the residual sulfate in artesian groundwater from Oman. These groundwater have several ppm of H_2S. Gypsum in the Cretaceous aquifer is the source of sulfate, with the likely electron donor being bitumen in the Umm Er Radhuma carbonate aquifer (Clark et al. 1995).

no longer detected by scent and becomes lethal at 300 ppmv. Hydrogen sulfide has a high solubility unless there are metals present in solution to form sulfides, such as FeS ($K_{FeS} = 10^{-3.92}$), ZnS ($K_{ZnS} = 10^{-9.05}$), or PbS ($K_{PbS} = 10^{-12.8}$).

Sulfate reduction, like denitrification, is accompanied by enrichment of ^{34}S and ^{18}O in the residual SO_4^{2-} in the groundwater. Bacteria preferentially reduced the isotopically light SO_4^{2-} to H_2S, leaving behind the ^{34}S- and ^{18}O-enriched SO_4^{2-} in the residual sulfate reservoir (Figure 6.16). A strong enrichment in $\delta^{34}S_{SO_4^{2-}}$ and $\delta^{18}O_{SO_4^{2-}}$ is a clear sign that SRBs are at work.

The enrichment for ^{18}O during sulfate reduction is less dramatic, due to exchange with ^{18}O of the water. Although oxygen exchange between sulfate and water at low temperatures essentially does not occur, it can during microbially mediated sulfate reduction (Fritz et al. 1989). For this reason, the Rayleigh enrichment seen in Figure 6.16 is less for ^{18}O than for ^{34}S.

In aquitards, bacterial sulfate reduction is evident by the formation of framboidal pyrite, which has the characteristic of clustered crystallites of pyrite (Figure 6.17).

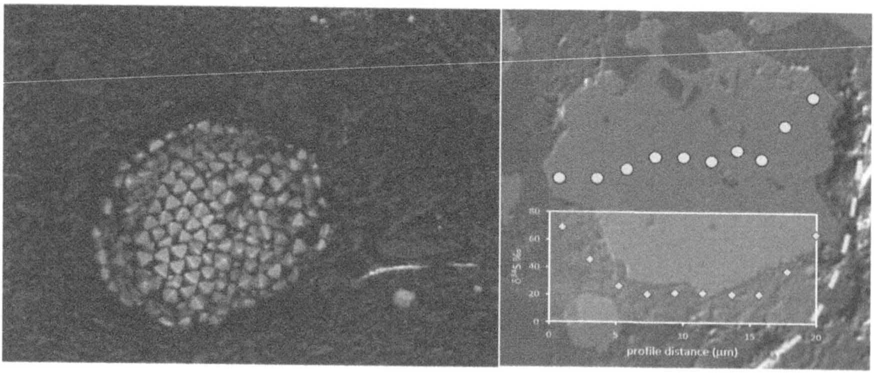

FIGURE 6.17 Framboidal pyrite formed in Ordovician shales of the Michigan Basin (Zhang 2018). Right: $\delta^{34}S$ profile across zone of nearby overgrowth with $\delta^{34}S$ measurements by secondary ion mass spectrometry (SIMS) showing enrichments up to 70‰ at rims due to Rayleigh enrichment as the porewater sulfate concentration is consumed (Jautzy et al. 2020). Scanning electron microscope (SEM) photo colorized using the Palette FM application.

Sulfate reduction is problematic in natural gas fields where abandoned well casings can corrode and connect normally isolated aquifers. Jackson et al. (2020) have identified such cases at a site in Southwestern Ontario where leaking well casings have connected deep gas-producing formations with overlying gypsiferous horizons that have high artesian pressure and with 300 ppm SO_4^{2-} with $\delta^{34}S_{SO4}$ enriched above that of the evaporites (Skuce et al. 2015) are now producing H_2S and HS^- (400 ppm) in these neutral pH waters according to the following reaction:

$$CH_4 + SO_4^{\ 2-} + H^+ \rightarrow H_2S + HCO_3^{\ -} + H_2O$$

The hydrogen sulfide accelerates casing corrosion, allowing flow into the shallow freshwater aquifer in the region (Figure 6.18, left). The recharge of this hydrogen sulfide groundwater into the freshwater aquifer has compromised the quality of a number of local potable water wells. The same reaction has been observed for groundwater proximal to an abandoned gas well near Lloydminster, Alberta (Van Stempvoort et al. 2005), with observed isotope trends in both DIC and SO_4^{2-} (Figure 6.18, right).

CO_2/CH_4: Methanogenesis

Near the bottom of the electromotive scale of aqueous redox reactions driven by excess organic carbon is methanogenesis. This is, in fact, a complicated series of

FIGURE 6.18 An abandoned gas well leaking CH_4, ubiquitous in regions of past production. Corrosion of abandoned gas wells can connect production zones with overlying evaporite horizons, provoking sulfate reduction and flow of sulfurous (H_2S and HS^-) groundwater into shallow freshwater aquifers. Right: Trends for isotopes of SO_4 and DIC during SO_4 reduction by CH_4 oxidation, from Van Stempvoort et al. (2005).

complementary reactions whereby different components of organic carbon function in the opposing roles of electron acceptor and electron donor, ultimately producing CO_2 and methane, CH_4. The overall reaction can be written simply as the oxidation and reduction of different forms of carbohydrate:

$$\text{Half-cell reaction: } CO_2 + 8e^- + 8H^+ \rightarrow CH_4 + 2H_2O \qquad \Delta G°_r = -130.58\,\text{kJ/mol}$$

$$\text{Methanogenesis: } 2CH_2O \rightarrow CH_4 + CO_2 \qquad \Delta G°_r = -316.28\,\text{kJ/mol}$$

Methanogenic reactions offer little energy to the bacteria and archaea that catalyze them, as much is left behind, stored as chemical energy in CH_4. Nonetheless, in the absence of the foregoing electron acceptors, from O_2 to SO_4^{2-}, microbial methanogenesis is viable and ubiquitous. Virtually all water-saturated environments host the microbial community capable of methanogenesis, including wetlands, tundra, landfills, and in the foregut of ungulates. Various fermentative bacteria degrade the complex organic carbohydrates (starch, cellulose, etc.) to simpler fatty acids, with by-products including acetate CH_3COOH, CO_2, and H_2.

$$C_6H_{12}O_6 \rightarrow CH_3CH_2CH_2COOH + 2CO_2 + 2H_2$$

Acetogenic bacteria thrive on fatty acid products to produce acetate and additional CO_2 and H_2:

$$CH_3CH_2CH_2COOH + 2H_2O \rightarrow 2CH_3COOH + 2H_2$$

Complementary archaeal reactions can then produce methane by either acetoclastic reaction or reduction of CO_2 with H_2 (Klass 1984):

$$\text{Acetate fermentation: } CH_3COOH \rightarrow CH_4 + CO_2 \qquad \Delta G°_r = -48.64\text{ kJ/mol}$$

$$CO_2 \text{ reduction: } CO_2 + 4H_2 \rightarrow CH_4 + 2H_2O \qquad \Delta G°_r = -130.58\text{ kJ/mol}$$

Biogenic methane occurs naturally in many Quaternary aquifers with buried organics of Holocene age. It can also migrate into groundwaters with leachates from unlined landfills. However, deeper-sourced methane of thermocatalytic origin can also migrate upward into shallow aquifers. Water supplies tapping such groundwaters can accumulate free methane, which presents an explosion hazard. The expanding development of shallow gas and shale gas increases this risk through losses from leaking well casings or by fracturing of the shale caps that normally protect the shallow groundwaters (Osborn et al. 2011).

Distinguishing the origin of methane in shallow groundwaters between natural biogenic sources and fugitive gas from deeper thermocatalytic sources can be discerned with isotopes. Archaeal methanogenesis is highly fractionating for ^{13}C with an enrichment factor $\varepsilon^{13}C_{CH_4-CH_2O}$ of −30‰ to −60‰. This imparts a strong depletion on microbial methane, with values typically less than −50‰ and as low as −90‰ (Figure 6.19 A). In contrast, thermogenic methane has only a modest depletion in $\delta^{13}C$ values due to fractionation at much higher temperatures.

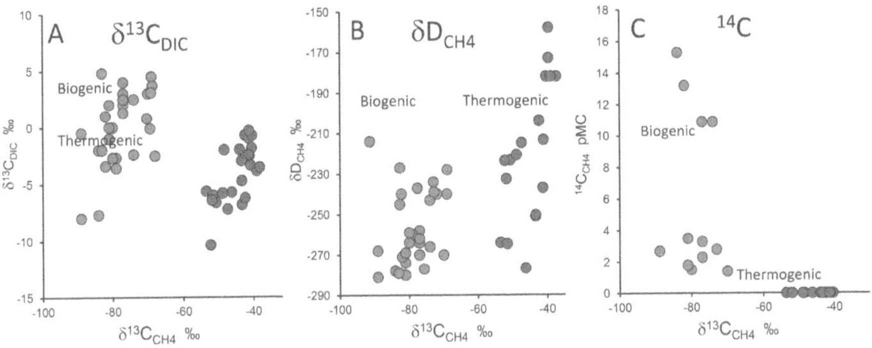

FIGURE 6.19 The $\delta^{13}C$ of CH_4 plotted with (A) $\delta^{13}C_{DIC}$, (B) δD_{CH4}, and (C) $^{14}C_{CH4}$ for biogenic methanogenesis in fresh water (data from Aravena et al. 1995; Coleman et al. 1988) and for thermogenic methane in sedimentary basins. Data from Barker and Pollock 1984.

DIC in methanogenic groundwater shows reciprocal values up to and greater than +5‰ to +10‰. This is due to CO_2 reduction (above), where the DIC substrate experiences a Rayleigh distillation during methanogenesis. Here, $^{12}C_{DIC}$ is favored for reduction with H_2 to methane, leaving the residual DIC enriched in ^{13}C. This strong enrichment in the residual DIC can be preserved in calcite cements and nodules and used to indicate a methanogenic environment during formation (Dimitrakopoulos and Muehlenbachs, 1987; Clark et al., 2004).

This enrichment is also observed for saline spring waters discharging methane on Anticosti Island in the Golfe du Saint-Laurent, Quebec (Daoust et al. 2023; Figure 6.20). The site hosts the hydrocarbon-rich Macasty shale, which was a target

T°C	23
pH	6.39
N_2	17.1 %
O_2	3.6 %
CO_2	4.83 %
DIC ppm C	74.9
$\delta^{13}C_{DIC}$	+9.75 ‰
$\delta^{13}C_{CO2}$	−2.1 ‰
$^{14}C_{DIC}$	3.52 pMC
CH_4	78.0 %
$\delta^{13}C_{CH4}$	−57.7 ‰
δD_{CH4}	−248 ‰
$^{14}C_{CH_4}$	1.59 pMC

FIGURE 6.20 Biogenic methane discharging from groundwater on Anticosti Island, Quebec. Shown is an inverted bottle for reactive gas sampling and gas-permeable passive samplers (see Figure 2.9) for noble gas analysis Daoust et al. 2023.

of exploration for unconventional resources. The spring waters are hypersaline (3× seawater) and degas CH_4 and CO_2, forming carbonate mounds (Figure 6.20). The enriched $\delta^{13}C_{DIC}$ suggests biogenic methanogenesis by CO_2 reduction; note that $\delta^{13}C_{CO2}$ is 10‰ depleted with respect to DIC due to fractionation between CO_2 and bicarbonate. The ^{14}C measured in both methane and DIC confirms a contribution from soil CO_2 in the recharge area. Thus, locally recharged groundwaters circulate through a local fault system to the Macasty shale, fostering biogenic methanogenesis. The occurrence of these springs in association with the surface expression of the local fault implies that this fault provides a pathway for deep circulation of meteoric waters.

The $\delta^{13}C$ of DIC associated with thermogenic methane generally has values up to only 0‰ due to exchange with carbonate minerals in the reservoir and is useful when used with $\delta^{13}C_{CH_4}$ to distinguish methane sources (Figure 6.19 B). The most diagnostic combination of isotopes is that of $\delta^{13}C$ with ^{14}C of the methane. Biogenic methane typically has a measurable radiocarbon activity due to production from buried organic carbon in surficial sediment (Quaternary) aquifers, or from soil DIC entrained during recharge (Figure 6.19 C). Thermogenic CH_4 is generated from ancient marine organic carbon and is ^{14}C-free. Radiocarbon, therefore is an important tool to unambiguously identify fugitive methane from deep production zones in shallow groundwaters.

Isotope fingerprinting with regional data was helpful to identify sources and processes of methane in groundwater in the oil and gas producing province of Alberta. Humez et al. (2016) use $\delta^{13}C_{CH_4}$ and δD_{CH_4} and δD_{H_2O} in groundwater to show that the methane is of natural, biogenic origin through CO_2 reduction (Figure 6.21).

H_2O/H_2: Reduction of Water

The lowest electromotive potential achieved in the redox evolution in groundwaters is the lower stability limit for water itself, where hydrogen gas is generated. The production of H_2 from the reduction of water is not associated with degradation of organics. Rather, it relies on electron donors with lower electromotive potential than carbon. Zero-valence iron is one potential electron donor and is discussed in

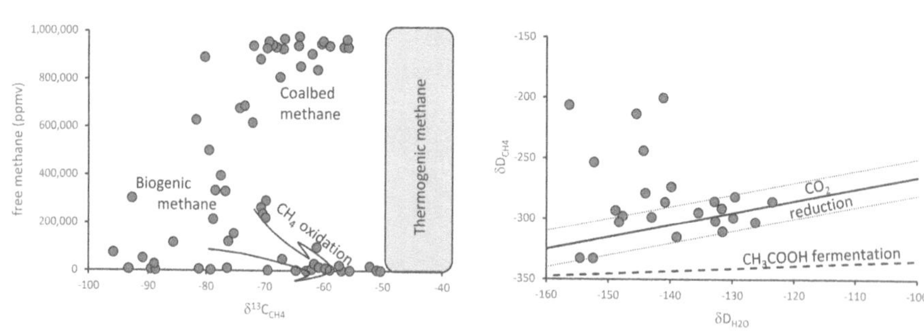

FIGURE 6.21 Isotope characterization of methane in Quaternary surficial clastic aquifers and shallow bedrock aquifers in gas producing regions of western Canada. Methane in groundwater is determined to be of biogenic origin, produced by CO_2 reduction (Humez et al. 2016).

Chapter 8 for its role in reductive dehalogenation of contaminants like trichloroethylene, but it does not occur naturally except in a few native iron-ore mines. In fact, there are few naturally occurring electron donors that exist below the electromotive potential of the H_2O/H_2 redox couple:

$$\text{Half-cell reaction} \qquad H_2O + e^- \rightarrow \tfrac{1}{2}H_2 + OH^- \qquad \Delta G^\circ_r = -14\,\text{kJ/mol}$$

A remarkable example of groundwater redox evolving as low as this is found in Oman, where hyperalkaline springs with pH over 12 are bubbling hydrogen gas (Barnes and O'Neil 1978; Neal and Stanger 1983; Figure 6.22). These springs occur in the ultramafic sequence of the Semail Ophiolite, the world's best exposure of oceanic crust, which was obducted onto the Arabian plate in late Cretaceous time. Extensive closed-system serpentinization of olivine and pyroxene by the deeply circulating groundwater produces hydroxide minerals including serpentine [$Mg_3Si_2O_5(OH)_4$], brucite [$Mg(OH)_2$], and ferrous iron hydroxide [$Fe(OH)_2$] according to the following:

$$5Mg_2SiO_{4\,ol} + Fe_2SiO_{4px} + 9H_2O \rightarrow 3Mg_3Si_2O_5(OH)_{4\,serpentine} + Mg(OH)_2 + 2Fe(OH)_2$$

FIGURE 6.22 Blue pools of Oman; hyperalkaline springs with pH > 12. Inset shows gas vents bubbling H_2.

pH	T°C	Na^+	K^+	Ca^{2+}	Mg^{2+}	Cl^-	SO_4^{2-}	OH^-	alk_C	P_{CO_2}	P_{H_2}
12.1	25	16.2	0.23	3.21	0.06	11.8	0.09	7.49	0.00 meq/L	$<10^{-12}$	>1 atm

The waters discharge with negligible carbonate alkalinity. The P_{CO_2} is more than eight orders of magnitude lower than the atmospheric P_{CO_2} of $10^{-3.4}$, such that CO_2 from the air reacts with the high OH^- to precipitate the white calcite seen in the pools. Photo, Lowry et al. 1986.

The hyperarid environment in Oman limits groundwater circulation with the result that the highly soluble brucite and ferrous iron hydroxide push the pH of these waters to such high levels.

The high pH of these groundwaters allows reduction of H_2O to produce H_2 via iron oxidation. From Figure 6.8, the steep slope of the ferric/ferrous iron line allows that at very high pH ferrous iron becomes a favorable electron donor for the reduction of water to H_2. The stable ferrous iron phase at this pH is $Fe(OH)_2$, and the ferric iron phase is Fe_2O_3, which react with water at depth according to the overall redox reaction:

$$2Fe(OH)_2 \rightarrow Fe_2O_3 + H_2O + H_2$$

From the two half-cell reactions, the *pe* corresponding to the pH for these waters can be determined and compared with the *pe* for reduction of water at pH 12.1.

$$2Fe(OH)_2 \rightarrow Fe_2O_3 + H_2O + 2H^+ + 2e^- \qquad \Delta G^\circ_r = -6.94 \text{ kJ/mol}$$

$$\log K = -\frac{-6.94}{5.708} = 1.2$$

$$\log K = 2\log a_e + 2\log a_{H^+}$$

$$1.2 = -2pe - 2pH$$

$$pe = -0.61 - pH$$

$$= -12.7$$

$$2H^+ + 2e^- \rightarrow H_2 \qquad \Delta G^\circ_r = 0 \text{ kJ/mol}$$

$$\log K = -\frac{0}{5.708} = 0$$

$$\log K = -2\log a_{H^+} - 2\log a_{e^-}$$

$$0 = -2pH - 2pe$$

$$pe = -pH$$

$$= -12.1$$

This shows that the iron oxidation reaction takes place at a lower *pe* level than the stability of water, so this reaction is electrochemically favorable. The high hydrostatic pressure at depth allows the hydrogen gas to accumulate. It then bubbles from the solution in the discharge area with the loss of hydrostatic pressure. Remarkable are the isotopes measured in the hydrogen gas (Fritz et al. 1992). The δD values were the lowest measurements made for natural hydrogen gas. Using the measured values given here and the relationship $\alpha D_{H_2O\text{-}H_2} = (\delta D_{H_2O} + 1000)/(\delta D_{H_2} + 1000)$ yields an astonishingly high value for deuterium fractionation. The estimated temperature of the low-temperature serpentinization reaction at depth can be calculated using the fractionation-temperature relationship of Bottinga (1969) for these three hyperalkaline springs. This gives temperatures between 25°C and 37°C, which are close to the mean annual air temperature of 30°C for this region.

	δD_{H_2O}	δD_{H_2}	$\alpha D_{H_2O\text{-}H_2}$	T°C
Al Khoud 1a	2.9‰	−721‰	3.59	32
Al Khoud 20a	5.2‰	−733‰	3.76	25
Nizwa 1b	2.9‰	−712‰	3.48	37

This gives temperatures between 25°C and 37°C, which are close to the mean annual air temperature of 30°C for this region.

SALINITY IN GROUNDWATER

Groundwater quality can be impacted by salinity from a range of sources (Figure 6.23). Groundwater recharged through soils typically has a bicarbonate geochemical facies with total dissolved solids (TDS) up to about 200–300 ppm. However, greater depth of circulation is generally accompanied by gains in TDS, dominated by chloride salinity. Chebotarev (1955) noted a sequence in the evolution of groundwater starting with a HCO_3^- facies from weathering in soils to a SO_4^{2-} facies and then Cl^- facies in deep crustal fluids. The Chebotarev sequence is a demonstration of the relative solubility of the carbonates, sulfates and halides, but does not fully address the origin of salinity in groundwater. Ultimately, most of the natural Cl^- salinity in groundwater is derived from marine sources, whether as marine aerosols, saline intrusion, dissolution of evaporite minerals in sedimentary strata or through infiltration of evaporated basin brines during the Phanerozoic. These sources have solute and isotope characteristics that differ from modern seawater (Table 6.1) and so can be used to trace the processes contributing salinity to groundwater.

Cl^- IN RAINWATER

A strong inland gradient is found for Cl^- and other ocean-derived solutes, with values up to 10 ppm decreasing to less than 1 ppm within 100 km of the coast. Continental rainfall has typically less than 1 ppm and as low as 0.5 ppm Cl^- (Junge and Werby 1958; Li 1992). This low concentration provides an important baseline to assess potential impacts of anthropogenic inputs from wastewater infiltration and from agriculture. Given the prevalence of agriculture and rural wastewater treatment systems, few exploited groundwaters remain that are not impacted.

One example, however, is groundwater from glacial outwash sand aquifers of Clearwater Township, Ontario, where pristine groundwater discharges from flowing artesian wells that are distal from the highlands bounding the till plains (Figure 6.24).

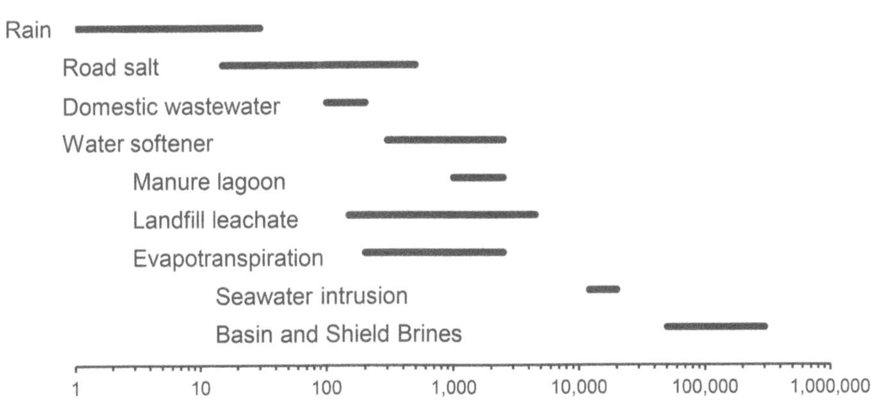

FIGURE 6.23 Sources and ranges for chloride in groundwaters. Data from Eriksson 1952; Howard and Haynes 1993; Krapac et al. 2002; Kinsley et al. 2005; Christensen et al. 2001; Clark et al. 2013.

TABLE 6.1
Geochemistry and Isotopes of Seawater

Anions	ppm	mmolal	Cations	ppm	mmolal
Cl^-	19,350	546	Na^+	10,760	468
SO_4^{2-}	2,710	28.3	Mg^{2+}	1,290	53.1
HCO_3^-	142	2.33	Ca^{2+}	411	10.3
Br^-	67	0.84	K^+	399	10.2
F^-	1.3	0.068	Sr^{2+}	8	0.091
I (I^- and IO_4^-)	0.045	0.00035	Li^+	0.18	0.026
NO_3^- (as N)	0.005–2	$0.8–300 \times 10^{-4}$	Ba^{2+}	0.02	0.00015
$H_3PO_{4(total)}$ (as P)	0.001–0.05	$1–50 \times 10^{-5}$	$H_4SiO°_4$ (as Si)	0.03–10	0.001–0.3
$H_3BO_{3(total)}$ (as B)	4.5	0.42	DOC	0.3–2 as C	0.02–0.2
Isotopes					
$\delta^{18}O$	0‰ VSMOW	$\delta^{13}C_{DIC}$	0‰ PDB	$\delta^{34}S_{SO4}$	21‰ CDT
δD	0‰ VSMOW	$\delta^{13}C_{DOC}$	−27‰ PDB	$\delta^{18}O_{SO4}$	9.5‰ VSMOW
δ^7Li	32.3‰ LSVEC	$\delta^{18}O_{PO4}$	19.7‰ VSMOW	$\delta^{37}Cl$	0‰ SMOC
$\delta^{11}B$	40.4‰ NBS 951	$^{87}Sr/^{86}Sr$	0.707924	$\delta^{81}Br$	0‰ SMOB

Sources: Bernat et al. 1972; Drever 1997; Tsunogai and Henmi 1971.

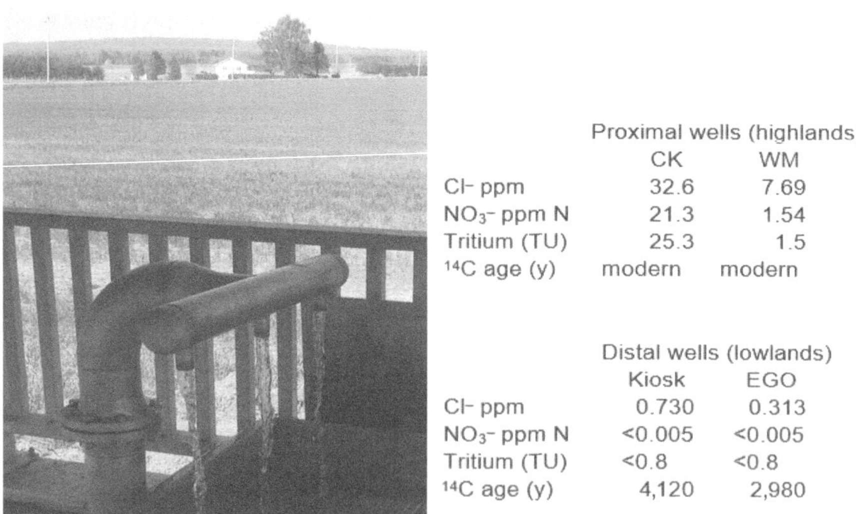

FIGURE 6.24 Elmvale artesian groundwater from distal well (Kiosk) in sand aquifer confined by overlying clay till. Recharge occurs in the highlands in background of photo.

Distal wells were recharged in the mid to late Holocene prior to human activity in the region and have Cl^- levels below 1 ppm. Proximal wells in the highlands, in contrast, have measurable tritium, which is a short-lived radioisotope that indicates recent recharge (see Groundwater Dating). Proximal wells have elevated Cl^- as well as measurable NO_3^-, both of which indicate anthropogenic contamination from the surface.

Aquitard Diffusion

Groundwater can acquire salinity through diffusion of solutes from adjacent aquitards along the flow path. Low-permeability clays and shales of marine origin often retain saline porewaters, which are a source of salinity to circulating freshwaters following uplift into a continental setting. The Milk River aquifer is a good example, where meteoric waters now flow through the Cretaceous Milk River Formation of marine sandstone with interbedded marine shales. Water quality deteriorates along the flow path to the north due to increases in Cl^-, Na^+, and the balance of conservative solutes of seawater origin. Proximal wells in and near the recharge area where the sandstone outcrops and along the Milk River channel have Cl^- near 1 mmol/L (35 ppm). Cl^- increases to over 60 mmol/L (over 2000 ppm) (Figure 6.25).

Crustal Brines

Hypersaline Ca–Na–Cl brines, with salinities often exceeding 200 g Cl^-, have been observed at depths greater than about 500 m in most crustal environments. Interest in these deep crustal settings for siting deep geological repositories generated considerable research on the origin and age of these fluids. Basin brines are found in large sedimentary basins and are routinely produced with oil and gas

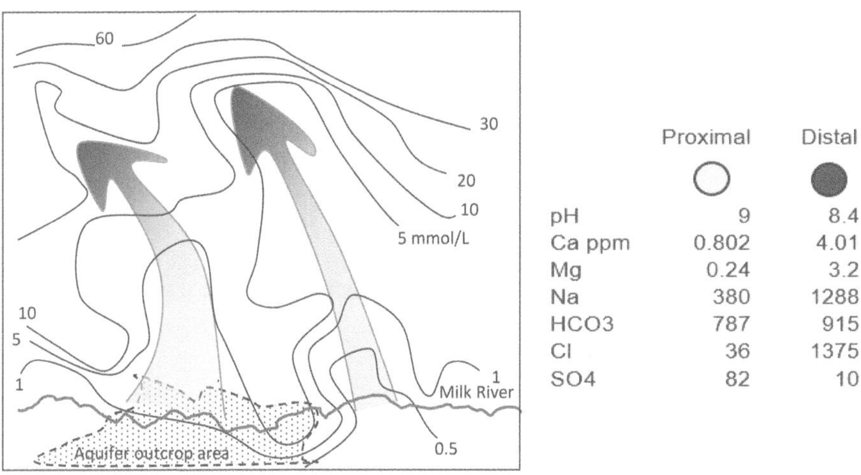

	Proximal	Distal
pH	9	8.4
Ca ppm	0.802	4.01
Mg	0.24	3.2
Na	380	1288
HCO3	787	915
Cl	36	1375
SO4	82	10

FIGURE 6.25 Chloride concentrations in the Milk River sandstone aquifer increase along the flow paths due to solute diffusion from confining aquitards of marine shales (Hendry and Schwartz 1990).

production. Shield brines are found in crystalline rocks and are routinely encountered during deep underground mining in such terrains. Geochemical and isotopic evidence supports an origin through strong evaporation in Paleozoic epicontinental seas, which produced high-density brines that infiltrated into the crust. These hypersaline crustal fluids are ubiquitous in both sedimentary basins and crystalline (shield) terrains, where they mix with deeply circulating meteoric waters (Clayton et al. 1966; Bottomley et al. 1994, 1999, 2005; Frape and Fritz 1982; Lowenstein et al. 2003).

The geochemical facies of the evaporating seawater evolves through a series of mineral precipitation reactions starting with carbonate precipitation, followed by gypsum saturation and then halite (Figure 6.26, left chart). Eventually Epsom salt [$MgSO_4·7H_2O$] removes residual sulfate, followed by K^+ as sylvite [KCl] as the geochemical facies evolves from Na–Cl to Mg–Cl, as observed for the Dead Sea with some 400 m of halite on its bed. Crustal brines typically did not reach that level of concentration before they infiltrated into the subsurface. The degree of evaporation is evident from the concentration of Br^-, which is conservative and does not form bromide minerals until much higher concentrations (Figure 6.26, right chart).

Infiltration of these high-density, halide-enriched brines into the subsurface led to a sequence of reactions that altered their geochemical facies, starting as evaporated brines with a Mg–Cl facies close to that of the Dead Sea. Penetration through carbonate strata consumed Mg^{2+} by dolomitization with the release of Ca^{2+} and engendered geochemical evolution toward that of the basin brine. Further evolution consumed K^+ by illitization of clay minerals, producing typical sedimentary basin brines dominated by an Na(Ca)–Cl facies. Deeper penetration of these brines into the basement crystalline rocks has typically exchanged Na^+ for Ca^{2+} by albitization of anorthite

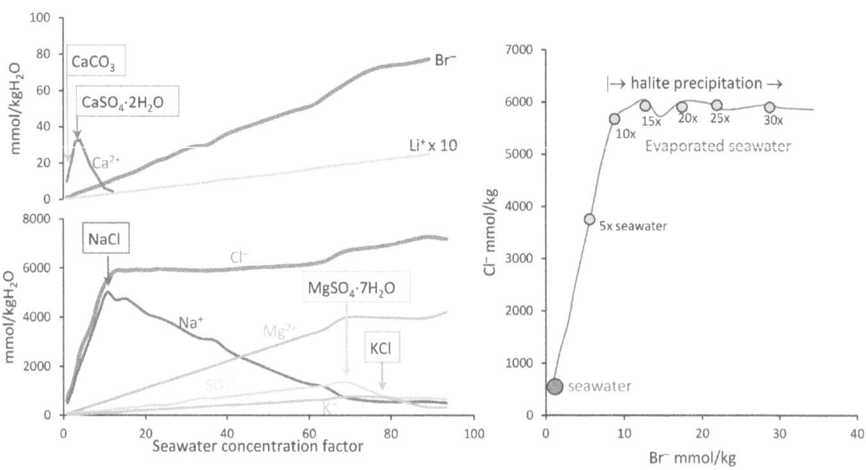

FIGURE 6.26 Evolution of seawater chemistry during evaporation and salt formation. Left: (top and bottom) in the order of calcium carbonate [$CaCO_3$], gypsum [$CaSO_4·2H_2O$], halite [NaCl], epsomite [$MgSO_4·7H_2O$], and sylvite [KCl]. Right: Evaporation pathway for Br^- and Cl^- correlation past halite saturation. Data from McCaffrey et al. 1987.

feldspar under conditions of very low water-to-rock ratios over long periods of time and elevated temperature (Bottomley et al. 1994, 2005; Greene et al. 2008).

Dolomitization: $2CaCO_3 + Mg^{2+} \rightarrow MgCa(CO_3)_2 + Ca^{2+}$

Illitization: $NaAl_2Si_4O_{10}(OH)_2 + K^+ \rightarrow KAl_2Si_4O_{10}(OH)_2 + Na^+$

Albitization of anorthite: $CaAl_2Si_2O_8 + 2Na^+ + 4H_4SiO_4 \rightarrow 2NaAlSi_3O_8 + 8H_2O + Ca^{2+}$

The bromide-to-chloride ratio is diagnostic of the source of salinity in such deep crustal brines. Progressive evaporation of seawater beyond the point of halite saturation depletes the brine in Na^+ with Cl^- held constant during halite precipitation (Figure 6.26, right chart). However, Br^- continues to increase as evaporation proceeds due to the much higher solubility of bromide minerals, and so the Br/Cl molar ratio evolves from the seawater value of 0.002 to as high as 0.01. Both solutes are considered to behave conservatively in the subsurface and so are reliable tracers of the origin of salinity.

The isotope signature of these brines is very diagnostic of their origin, where on a $\delta^{18}O$-δD diagram, shield brines in crystalline rocks plot above the meteoric waterline with strong deuterium enrichment, while the basin brines plot below, with strong ^{18}O enrichment (Figure 6.27). While the $\delta^{18}O$ enrichment can be attributed to the strong primary signal of evaporation, the deuterium enrichment of shield brines is a secondary effect. Reardon and Fritz (1979) attributed the unusual isotopic composition of shield brines to hydration of primary silicates under closed system conditions. Partitioning of ^{18}O differs from D during hydration reactions. The enrichment factor for the $\delta^{18}O$ shows an enrichment of the clay relative to the water ($\varepsilon^{18}O_{kaolinite-water} = 34‰$) at 25°C, whereas clays are depleted in deuterium over water ($\varepsilon D_{kaolinite-water} = -32‰$). This reversal in fractionation is similarly observed for gypsum (Table 4.2) and occurs during reactions such as alteration of albite or hydration of anhydrite:

$$CaAl_2Si_2O_{8\,anorthite} + 3H_2O \rightarrow Ca^{2+} + Al_2Si_2O_5(OH)_{4\,kaolinite} + 2OH^-$$

$$CaSO_{4\,anhydrite} + 2H_2O \rightarrow CaSO_4 \cdot 2H_2O_{gypsum}$$

Under conditions of low water-to-rock ratios, the residual water hosting such reactions becomes measurably depleted in ^{18}O but enriched in D, such that these waters plot above the meteoric water line (MWL) with a strong deuterium excess (Figure 6.27).

In contrast to these evolved evaporitic brines are high-salinity groundwaters from the dissolution of the evaporite formations of these evaporitic basins by deeply circulating meteoric waters in a continental setting. Here, the isotopic signature of the meteoric waters is largely preserved, although hydration of anhydrite to gypsum can impart some isotope effects (Figure 6.28).

The example of brines in the Silurian and Ordovician aquitards at a site in Southern Ontario illustrates these two sources of salinity (Figure 6.29). Porewater profiles show a downward trend of increased Cl^- salinity, with an excursion near 300 m depth attributed to flushing by glacial meltwater at the end of the last

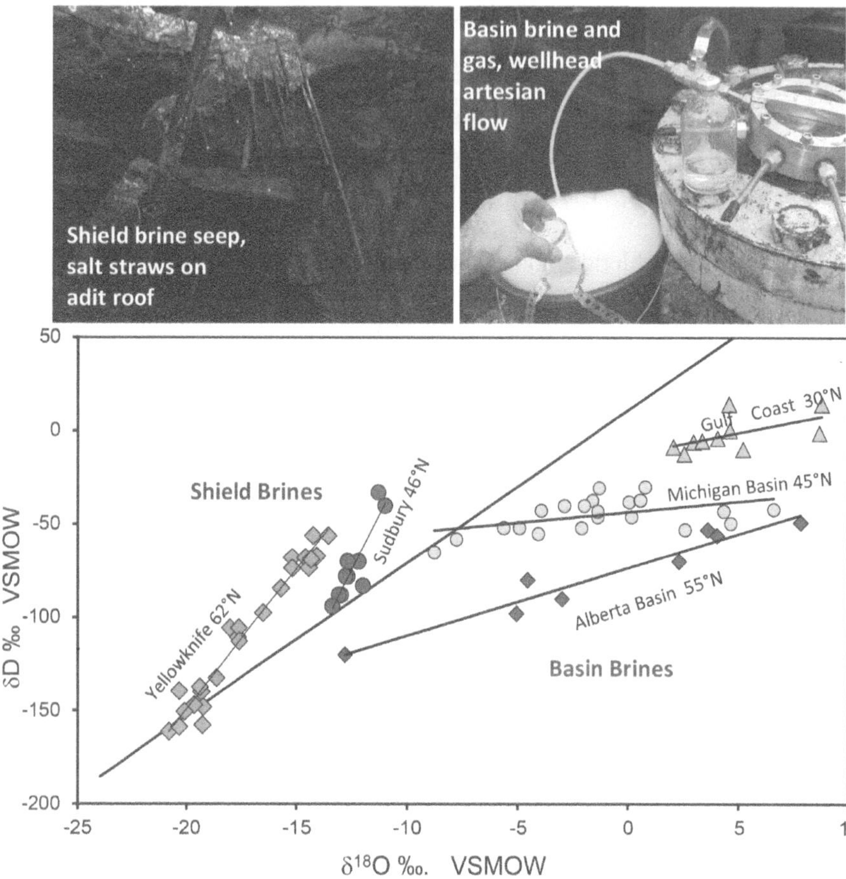

FIGURE 6.27 Stable isotopes of crustal brines. Shield brine crystalline rocks are deuterium-enriched due to feldspar alteration (Douglas et al. 2000; Fritz and Frape 1982). Basin brine in shale and carbonate rock (Clayton et al. 1966; Intera Engineering Ltd. 2011). Left photo: Shield brine salt stalagmite straws growing from the adit roof in the Con Mine. Right photo: Gas diffusion sampler and effervescing brine flowing from 850-m depth in Cambrian Sandstone, Bruce Nuclear site (Intera Engineering Ltd. 2011).

glaciation. Salinity in the Silurian section is characterized by an Na/Cl ratio near 1 with low Br/Cl, confirming a source from dissolution of halite by meteoric groundwater. This transitions through the lower Silurian into a very high salinity, Br-enriched and Na-depleted brine concluded to be remnant brine from seawater that was evaporated beyond halite saturation and that infiltrated the section in Silurian time (Intera Engineering Ltd. 2011). The porewater $\delta^{18}O$ profile is consistent with meteoric waters in the Silurian section and an evolved basin brine in the Upper Ordovician section. The Lower Ordovician section has been affected by more recent fluid flow in the underlying Cambrian sandstone, which has shifted the profile (Clark et al. 2013).

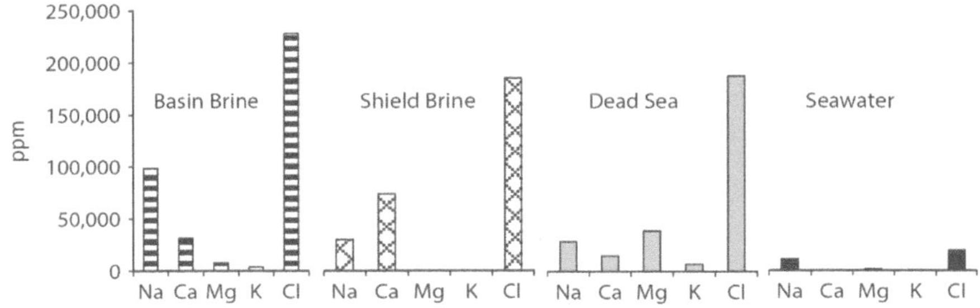

FIGURE 6.28 Major ion geochemistry of highly evaporated basin brine (Intera Engineering Ltd. 2011), crystalline shield brine (Douglas et al. 2000), Dead Sea water, and modern seawater.

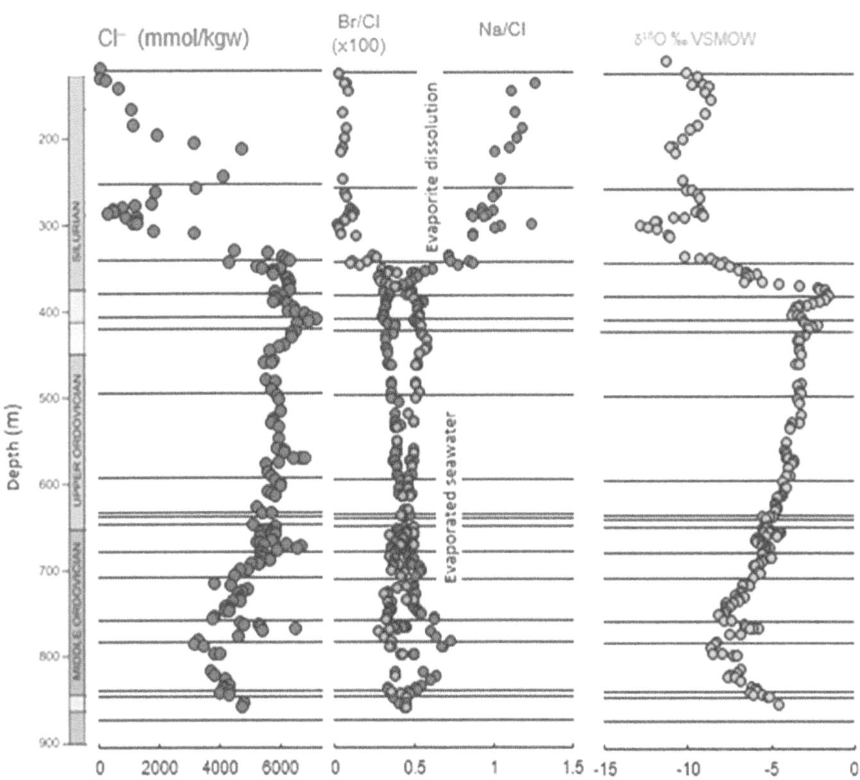

FIGURE 6.29 Cl, Na/Cl, and Br/Cl profiles in porewaters at the Bruce site investigated as a potential site for a nuclear waste repository near Kincardine, Ontario (modified from Clark et al. 2013). The upper, Silurian section is characterized by low Br/Cl and Na/Cl values close to unity, suggesting a source of salinity from the dissolution of halite. These Silurian formations host salt deposits near Windsor, Ontario. The deeper, Ordovician section has high Br/Cl and low Na/Cl consistent with seawater evaporation beyond the point of halite saturation, as shown in Figure 6.26.

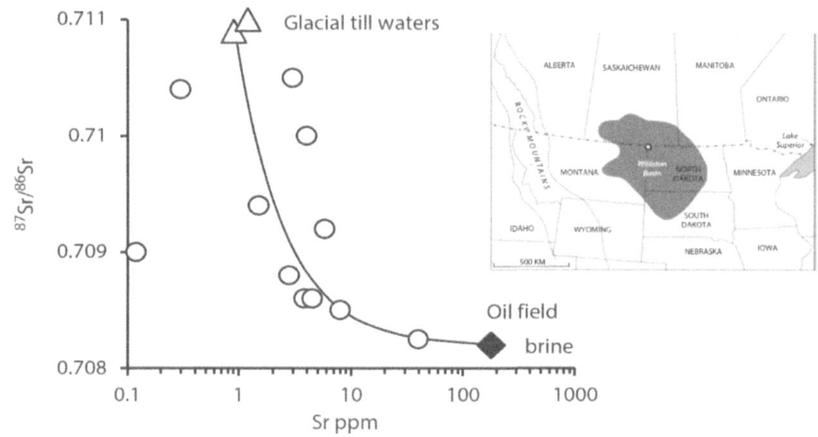

FIGURE 6.30 Use of $^{87}Sr/^{86}Sr$ to quantify salinity contributions from deep brines (black diamond) to shallow groundwaters (open discs) in the Williston Basin, Montana. (Modified from Peterman et al. 2010.)

Strontium Isotopes and Salinity

The $^{87}Sr/^{86}Sr$ ratio can be used to identify salinity from seawater brines. An example is in oil-producing regions, where improperly abandoned legacy wells leak deep brines into shallow aquifers. Peterman et al. (2010) use $^{87}Sr/^{86}Sr$ to calculate the amount of oil field brine from deep wells impacting shallow groundwater in the Williston Basin of northeastern Montana (Figure 6.30). In this study, shallow meteoric waters have Sr^{2+} with a radiogenic $^{87}Sr/^{86}Sr$ signature (see Chapter 5) from weathering of the Canadian Shield–derived glacial till, whereas the basin brines have a marine ratio. Their measurements show a range of mixtures of these two sources. While the Sr^{2+} concentration can be affected by sorption and exchange reactions, the $^{87}Sr/^{86}Sr$ ratio is conservative.

Seawater Intrusion in Coastal Aquifers

Seawater intrusion into near-shore aquifers due to over-pumping is common in many coastal regions relying on groundwater for irrigation. Such salinity can be differentiated from crustal brines through their isotope and halide signature in the saline water, which is conservative during mixing. Salinity gained through dissolution of evaporite minerals will not have a measurable effect on the stable isotope contents of the impacted groundwaters (Figure 6.31).

Sources of groundwater salinity can also be traced on the basis of the $\delta^{37}Cl$ and, to a lesser extent, with $\delta^{81}Br$ in dissolved halides. Isotope fractionation for halides is limited to mineral reactions, such as evaporite dissolution and precipitation. Unlike C, N, and S, Cl^- and Br^- do not participate in bacterially mediated redox reactions, which typically have significant fractionation. Halides do fractionate during diffusion through low-permeability materials. Further limiting fractionation are the lower

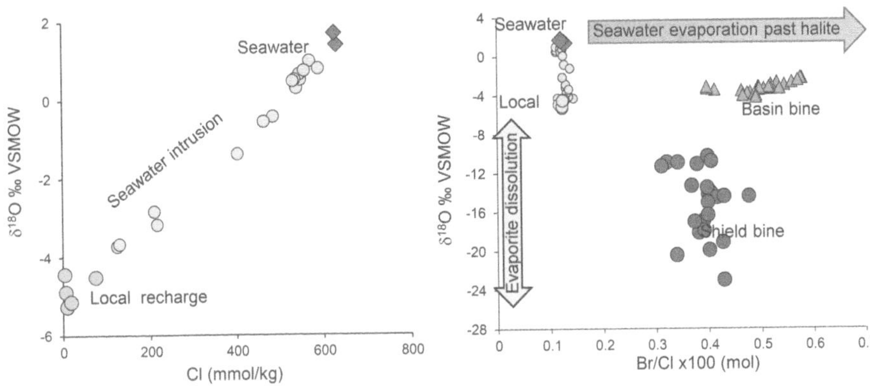

FIGURE 6.31 Left: Salinization by seawater intrusion is clearly identified by the mixing line between local seawater and locally recharged groundwater (modified from Sivan et al. 2005). Right: The Br/Cl ratio is diagnostic in distinguishing seawater intrusion (Br/Cl similar to modern seawater) from crustal brines (Br/Cl ratio enriched by seawater evaporation beyond halite saturation). Michigan Basin brine is shown as an example (Clark et al. 2013). The Canadian Shield brines shown have a broad range for $\delta^{18}O$, reflecting exchange with rock and mixing with regional meteoric waters (Bottomley et al. 2005; Frape and Fritz 1982). Salinity from evaporite (halite) dissolution has a low Br/Cl ratio, less than seawater, and with a range in $\delta^{18}O$ reflecting dissolution by meteoric water circulation.

mass effects provided by these isotopes. For $^{37}Cl/^{35}Cl$ this is less than 6% compared with 12.5% for $^{18}O/^{16}O$ and is only 2.5% for $^{81}Br/^{79}Br$. Stotler et al. (2010) show that $\delta^{37}C$ for Shield brines in Canada and Scandinavia vary within a range of −1‰ and +2‰ SMOC (Standard Mean Ocean Chloride). Values for $\delta^{81}Br$ fall within a range of 0‰–2‰ SMOB (Standard Marine Ocean Bromide). Cl^- in basin brines are relatively depleted in ^{37}Cl due to precipitation of marine halite into evaporite deposits spanning the Phanerozoic (Cambrian to Tertiary), which have marginally positive $\delta^{37}Cl$ values ranging from 0.0‰ to 0.9‰ SMOC (Eastoe et al. 2007) due largely to fractionation during crystallization.

Road Salt in Urban Aquifers

Urban hydrogeology has become an important field in view of the rapid pace of development in rural regions. Considerable focus has been made on the use of deicing chemicals, which are greatly dominated by halite. Infiltration into the subsurface is, of course, aggravated by the capacity of the chloride brine runoff to penetrate frozen soil and represents perhaps the most significant single threat to groundwater quality in urban regions (Howard et al. 1993). The increases in Na^+ and Cl^- are not the only threats to water quality. Bäckström et al. (2004) show releases of sorbed Cd, Cu, Pb, and Zn due to cation exchange in soil water following road salting.

A good example of road salt impact is the Waterloo moraine, which supplies the twin cities of Kitchener and Waterloo, Ontario, with groundwater from a series of well fields. Monitoring of groundwater quality at the Greenbrook well field over

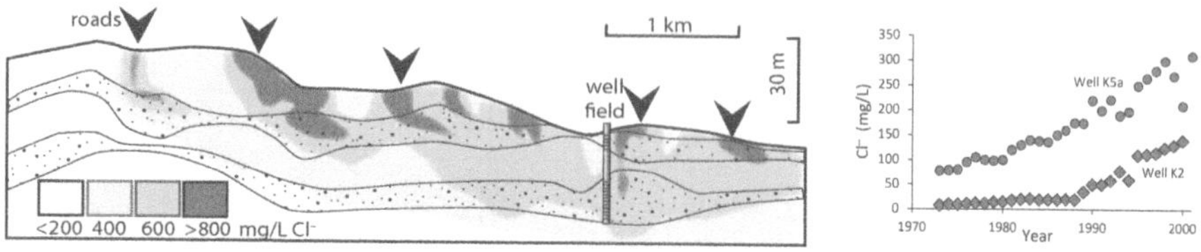

FIGURE 6.32 Increased Cl^- in the urban water supply well and modeled inputs from regional road deicing salt (after Bester et al. 2006). Road crossings on the section are shown with downward arrowheads.

the past several decades shows a steady increase in Cl^- from levels below 50 mg/L back in the early 1970s to levels above the drinking water limit by 2000 (Bester et al. 2006; Figure 6.32). Further, the steady increase in Cl^- correlates with increased urbanization, including the addition of new local and arterial roads that are salted in winter months. The impact of these increased road-salt loadings on the aquifer was modeled in a three-dimensional grid that captured the complex architecture of the glacial outwash sediments (Bester et al. 2006), using the measured increases of Cl^- from a series of wells for calibration. Figure 6.32 shows infiltration of Cl^- from the main arterial network. Projections from the study allowed testing of mitigation strategies and a modification of the protocols for road-salt applications.

Mitigation of the impacts of salinity buildup in urban aquifers requires reliable identification of the source of salinity. An increase in Cl^- in groundwaters cannot necessarily be attributed to road salt. Domestic septic infiltration beds are typically a source of Na^+ and Cl^-, largely from the prevalence of salt in food. Landfill leachate is also enriched in Na^+ and Cl^- from domestic waste. Finally, deep crystalline and sedimentary basin settings typically host hypersaline brines mixed to varying degrees with meteoric waters. However, each of these sources has additional geochemical components that can be useful in discriminating the source of salinity. Although Na/Cl ratios can be distorted by cation exchange along the flow path, Br^- and I^- are good complements to Cl^-. The Br/Cl ratio is elevated in bedrock brine but depleted in salt (Figure 6.33). Nitrogen species and DOC are useful in identifying domestic

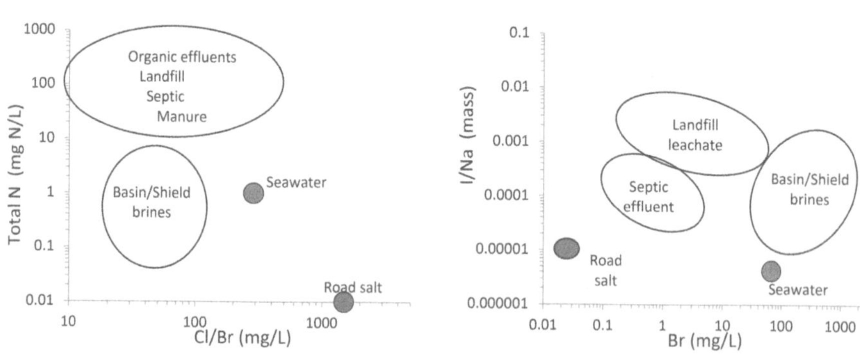

FIGURE 6.33 Distinguishing sources of salinity in groundwater with halides and nitrogen species. Modified from Panno et al. 2006, with data from Bottomley et al. 2002, 2005.

wastewater and landfill leachate (Panno et al. 2006), as they have significant enrichments in these sources.

Sources of Sulfate in Groundwater

The Chebotarev model for the evolution of geochemical facies in groundwater suggests that between the shallow low-salinity bicarbonate groundwaters and the deep chloride brines is an intermediate facies dominated by the solubility of sulfate minerals like gypsum, which provide salinities over 2000 ppm TDS and have an impact on water quality.

Sulfate in groundwater has the capacity to compromise water quality both by the elevated salinity it can contribute (drinking water limit is 250 ppm) and by the potential for reduction to toxic H_2S (see above). Sources of sulfate in groundwater include the dissolution of gypsum and anhydrite:

$$CaSO_4 \cdot 2H_2O \rightarrow Ca^{2+} + SO_4^{2-} + 2H_2O \quad K_{gyp} = 10^{-4.58}$$
$$CaSO_4 \rightarrow Ca^{2+} + SO_4^{2-} \quad K_{anh} = 10^{-4.36}$$

Gypsum dissolution can generate some 1500 ppm SO_4^{2-} in solution. Without the solubility control of Ca^{2+}, for example, in acid mine drainage, SO_4^{2-}concentrations can increase to the g/L range. Other sulfate minerals, such as celestite ($SrSO_4$; $K_{cel} = 10^{-6.63}$) and barite ($BaSO_4$; $K_{bar} = 10^{-9.97}$) are highly insoluble and maintain SO_4^{2-} concentrations below 100 ppm. The other principal source of sulfate in groundwater is from the oxidation of pyrite:

$$FeS_2 + 3½O_2 + H_2O \rightarrow Fe^{2+} + 2SO_4^{2-} + 2H^+$$

This proceeds naturally in soils developed on newly exposed materials, such as glacial till which may have a component of pyrite that can be oxidized and contribute sulfate to recharging groundwater. This also takes place, although perhaps more slowly, during the weathering of sulfide-bearing bedrock, such as black shales and ore bodies. Exposure of such materials to atmospheric oxygen as a consequence of mining or road construction, as discussed above, is a mechanism to release sulfate to the environment.

Determining the sources of sulfate and sulfide in groundwater is helped by measuring their isotopes, $\delta^{34}S_{SO_4^{2-}}$, $\delta^{34}S_{H_2S}$, and $\delta^{18}O_{SO_4^{2-}}$. The $\delta^{34}S$ of sulfate can be simplified into that originating from marine sulfate, which is enriched in ^{34}S, and that originating from oxidation of pyrite and other reduced sources such as volcanic sulfur, which are depleted in ^{34}S. Today, seawater sulfate has a $\delta^{34}S$ value of 21‰ CDT (Canyon Diablo Troilite) and a $\delta^{18}O$ value of 9.5‰ VSMOW (Vienna Standard Mean Ocean Water). In the past, $\delta^{34}S$ seawater varied between about 10‰ and 30‰, essentially controlled by the relative fluxes of carbon and sulfur through oxidation and reduction (Veizer et al. 1980; Figure 6.34). Oxygen-18 followed a more constrained range, averaging close to 15‰ during most of the Phanerozoic with excursions up to 16‰ during the Carboniferous and down to 10‰ during the Permian. Values for

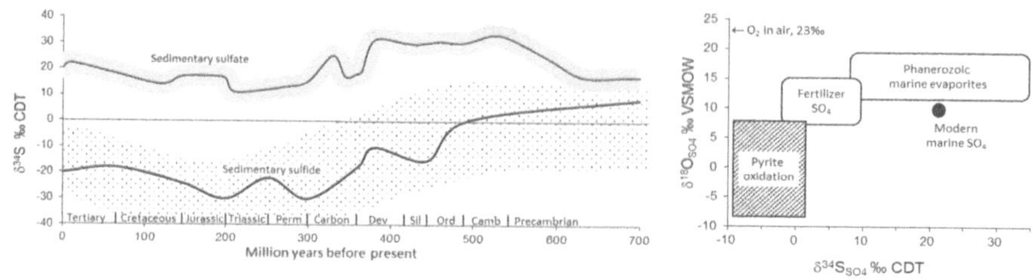

FIGURE 6.34 Left: Ranges in $\delta^{34}S$ in sedimentary (marine) sulfate and diagenetic sulfide over the Phanerozoic era. Right: $\delta^{34}S$–$\delta^{18}O$ composition of major sources of sulfate in groundwater. Data from Claypool et al. 1980; Veizer et al. 1980; Turchyn and Schrag 2006; Seal 2006.

marine $\delta^{18}O_{SO_4^{2-}}$ are largely controlled by continental oxidation of sulfides with varying contributions of ^{18}O from the water ($\delta^{18}O \approx -5$ to -20‰) and from atmospheric O_2 ($\delta^{18}O = +23.5$‰).

The difference between the two forms is largely due to the fractionation that accompanies sulfate reduction. Bacterial reduction of sulfate in marine sediments is accompanied by a fractionation, $\varepsilon^{34}S_{SO_4^{2-}-H_2S} \cong 20$‰–40‰, and so sedimentary sulfide has the low values (Figure 6.34). Sulfide in coal and sulfide ore is similarly depleted, with $\delta^{34}S$ values less than 10‰. Mantle-derived sulfide (volcanic H_2S) and meteorites also have $\delta^{34}S$ values near or less than 0‰.

Industrial emissions are characterized by $\delta^{34}S$ below about 10‰ and an elevated $\delta^{18}O$ from atmospheric O_2 in groundwater. Proemse et al. (2013) use this co-isotope signature to trace fallout from the oil sands in western Canada. Einsiedl and Mayer (2005) identify industrial fallout as being the dominant source of SO_4^{2-} in karst groundwaters in southern Germany, where concentrations are attenuated in older groundwaters by sulfate reduction based on the oxidation of organic contaminants accompanying this anthropogenic source.

Lithium, δ^7Li, and Salinity

The lightest alkali element is a lithophilic and highly soluble cation with high solubility due to a high hydration number of 6. Lithium also does not participate in redox or biological reactions. These characteristics make it a relatively conservative solute. Lithium is found in igneous rocks, particularly pegmatites, and in minerals such as lepidolite ($K[Li,Al,Rb]_3[Al,Si]_4O_{10}[F,OH]_2$) and spodumene ($LiAl[SiO_3]_2$), but more abundantly in brines associated with evaporite deposits (playas and salars). Its broad range in isotope distribution between seawater and igneous rocks makes it a good tracer of origin.

Li^+ concentration in seawater is only 0.18 ppm, with δ^7Li of +32.3‰. Studies show that through evaporation, its isotope ratio is conservative. In contrast, igneous rocks have δ^7Li values close to the LSVEC (lithium carbonate) standard. The dominantly Na–Cl brines of the groundwater-fed salar or brine lakes in the Atacama Desert of northern Chile, for example, contain up to 4 g/L Li^+ (Bradley et al. 2013) originating from hydrothermal alteration of the andesite rocks of the Andes. These high

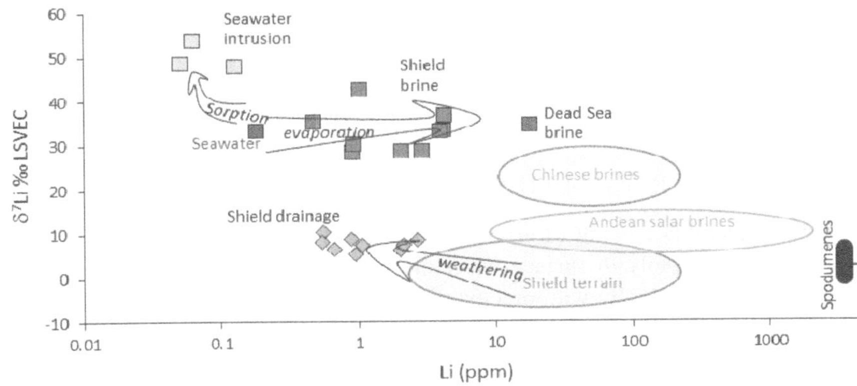

FIGURE 6.35 Lithium and δ^7Li in various fresh and saline waters and in rocks (MORB = Midoceanic Ridge basalt). Sources of data: Svec and Anderson 1965; Chan and Edmond 1988; Bottomley et al. 2003; Hogan and Blum 2003; Lemarchand et al. 2010; Araoka et al. 2013; Desaulty et al. 2022.

concentrations are augmented by further evaporation in surface lagoons, and the highly concentrated Li^+ is then recovered for commercial use by alkaline precipitation as Li_2CO_3. Strong evaporation of brines in playas and salars can precipitate Li^+ in a range of phosphate and sulfate evaporite minerals. In contrast to these igneous origins for lithium are brines derived from the evaporation of seawater (Figure 6.35), such as the Dead Sea brine, which has concentrated Li^+ to 18 ppm, representing a hundredfold increase over seawater (Epstein et al. 1981). Brines derived from seawater evaporation can then be identified on the basis of δ^7Li, which is conservative during simple concentration. Bottomley (1999, 2003) uses δ^7Li to trace the source of deep crustal brines to an evaporated seawater source.

Weathering of granitic bedrock is a source of Li^+ in groundwaters, although fractionation by sorption can occur along the flow path. Lemarchand et al. (2010) show this for drainage in granitic terrain with a highly weathered saprolite developed at the surface (Figure 6.35).

7 Groundwater Dating

GROUNDWATER *AGE* AND MEAN RESIDENCE TIME

Of the considerable amount of groundwater recharge that occurs on a daily basis, most circulates rapidly to surface water drainage systems such that storm hydrographs correspond closely with precipitation. Phreatic aquifers can respond rapidly to recharge and have groundwater with mean residence times on the order of months to a few years (Figure 7.1). Deep and confined aquifers host groundwaters that may circulate over hundreds to thousands of years, often recharged under different climate conditions.

Groundwater age relates to circulation and renewability, critical for water resource planning. Groundwater resources identified as being modern are regularly recharged

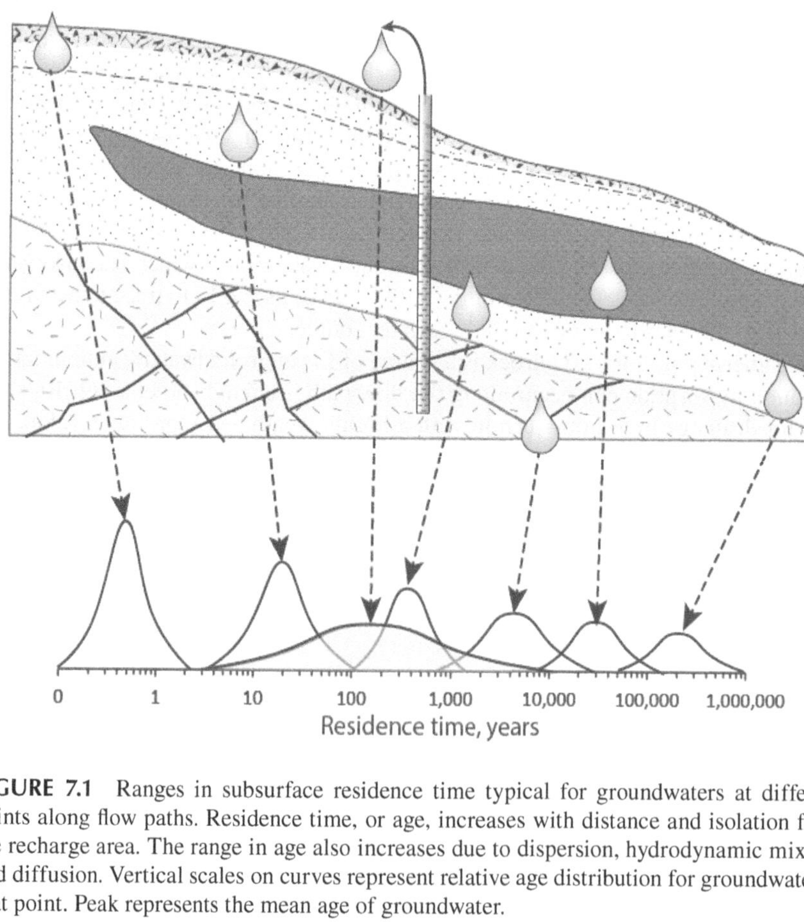

FIGURE 7.1 Ranges in subsurface residence time typical for groundwaters at different points along flow paths. Residence time, or age, increases with distance and isolation from the recharge area. The range in age also increases due to dispersion, hydrodynamic mixing, and diffusion. Vertical scales on curves represent relative age distribution for groundwater at that point. Peak represents the mean age of groundwater.

DOI: 10.1201/9781003353836-7

and can be sustainably developed, whereas exploiting fossil groundwater is essentially mining the resource. Modern groundwater will have more direct pathways from the surface and so is vulnerable to contamination from surface activities in the recharge area.

Hydrodynamic dispersion and differential pathways mix recharge water from different events such that groundwater below the water table will integrate a range of ages, the mean of which provides a mean residence time. A range of tracers, both anthropogenic and natural, can be used to constrain groundwater age. Figure 7.2 shows the effective age range for various methods, with details given in Table 7.1. Phillips and Castro (2003) provide a detailed overview of many of these methods. Note that many are non-routine and experimental, and so their use is often restricted to research studies.

Modern groundwater recharge can be identified by solutes such as chloride and nitrate at levels above regional background. As discussed in Chapter 6, Cl^- concentrations above about 1 ppm in shallow groundwater for most continental settings are usually derived from septic systems, agriculture, and road salt. Similarly, NO_3^- is generally below 1 ppm in non-impacted groundwater. Other indicators of modern recharge include a host of recalcitrant pharmaceuticals and fluorinated compounds, which at trace concentrations are clear indicators of a component of modern recharge. These are, however, qualitative indicators with variable input functions and provide no age constraint.

Emissions of certain anthropogenic components to the atmosphere have known atmospheric scenarios and so provide quantitative tracers of recent groundwater recharge. Several tracers fall into this category, including chlorofluorocarbons (CFCs), sulfur hexafluoride, SF_6, and ^{85}Kr. Two other cosmogenic isotopes, ^{35}S ($T_{½}$ = 87.5 d) and ^{22}Na ($T_{½}$ = 2.6 y), have interesting half-lives for very young systems. Improving analytical methods are increasing interest in them (Schubert et al.

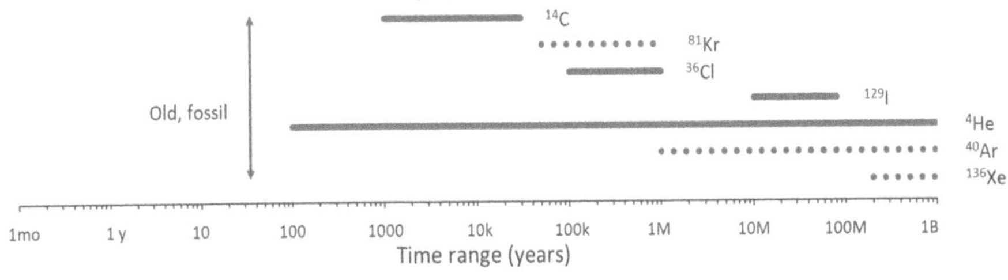

FIGURE 7.2 Age range for routine, non-routine radioisotopes, and anthropogenic gases in groundwater dating.

TABLE 7.1
Tracers for Age Dating Groundwater

Tracer	Half-Life	Source (Target Nuclide)	Production
		Modern Groundwater	
^{35}S	87.5 d	Cosmogenic (^{40}Ar)	Steady state
^{22}Na	2.6 y	Cosmogenic (^{22}Ne)	100–300 mBq/m^2/y
T–^{3}He	12.32 y	Cosmogenic (^{14}N)	5–15 TU
^{85}Kr	10.74 y	U-fission, fuel reprocessing	1.5 Bq/m^3, increasing
SF_6	Stable	Electrical insulator emissions	7 pptv, increasing
CFC	Stable	Refrigerant emissions, aerosols	CFC-12: 538 ppt in 2000
		Sub-Modern Groundwater	
^{39}Ar	269 y	Cosmogenic (^{40}Ar)	16.7 mBq/m^3
^{32}Si	~172 y	Cosmogenic (^{40}Ar)	2 atoms/m^2/s
		Old Groundwater/Brines	
^{14}C	5730 y	Cosmogenic (^{14}N)	100 pMC, 0.226 Bq/g C
^{81}Kr	229,000 y	Cosmogenic (^{80}Kr)	1.8 μBq/m^3
^{36}Cl	301,000 y	Cosmogenic (^{35}Cl)	< 50 atoms/m^2/s
		Subsurface, (^{35}Cl)	$^{36}Cl/Cl > 500 \times 10^{-15}$
^{129}I	15.8 My	Cosmogenic (^{129}Xe)	100,000 atoms/L
		Subsurface (^{238}U sp. fission)	$< 500 \times 10^6$ atoms/L
^4He	Stable	Subsurface (U and Th α-decay)	5.24×10^{-6} ppmv
^{40}Ar	Stable	Subsurface (^{40}K)	1.17% of K
^{136}Xe	Stable	Subsurface (^{238}U sp. fission)	8.9% of Xe_{atm}

2016; Kaste et al. 2016). However, for modern groundwater recharged in the past several decades, tritium remains the most commonly used tracer.

DATING MODERN GROUNDWATER WITH TRITIUM

Tritium has a number of attributes that make it an important tracer for age-dating groundwater. It is a component of H_2O and so a tritium age is of water itself and not its solutes. T has a short half-life, so its detection in water is a clear indication of at least some component of modern recharge. Tritium is easy to sample and can be analyzed in many commercial laboratories worldwide. It also has a daughter, ^3He, that can be analyzed to add greater precision to age calculations. Finally, the thermonuclear peak produced in the 1960s has attenuated, so now levels in precipitation have returned to background.

Natural Production of Tritium in the Atmosphere

Cosmogenic or natural tritium production occurs in the upper atmosphere from the bombardment of nitrogen by the flux of secondary neutrons generated in the stratosphere by cosmic radiation, then combining with stratospheric oxygen and hydroxyl ions to form water:

$$^{14}N + {}^{1}n \rightarrow {}^{12}C + T \rightarrow HTO$$

Tritium is transferred to meteoric water vapor by seasonal transfers from the stratosphere to the troposphere through mixing at the tropopause with seasonal migration of the Hadley cells. This produces a characteristic "spring peak" for tritium in precipitation, as observed in monitoring records for Vienna and two southern hemisphere sites at Kaitoke, NZ and Vernadsky, Antarctica (Figure 7.3).

The cosmic ray flux and tritium production varies with geomagnetic latitude with elevated production at higher latitudes. This latitudinal variability and the effect of dilution by coastal water vapor from the tritium-free oceans cause some variability in the distribution of global tritium in precipitation. The low latitude oceanic station Quezon City at 15°N in the Philippines has stabilized at tritium values less than 2 TU, whereas a high latitude, inland station such as Vienna, (48°N), is at 10 TU. Stations near centers of nuclear activity, such as Ottawa, Canada, are generally above natural background for that latitude. Stations remote from nuclear activities, such as Kaitoke, New Zealand, and Vernadsky, Antarctica (Figure 7.3) have harmonic signals that suggest an influence of solar cycle 24 on tritium production by attenuation of cosmic radiation during increased solar activity. Elevated tritium is apparent at these stations during the solar minima in 1995 and 2009, and lower values correlate with the solar maximum near the year 2000. Regional and local records for tritium in precipitation are available from the IAEA GNIP monitoring network through Wiser (Water Isotope System for Electronic Retrieval).

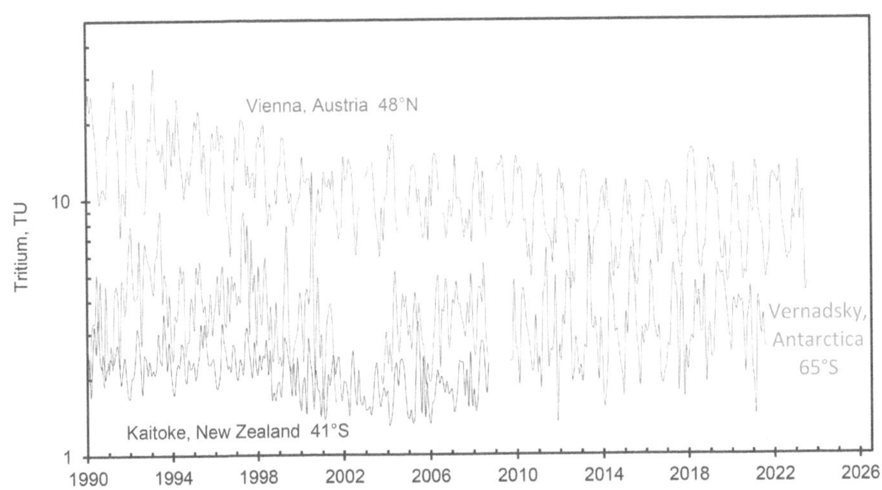

FIGURE 7.3 Trends for global tritium in precipitation since 1990. (Data from IAEA/WMO, the GNIP Database is accessible at: https://www.iaea.org/services/networks/gnip.) High latitude and high elevation as well as distance from coastal waters contribute to increased natural tritium production.

The peak levels in early 1960s precipitation from atmospheric testing of hydrogen bombs, provided a time horizon for dating groundwater in the ensuing decades. The first test of a nuclear weapon was the *Trinity* plutonium fission device detonated in the desert of New Mexico on July 16, 1945. Six years later, hydrogen fusion was demonstrated on May 9, 1951, with a fission device code-named *George*, which fused a small core of deuterium to helium. On November 1, 1952, the first hydrogen bomb, *Ivy Mike*, was tested. The 10.4 megaton explosion on the Enewetak Atoll in the Marshall Islands of the western Pacific Ocean, generated a fireball over 3 miles wide. The ensuing decade of American and Soviet atmospheric tests ended in 1962 with a final release of over 140 megatons of energy just prior to the 1963 Soviet–American atmospheric test-ban treaty, brought subsequent testing of nuclear weapons underground. The neutron flux associated with these thermonuclear tests activated ^{14}N and other nuclides in the atmosphere, producing tritium as well as ^{14}C, ^{36}Cl, and other radionuclides. The resulting spike of tritium in precipitation was over 5000 TU in the northern hemisphere in the spring of 1963 (Figure 7.4).

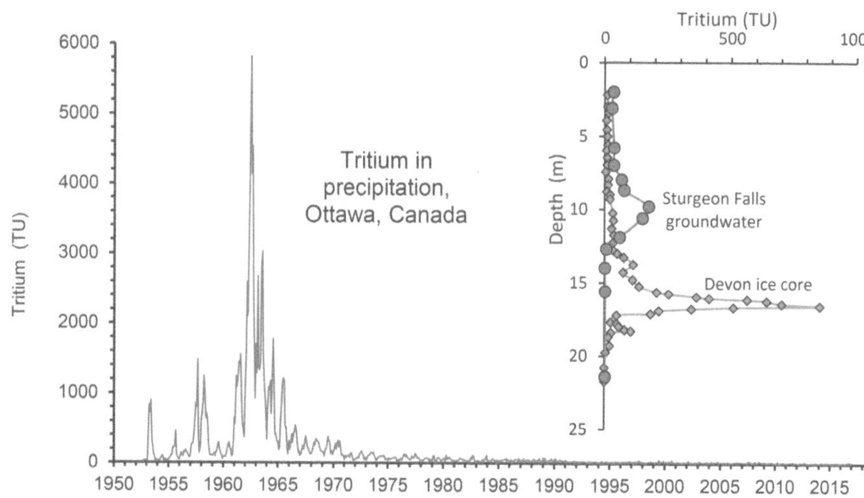

FIGURE 7.4 The production of anthropogenic tritium as a consequence of the high neutron flux generated by above-ground testing of thermonuclear (hydrogen-fusion) weapons, through spallation reactions on ^{14}N in the atmosphere. The Ottawa record is the longest continuously-monitored tritium record for precipitation, begun in 1951 by Dr. Bob Brown, a radiochemist with Atomic Energy of Canada Limited, Chalk River, Ontario. Available through IAEA/WMO (GNIP). The limited test ban treaty of 1963 between the USSR and USA began an era of recovery to background levels. Soviet tests in Kazakhstan (48°N) and American tests in the Marshal Islands (7°N) produced higher atmospheric tritium for the northern latitudes. Following the test-ban treaty, thermonuclear tritium has been lost from the atmosphere by a high rate of mixing in the oceans and decay and has returned to natural, pre-thermonuclear levels maintained by cosmogenic production. The 1963 "bomb peak" can be observed in older groundwater and in ice cores (inset: Sturgeon Falls data from Renaud et al. 2005, Robertson and Cherry 1989; Devon data from Clark et al. 2007). Note the highly attenuated peak in the groundwater due to hydrodynamic dispersion and decay, whereas the glacier ice experiences only decay.

Very few measurements exist of natural, pre-bomb tritium in precipitation. Precipitation near Ottawa had about 10 to 15 TU before 1951, as determined from archived samples and vintage wines bottled in the 1940s (Brown 1961).

The 1963 peak was a useful time horizon used in many hydrogeological studies to determine recharge and flow velocities, such as the case for the vertically recharged groundwater in the Sturgeon Falls deltaic sand aquifer (Robertson and Cherry 1989; Figure 7.4). As the 1963 tritium spike recedes into the past, its use as a time horizon for dating groundwaters is lost.

Groundwater Dating by Tritium Decay

Subsurface production of tritium is negligible, occurring only in U-enriched bedrock through fission of ^6Li. For example, Andrews et al. (1989) calculate a concentration of less than 1 TU in the Stripa granite, Sweden, which has a high uranium content of 44 ppm. Cosmogenic production is therefore the only significant contributor of tritium in groundwater. As this decays, the tritium remaining after some time, T_t, can be used to calculate age since recharge using the exponential decay equation:

$$T_t = T_0 e^{-\lambda t}$$

The decay constant, λ, is equal to ln(2) divided by the half-life, $T_{½} = 12.32$ years:

$$t = -\frac{1}{\lambda} \times \ln\left(\frac{T_t}{T_0}\right)$$

$$t = -17.77 \times \ln\left(\frac{T_t}{T_0}\right)$$

Such an age presumes that the groundwater has followed a simple piston flow and the initial tritium concentration at the time of recharge, T_0, is known. More likely, recharge involves the integration of precipitation with variable tritium concentrations over a substantial period. Consider the two wells in Figure 7.5. The phreatic aquifer gains recharge along its full length, such that groundwater sampled down gradient integrates a greater range of ages than the groundwater in the confined aquifer.

More realistic are lumped parameter models that combine cumulative tritium input over a specified period with dispersion, mixing, and decay (Maloszewski and Zuber 1982, 1996). Doney et al. (1992) parameterize the bomb contribution to annual precipitation globally to determine atmospheric scenarios for bomb tritium at specific locations. The United States Geological Survey (USGS) has developed this approach into an accessible Excel-based program, Tracer LPM, to model tritium and other atmospheric tracers of groundwater age (Jurgens et al. 2012).

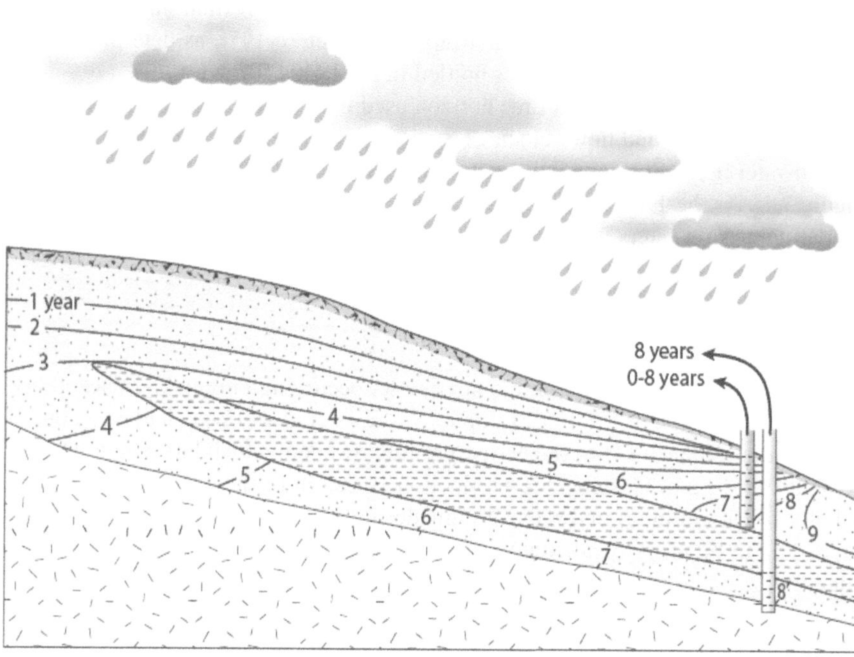

FIGURE 7.5 Schematic of groundwater age for exponential mixing in a phreatic aquifer and piston-flow in a confined aquifer. Note that the simple piston-flow hydrodynamic dispersion would broaden the range of ages in the sample. Groundwater sampled from the well in the phreatic aquifer has recharge along its length and so integrates precipitation over a longer period of time, giving a broader range of ages, whereas the confined aquifer well samples groundwater recharged over a well-constrained period of time.

Groundwater Dating with Tritium – ^{3}He

Initial tritium is poorly known in most groundwater systems due to seasonal and spatial variability. However, knowledge of T_0 is not required if the accumulated daughter, tritigenic ^3He_T, is also measured in the groundwater at the time of sampling. Measuring tritigenic ^3He_T can allow much more precise age calculations, often with resolution to months. An increasing number of noble gas laboratories are now available to provide measurements of ^{3}He in groundwater to complement enriched tritium measurements for this now routine dating scheme.

The T–^{3}He dating method combines the two exponential functions for loss of parent and gain of daughter presented in Chapter 3, Figure 3.10:

$$\text{Loss of initial tritium} \qquad T_t = T_0 e^{-\lambda t}$$

$$\text{Gain of tritigenic } ^3He_T \qquad ^3He_T = T_0\left(1 - e^{-\lambda t}\right)$$

By rearranging the tritium decay equation and substituting into the ^3He ingrowth equation, knowledge of the tritium input concentration, T_0, is removed, and time, t, can be calculated from the measured tritium and ^3He concentrations:

$$^3He_T = T_t\left(e^{\lambda t} - 1\right)$$

$$t = 17.77 \times \ln\left(\frac{^3He_T}{T_t} + 1\right)$$

However, the measured ^3He concentration includes atmospheric ^3He that has accompanied the groundwater during recharge and must be subtracted to give tritogenic ^3He_T. The atmospheric ^3He includes both a dissolved component and an excess air component that can be entrained as microbubbles during recharge.

$$^3He_T = {^3He}_{measured} - \left({^3He}_{atm-diss} + {^3He}_{excess\ air}\right)$$

Atmospheric He has a partial pressure of 5.24×10^{-6} and a dissolved concentration that ranges from $5.0 \times 10^{-8}\ cc_{STP}/cc_{H_2O}$ at 0°C to $4.58 \times 10^{-8}\ cc_{STP}/cc_{H_2O}$ at 30°C (Figure 7.6) and has a $^3He/^4He$ ratio in air-equilibrated water of 1.36×10^{-6}. This ratio is slightly less than the atmospheric ratio of 1.38×10^{-6} due to fractionation during dissolution, which favors ^4He in water. Note that the abundance of ^3He is only 1.38 ppm, and so measurements of total helium are essentially a measurement of ^4He. In calculations below, ^4He is taken to be total helium for solubility considerations,

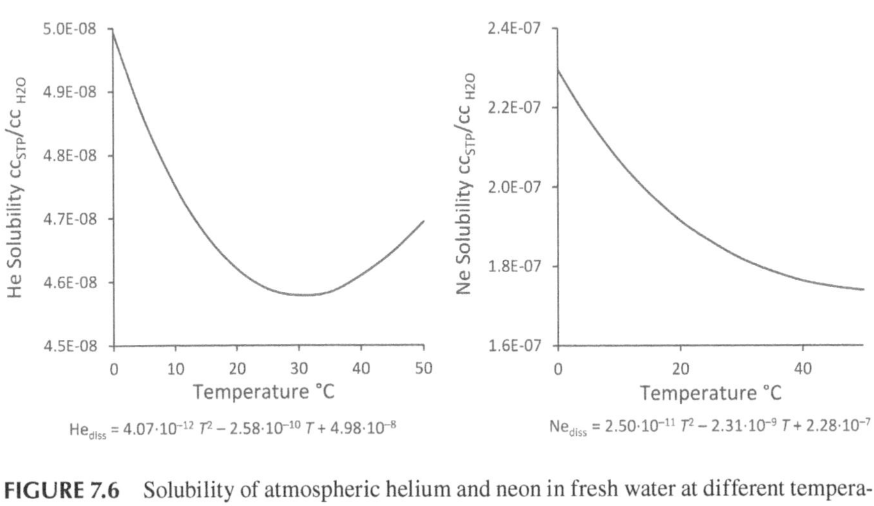

FIGURE 7.6 Solubility of atmospheric helium and neon in fresh water at different temperatures. Equations use temperature in degrees Celsius.

and ^3He concentrations are then determined by multiplying He (cc_{STP}/cc_{H_2O}) by the $^3He/^4He$ ratio.

Contributions of ^3He from excess air are normally determined by the excess Ne in the sample, for which its solubility is less sensitive to temperature than for He. If Ne is not measured, the dissolved plus excess air components can be calculated from the measurement of total He. A sample solved example is given with problems for this chapter, available online.

The tritium–helium dating method assumes that there is no loss of ^3He from shallow groundwater to the phreatic zone and atmosphere. This is unlikely if flow is sufficiently deep below the water table. Loss of ^3He produces a bias to erroneously young groundwater ages. This may be the case for groundwater less than a couple of meters below the water table. Also, groundwater that are deep and close to crystalline bedrock may gain geogenic ^4He and ^3He, producing erroneously old ages. Helium-4 is produced in the subsurface by alpha decay of uranium and thorium, and ^3He is produced by fission of ^6Li. Fission of ^6Li occurs with the secondary neutron flux in the subsurface from spontaneous fission of ^{238}U and impact of α particles on light nuclei (Figure 3.10):

$$^6\text{Li} + \text{n} \rightarrow {}^4\text{He} + {}^3\text{He}_{\text{geo}}$$

The contribution of $^3He_{geo}$ to groundwater can be determined by calculating any helium in excess of the excess air component from recharge as determined with Ne and the $^3He/^4He$ ratio for geogenic helium. This ratio is generally taken to be 2×10^{-8} for shield terrains, given the greater production of He from alpha decay than from lithium fission (Bottomley et al. 1984).

$$\text{He}_{\text{geo}} = \text{He}_{\text{meas}} - \text{He}_{\text{diss}} - \text{He}_{\text{Ne-excess}}$$
$$^3\text{He}_{\text{geo}} = \text{He}_{\text{geo}} \times ({}^3\text{He}/{}^4\text{He})_{\text{geo}}$$

A good case study using T–^3He to determine groundwater ages is the site of a trichloroethylene (TCE) plume at Valcartier, Quebec (Murphy et al. 2010). Groundwater ages were required to determine the velocity of the TCE plume to wells and river water down-gradient from the source. The aquifer is composed of up to 10 to 20 m of clean glaciofluvial outwash sands overlying a silt aquitard with underlying proglacial sands and gravels. The bedrock is granite gneiss of the Canadian Shield.

The measured ^3He concentrations together with the geogenic ^3He calculated from the Ne and ^4He concentrations are given in Table 7.2. The T–^3He ages, corrected for atmospheric helium incorporated during recharge and geogenic contributions in the subsurface, show consistent increases with depth and along the flow path (Figure 7.7). Groundwater at the down-gradient end of the aquifer have mean ages of roughly 30 years with little vertical variation, reflecting upward flow and mixing over basal topography. These ages were used to constrain aquifer parameters in the particle tracking model (Figure 7.7).

TABLE 7.2
Tritium and Helium Concentrations with Calculated T–^{3}He Ages for Groundwater from a Glaciofluvial Outwash Aquifer in Valcartier, Quebec

	Piezometer Nest P-125-126 (Figure 7.7)			
Piezometer depth (m)	3.08	6.11	11.68	24.56
T (°C)	15.5	15.5	15.5	15.5
Tritium (TU)	14.1	13.5	12.7	11.7
$He_{measured}\left(cc_{STP}/cc_{H_2O}\right)$	8.80E–08	9.17E–08	8.31E–08	1.30E–07
$^3He/^4He_{measured}$	1.17E–06	1.41E–06	1.45E–06	8.96E–07
$^3He_{measured}\left(cc_{STP}/cc_{H_2O}\right)$	1.27E–13	1.31E–13	1.21E–13	1.26E–13
$^{20}Ne_{measured}\left(cc_{STP}/cc_{H_2O}\right)$	3.54E–07	3.44E–07	2.88E–07	2.77E–07
$^{20}Ne_{excess\ air}\left(cc_{STP}/cc_{H_2O}\right)$	1.37E–07	1.27E–07	7.09E–08	5.99E–08
$^4He_{air\ equilibrated}\left(cc_{STP}/cc_{H_2O}\right)$	4.68E–08	4.68E–08	4.68E–08	4.68E–08
$He_{air\ equilibrated}\left(cc_{STP}/cc_{H_2O}\right)$	6.36E–14	6.36E–14	6.36E–14	6.36E–14
$He_{excess\ air}\left(cc_{STP}/cc_{H_2O}\right)$	3.94E–08	3.65E–08	2.04E–08	1.73E–08
$^3He_{excess\ air}\left(cc_{STP}/cc_{H_2O}\right)$	6.19E–14	5.79E–14	3.57E–14	3.13E–14
$He_{geogenic}\left(cc_{STP}/cc_{H_2O}\right)$	1.66E–08	3.88E–09	1.06E–08	7.11E–08
$He_{geogenic}\left(cc_{STP}/cc_{H_2O}\right)$	3.33E–16	7.75E–17	2.12E–16	1.42E–15
$^3He_{tritogenic}\left(cc_{STP}/cc_{H_2O}\right)$	1.15E–15	9.38E–15	2.15E–14	2.97E–14
$^3He_{tritogenic}$ TU	0.46	3.77	8.65	11.9
Age (years)	0.6	4.4	9.2	12.5

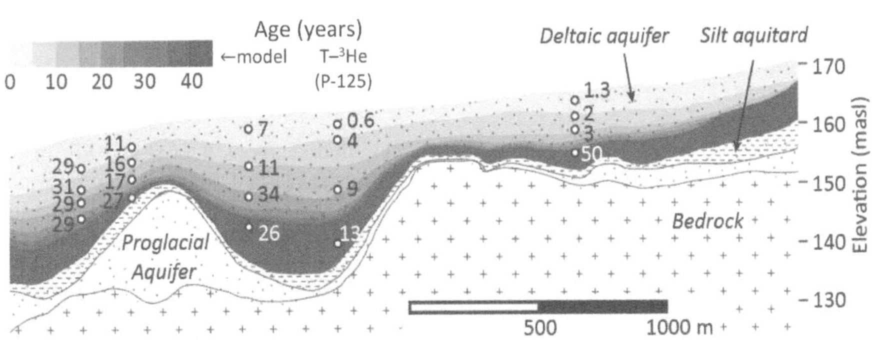

FIGURE 7.7 Cross-section of a glaciofluvial sand aquifer contaminated by trichloroethylene (TCE) in Valcartier, Quebec. Sampling points of groundwater samples were taken in vertical profiles at five locations along the flow path, with measured T–^{3}He ages in comparison with particle track ages. After Murphy et al. 2010.

ANTHROPOGENIC AGE TRACERS

Fluorinated gases, including CFCs and SF_6, manufactured since the mid-20th century have generated growing atmospheric scenarios that can be retained in groundwater and so are useful for dating modern groundwater. Although the 1989 ban on CFCs has led to a decrease in their atmospheric concentrations, their hydrofluorocarbon (HFC) replacements and SF_6 continue to grow in atmospheric concentration. CFCs and SF_6 are recalcitrant in oxidizing groundwater. HFCs degrade to persistent poly-fluoroalkyl substances. A short-lived isotope of krypton, ^{85}Kr, adds to the short-list of anthropogenic age tracers that can be used to provide groundwater ages.

CFCs AND SF_6

CFC compounds, used in air conditioners and as aerosol propellants since the 1940s, steadily accumulated in the atmosphere for decades. With their increasing atmospheric concentrations and conservative behavior in aerobic groundwater, CFCs became an important tool for dating (IAEA 2006). Following discovery of their role in the photochemical degradation of stratospheric ozone, the 1989 Montreal Protocol greatly reduced global emissions such that atmospheric concentrations leveled off in the 1990s and are now declining (Figure 7.8). Like tritium, this has now introduced

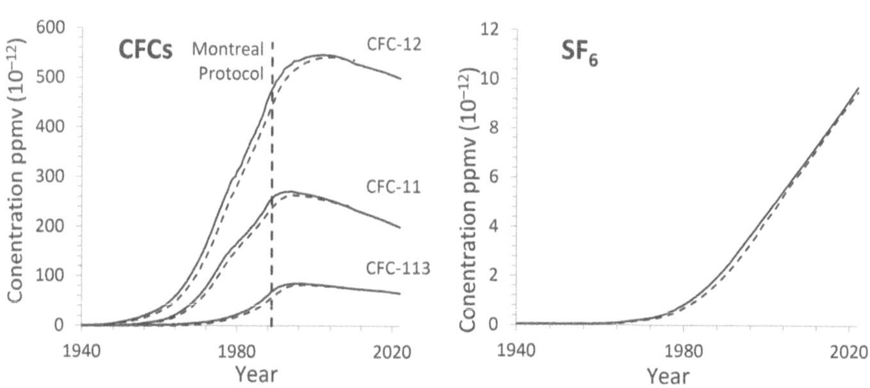

FIGURE 7.8 Atmospheric concentrations of chlorofluorocarbons (CFCs) and SF_6 in the northern hemisphere (solid lines) and southern hemisphere (dashed lines). (Data from water.usgs.gov/lab/software/air_curve/.) The air-equilibrated water concentrations depend on water temperature. For low-salinity waters, the temperature-dependent Henry's constant values can be calculated from the following equations (T in degree Celsius) based on data in IAEA (2006) for CFCs and in Sander (1999) for SF_6:

$$K_{\text{CFC-11}} = -6.22\times10^{-7}T^3 + 6.15\times10^{-5}T^2 - 2.27\times10^{-3}T + 3.85\times10^{-2}$$

$$K_{\text{CFC-12}} = -1.28\times10^{-7}T^3 + 1.30\times10^{-5}T^2 - 5.03\times10^{-4}T + 9.37\times10^{-3}$$

$$K_{\text{CFC-113}} = -2.28\times10^{-7}T^3 + 2.21\times10^{-5}T^2 - 7.95\times10^{-4}T + 1.25\times10^{-2}$$

$$K_{SF_6} = -1.95\times10^{-9}T^3 + 2.60\times10^{-7}T^2 - 1.57\times10^{-5}T + 5.01\times10^{-4}$$

ambiguity into the interpretation of CFC ages. Other complications include their temperature-dependent solubilities, elevation, and effect of salinity. Warner and Weiss (1985) and Sanders (1999) determined Henry's Constants for CFCs, showing that they vary by almost an order of magnitude between 0°C and 40°C.

The ambiguity from declining atmospheric CFC signals is not found for SF_6. This nontoxic, inert gas is used as an insulator in electrical equipment, such as high-voltage supplies. It has been accumulating in the atmosphere since the 1950s, with an ingrowth function that has remained essentially linear over the past two decades (Darling et al. 2012; Chambers et al. 2019). SF_6 has an even lower atmospheric concentration and lower Henry's constant (0.00024 at 25°C), such that aqueous concentrations are less than about 3 fmol/L (3×10^{-15} mol/L). There is evidence, however, that the atmospheric concentrations reconstructed at the hemispheric scale are exceeded in the regions around major urban centers, requiring special consideration of the input function used in such areas (Santella et al. 2008). Despite these constraints on the input function and measurement, SF_6 has become a complementary tracer to CFCs and is now routinely analyzed by many laboratories.

Like SF_6, the radionuclide ^{85}Kr released from nuclear fuel reprocessing has a steadily increasing atmospheric concentration. Unlike SF_6, it decays, with a half-life of 10.76 years, by β emission to common ^{85}Rb. Back-extrapolation from a measured value along a decay line to the input function provides an unambiguous age that can be quite accurate. However, its use is not routine due to the large samples required (thousands of liters) to strip enough gas for radiometric counting and the potential for atmospheric contamination.

The use of the dissolved concentration of atmospheric gases, such as CFCs, SF_6, and ^{85}Kr, has an additional complication from the physical incorporation of microbubbles of air during recharge at the water table. Entrained with the influx of groundwater following a recharge, this excess air can add a significant fraction of gas, which is then dissolved as the groundwater moves downward to higher pressures below the water table. This can be corrected for using the solubility and concentration of a gas with a known and stable concentration, such as Ne, Ar, or N_2.

ANTHROPOGENIC ^{85}Kr

The reprocessing of spent nuclear fuel to recover uranium and plutonium and remove fission products releases radioisotopes of noble gases to the atmosphere. The short-lived noble gases, including ^{133}Xe ($T_{½}$ = 5 days) and ^{37}Ar ($T_{½}$ = 35 days), readily decay to stable nuclides. However, the longer-lived krypton fission product, ^{85}Kr ($T_{½}$ = 10.74 years), has been steadily accumulating in the atmosphere, although with a decreased rate in the past decade (Bollhöfer et al. 2019) (Figure 7.9). As a noble gas, it provides a geochemically conservative tracer of modern recharge.

In a study of groundwater in an aquifer of pyroclastic material in the Kumamoto volcanic region of Japan, Momoshima et al. (2011) and subsequently Kagabu et al. (2017) use ^{85}Kr to date groundwater along the hydrodynamic flow path of this aquifer with declining water levels, which supplies a population of nearly one million. They successfully use ^{85}Kr in this urban region where CFCs and SF_6 tracers were

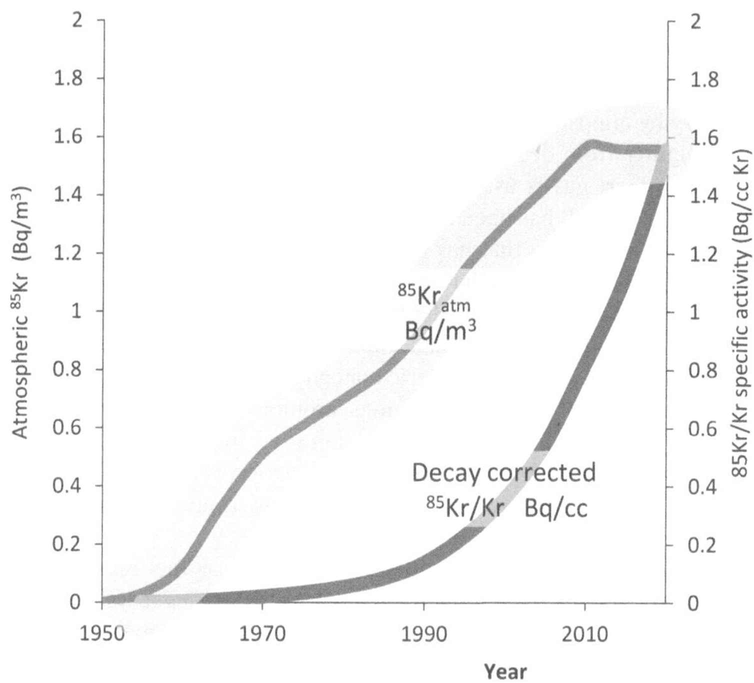

FIGURE 7.9 Global atmospheric ^{85}Kr. Gray represents variability from nuclear reprocessing activities. The ^{85}Kr/Kr specific activity ratio in the atmosphere, corrected for decay to 2020 for comparison with ^{85}Kr/Kr in groundwater. Data source: Kagabu et al. 2017; Momoshima et al. 2011.

compromised by local emissions. Their method is based on scintillation counting, requiring recovery of gas from 10,000 L of groundwater using an in-field membrane gas separator. Contamination from the atmosphere is a concern for such tracers, requiring closed system extraction methods in the field. Measurements are made on the total krypton separated from the extracted gases in the groundwater and expressed as the ^{85}Kr activity in Bq normalized to the total stable Kr extracted. These were then simply compared to the decay-corrected ^{85}Kr/Kr atmospheric scenario (Figure 7.9) to give groundwater ages of 8.2 ± 0.7 years and 20.7 ± 0.6 years at the respective sites, which compared favorably with tritium and hydrodynamic ages.

Recently, new atom-counting techniques using laser trapping and cooling of Kr (Atom Trap Trace Analysis [ATTA]; Sturchio et al., 2004; Jiang et al., 2012) has reduced the sample requirement from over 200 μL of Kr to less than 0.5 μL standard temperature and pressure (STP). This greatly reduces the amount of groundwater from which the Kr sample must be stripped from solution.

Thermonuclear ^{14}C and ^{36}Cl

In addition to a tritium peak, atmospheric testing of nuclear devices generated peaks for levels of ^{36}Cl and $^{14}CO_2$. Early testing with shipborne devices in late 1950 were

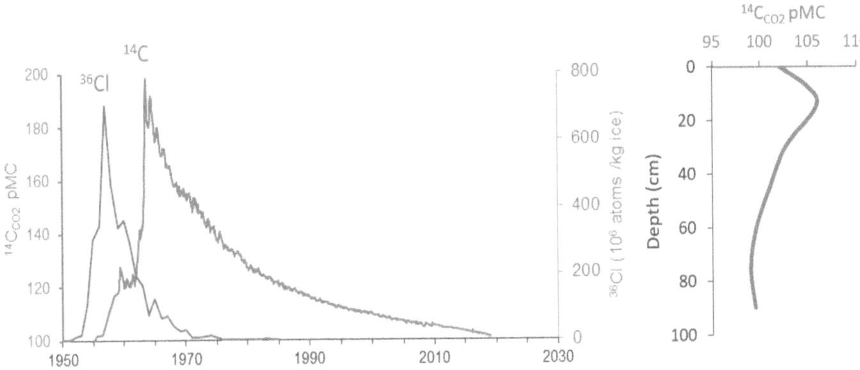

FIGURE 7.10 Left: Anthropogenic ^{36}Cl and ^{14}C in the atmosphere following thermonuclear bomb testing. The ^{36}Cl peak follows detonations at sea between 1954 and 1958, as observed in fallout recovered from the Dye-3 ice core in Greenland (Synal et al. 1990). The thermonuclear radiocarbon record (Nydal and Lövseth 1983; Hua et al. 2013, with data before 1962 and after 1980 from Jurgens et al. 2012 and Zal 2023) is similar to that of a bomb tritium peak, as both are from neutron activation of atmospheric ^{14}N. Right: Radiocarbon measured in soil CO_2 respired from organics in soil core (Zal 2023).

detonated at sea level, resulting in the neutron activation of ^{35}Cl in seawater and subsequent injection into the upper atmosphere by the ensuing fireball. Synal et al. (1990) recovered the atmospheric fallout pattern over this period in an ice core from the Greenland ice sheet near the Dye-3 site (Figure 7.10). The high neutron flux from upper atmospheric tests activated ^{14}N to produce ^{14}C, raising $^{14}CO_2$ activities to over 200 percent modern carbon (pMC) (Figure 7.10, left chart). Like tritium, the ^{36}Cl has recovered to natural production levels through exchange with the oceans and has largely been flushed through shallow flow systems. This has not been the case for thermonuclear radiocarbon, which has attenuated over a longer period of time through exchanges with the biosphere and by additions of ^{14}C-free CO_2 from the combustion of fossil fuels and cement production.

Biomass that has been growing over the past six decades retains these high radiocarbon activities and so serves as a useful tracer of carbon from the soil zone. Figure 7.10 (right chart) shows the ^{14}C content of soil CO_2 respired from soil cores from a farm field in Eastern Ontario. The enrichments, peaking at 15 cm depth, are from organics grown during the thermonuclear peak shown in the left chart. The modest enrichment is due to the considerable attenuation by the mix of different-aged carbon in the soil. Values below 100 pMC at greater depths reflect organics grown prior to nuclear weapons testing, before about 1950.

DATING SUBMODERN GROUNDWATER

Anthropogenic tracers and natural tritium provide very good coverage for dating groundwater recharged in the past five to six decades. Radiocarbon, discussed in the following section, remains the most useful dating tool for groundwater recharged during the Holocene and late Pleistocene. However, uncertainties in estimating the initial ^{14}C activity restrict applications to groundwater older than about 1000 years. Few radionuclides have half-lives that are appropriate for dating the submodern range of about 50 to 1000 years before present.

Atmospheric ^{39}Ar

Argon-39 is an atmospherically produced radioisotope with a half-life of 269 years, is not reactive in the subsurface, and so is a good tool for estimating groundwater age in the submodern range. Atmospheric production of ^{39}Ar is by neutron impingement on stable ^{40}Ar. It decays to stable ^{39}K by beta emission:

$$^{40}Ar + n \rightarrow {}^{39}Ar + 2n$$
$$^{39}Ar \rightarrow {}^{39}K + \beta$$

The atmospheric production and decay maintains an atmospheric abundance ratio for $^{39}Ar/^{40}Ar$ of 8.2×10^{-16} or 0.0018 Bq per L Ar at STP (Benneti et al. 2007). However, subsurface production can contribute to this through the activation of ^{39}K by neutrons that are continually generated by α particles from U and Th decay impinging on light nuclei:

$$^{39}K + n \rightarrow {}^{39}Ar + p$$

Loosli and Oeschger (1980) suggest that high production in granitic terrain and K-rich sediments can generate up to fourfold increases over atmospheric ^{39}Ar. In contrast, production in K- and U/Th-poor aquifers, such as carbonates, would have much less impact on calculated ages.

Until recently, a handicap was the great volumes of water that were required to be stripped of Ar, out of contact with the modern atmosphere, for a radiometric analysis (Loosli et al. 2000). Atmospheric production of ^{39}Ar maintains an activity of 0.016 Bq/m^3 in air. Dissolved in water using the Bunsen coefficient of 0.0313 cc_{STP} / cc_{H_2O} for Ar gives an activity of ^{39}Ar in air-equilibrated water of 5.25×10^{-7} Bq/L. This amounts to only about 6000 atoms of ^{39}Ar per liter, requiring extraction from several cubic meters of water for measurement by decay counting (Loosli and Oeschger 1980). A new method for counting atoms rather than decay events has greatly reduced the required sample size. ATTA using a laser trap and atom counting requires the argon stripped from only a few liters of water for reliable measurement (Welte et al. 2010; Jiang et al. 2011; Reichel et al., 2013). This greatly expands the use of the method for the important submodern groundwater age range.

Reichel et al. (2013) report the first ^{39}Ar dating of groundwater by ATTA, using membrane separation of gas from groundwater in the field followed by cryogenic separation of Ar in the laboratory. This was carried out on confined groundwater from the upper Rhine Graben of southwestern Germany, yielding a subsurface residence time of 360 ± 68 years, which matched with hydrodynamic models for groundwater flow in this highly exploited aquifer. Musy et al. (2023) show that cosmic muon capture by potassium in the recharge area [^{39}K ($\mu - \gamma$) ^{39}Ar] provides additional ^{39}Ar during recharge, producing over-modern waters.

The subsurface production of ^{39}Ar in environments with high ^{39}K and high secondary neutron flux from alpha decay of U and Th remains an obstacle to dating the submodern range. However, like other geogenic radionuclides, such as ^{129}I and ^{36}Cl, dating by ingrowth toward secular equilibrium can be used to extend its useful dating range (Yokochi et al. 2012; Lu et al. 2013).

Cosmogenic ^{32}Si

Silicon-32, a cosmogenic nuclide with a half-life of about 170 years, offers potential to date groundwater in the important submodern range of 50 to 1000 years (Fröhlich et al. 1987). It is produced by cosmic ray spallation of ^{40}Ar (n – 2α) with an average rate of only two atoms per m^2 per second in the atmosphere, followed by wet or dry fallout to the surface. It has a low activity in recharge waters of less than about 0.0001 Bq/L, which complicates analysis. More restricting for routine use is its high reactivity in soils, where most is lost through exchange with stable silica on minerals in soils (Morgenstern 2000). Fifield and Morgenstern (2009) show improvements for analysis by AMS that suggest ^{32}Si may find applications in Quaternary dating. However, dilution with stable Si in the subsurface combined with uncertainties with the atmospheric input as required for calculating the decay period make ^{32}Si a poorly constrained geochronometer for groundwater.

RADIOCARBON DATING OLD GROUNDWATER

For groundwater beyond about 1000 years old, radiocarbon remains the most useful and routine dating tool. Its 5730-year half-life is well suited to dating groundwater recharge during the Holocene and late Pleistocene, a period with significant variations in climate associated with the melting of continental ice sheets in high-latitude regions and shifting pluvial regimes at lower latitudes. Further, sampling is straightforward with minimal risk of atmospheric contamination and, with recent developments in accelerator mass spectrometry (AMS), analysis is rapid and inexpensive.

Basis of Radiocarbon Dating Groundwater

Radiocarbon is produced in the stratosphere through the activation of ^{14}N by secondary neutrons produced when high energy cosmic ray particles impinge on other atmospheric gases. The ^{14}C nuclide is then oxidized to $^{14}CO_2$ and mixes with the reservoir of modern atmospheric CO_2, with a current mixing ratio close to 1.18×10^{-12}. This low concentration defines the reference standard, Modern Carbon (MC), which has an activity of 0.226 Bq/g C. Measurements are referenced to MC, either as a percent (pMC) or fraction (*f*MC), or in permil units ($\Delta^{14}C = (fMC - 1) \times 1000$), as seen in Table 7.3 for a number of examples including the standard (MC), the 1963 bomb peak, and from the mid and early Holocene.

TABLE 7.3
Radiocarbon Standard and Units of Measurement for a Variety of Times

Sample	Year	$a^{14}C$ Bq/g C	$^{14}C/^{12}C$ Ratio	pMC	*f*MC	$\Delta^{14}C$ ‰ MC
Modern Carbon, MC	1950	0.226	1.18E-12	100	1	0
Bomb Peak	1963	0.452	2.36E-12	200	2	1000
y2k	2000	0.249	1.30E-12	110	1.1	100
Today	2026	0.228	1.19E-12	101	1.01	10
Mid-Holocene	5000 yBP	0.123	6.44E-13	55	0.55	–454
Early Holocene	10,000 yBP	0.067	3.52E-13	30	0.30	–702

Like stable ^{13}C, ^{14}C fractionates during reactions like photosynthesis, and so for consistent dating, all data are normalized to an equivalent $\delta^{13}C$ value of –25‰. Atmospheric $^{14}CO_2$ cycles into the marine system as dissolved inorganic carbon (DIC) and into terrestrial vegetation by photosynthesis. Surface production of radiocarbon from neutron impact on ^{17}O can be used for surface dating (Gosse et al. 2006) but has a negligible contribution to ^{14}C in groundwater (Andrews et al. 1989).

Radioactive decay of ^{14}C produces the stable daughter nuclide ^{14}N, which is indistinguishable from atmospheric and subsurface nitrogen,

$$^{14}C \rightarrow {}^{14}N + \beta$$

and so the dating method is based on the loss of the parent, according to the decay equation rearranged to yield time,

$$a_t^{14}C = a_0^{14}C \times e^{-\lambda t}$$

$$-\lambda t = \ln\left(\frac{a_t^{14}C}{a_0^{14}C}\right)$$

and recalling $\lambda = \ln 2/T_{½}$

$$t = -8267 \times \ln\left(\frac{a_t^{14}C}{a_0^{14}C}\right)$$

This equation assumes that the initial concentration of the parent is known and that the system is closed to subsequent gains or losses of the parent, except through radioactive decay.

The activity of atmospheric $^{14}CO_2$ has not been constant in the past, with fluctuations of up to 10% during the Holocene (Figure 7.11). These are attributed to secular variations in the heliomagnetosphere from solar activity and changes in the geomagnetosphere, both of which attenuate the cosmic ray flux and therefore the production of ^{14}C (Solanki et al. 2004). Higher activities of 20% to 30% in the late Pleistocene are due to cosmogenic ^{14}C production mixing with a much smaller reservoir of atmospheric CO_2 of some 180 ppmv. For precise age dating, the secular

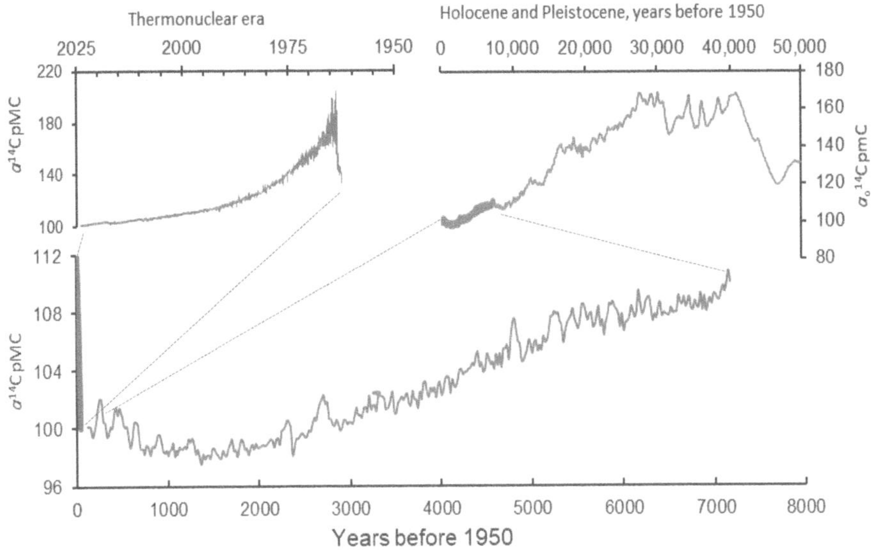

FIGURE 7.11 ^{14}C activity of atmospheric CO_2 during the Holocene and late Pleistocene epochs (data from Bayari 2002) based on tree rings (Holocene) and shallow marine corals (Pleistocene), based on their U/Th age. Thermonuclear era data based on direct measurements of $a^{14}CO_2$.

variations of $a^{14}CO_{2\text{-atm}}$ are incorporated into radiocarbon ages modeled with internationally recognized calibration programs, such as InterCal09 (Reimer et al. 2009), with dating possible to beyond 50,000 years before present. For groundwater dating, the effective range is much shorter, up to about 25,000 or 30,000 years, due to carbonate geochemical reactions that dilute the soil-derived ^{14}C and limit precision.

Radiocarbon dating of groundwater is based on dating the DIC and/or the dissolved organic carbon (DOC) in groundwater. Atmospheric $^{14}CO_2$ is incorporated into soils through photosynthesis, root respiration, and biodegradation. Infiltrating groundwater gains ^{14}C-active DIC and DOC by dissolving both soil CO_2 and soil organics. The long half-life of radiocarbon and comparatively rapid cycling of carbon in soils is such that the DIC and DOC dissolved from soil during infiltration are considered to have the same activity as MC at that time (Blume et al. 2022).

Following infiltration, both $a^{14}C_{DIC}$ and $a^{14}C_{DOC}$ can experience dilution from geochemical processes. Additions of ^{14}C-free DIC from weathering carbonates can occur. Similarly, groundwater can gain old DOC from buried organics in the aquifer. DIC is the most significant pool of carbon in groundwater and the most accessible for dating. It is essential to correct for DIC dilution and determine a value for the initial ^{14}C activity of DIC, $a_o^{14}C_{DIC}$, after which losses of radiocarbon are due to decay alone and which can be compared with $a_t^{14}C_{DIC}$ measured in groundwater sampled along the flow path to determine age, t.

Weathering and Initial $a^{14}_{o}C_{DIC}$

Aerobic respiration from the degradation of modern vegetation combined with root respiration in the rhizosphere maintains a large reservoir of $^{14}CO_2$ in the soil zone with modern ^{14}C activity (or slightly higher, see right chart in Figure 7.10) and $\delta^{13}C$ close to –25‰. This is dissolved by infiltrating precipitation and generates carbonic acid for weathering, as discussed in Chapter 5. However, weathering of old carbonate rocks and fracture minerals contribute ^{14}C-free DIC that dilutes this initial $^{14}C_{DIC}$, imparting an artificial age to this modern recharge.

If open system conditions prevail during carbonate weathering, then the DIC will have a modern ^{14}C activity. However, if carbonate weathering takes place under closed system conditions, $a^{14}_{o}C$ will be closer to 50 pMC due to the dilution with ^{14}C-free carbonate from the marine limestone matrix or fracture minerals:

Open System Conditions

$$^{14}CO_2 + H_2O + CaCO_3 \rightarrow Ca^{2+} + H^{14}CO_3^- + HCO_3^- \quad \text{calcite dissolution}$$
$$^{14}CO_2 + HCO_3^- \Leftrightarrow CO_2 + H^{14}CO_3^- \quad \text{exchange with soil } ^{14}C_{DIC}$$
$$^{14}CO_2 + H_2O + CaCO_3 \Leftrightarrow Ca^{2+} + 2H^{14}CO_3^- \quad \boldsymbol{a^{14}_{o}C_{DIC} = 100\,pMC}$$

Closed System Conditions

$$^{14}CO_2 + H_2O + CaCO_3 \rightarrow Ca^{2+} + H^{14}CO_3^- + HCO_3^- \quad \boldsymbol{a^{14}_{o}C_{DIC} = 50\,pMC}$$

The degree of openness of the system during mineral weathering then controls the initial radiocarbon activity of the DIC. As seen in Chapter 5, open system conditions often prevail, and so groundwater recharges with close to 100 pMC. This is not always the case. For example, in karstic regions, rapid recharge can minimize the exchange with soil CO_2, and so $a^{14}_{o}C_{DIC}$ is less than 100 pMC. The impact of this loss is an apparent age of groundwater that is greater than their true age and must be corrected by the determination of a radiocarbon dilution factor, q, which represents the fractional reduction of $a^{14}_{o}C_{DIC}$ to some value less than 100 pMC.

$q = 1$	No dilution of ^{14}C by reaction, typical of crystalline terrains
$q \cong 0.75$	Minor dilution of initial ^{14}C from partial closed system weathering
$q \cong 0.5$	Closed system carbonate weathering and exchange
$q < 0.5$	Extensive carbonate exchange and contributions from other geogenic sources

The correction for dilution of $a^{14}_{o}C_{DIC}$ during recharge applied to the atmospheric radiocarbon activity at the time of recharge, $a^{14}_{o}C_{atm}$, to produce a value for the activity of DIC when the clock starts and radioactive decay alone dominates. This activity is $a^{14}_{o}C_{DIC}$,

$$a^{14}_{o}C_{DIC} = q \times a^{14}_{o}C_{atm}$$

The value for $a_o^{14}C_{atm}$ is generally taken to be 100 pMC, although refinements can be made using a value from Figure 7.11 which is close to the age of the groundwater. Accordingly an early Holocene age using $a_o^{14}C_{atm} = 100\,\text{pMC}$ can be refined by recalculating using $a_o^{14}C_{atm} = 108\,\text{pMC}$, in an iterative determination using the decay equation:

$$t = -8267 \times \ln\left(\frac{a_t^{14}C_{DIC}}{a_o^{14}C_{DIC}}\right)$$

For example, groundwater with $a_t^{14}C_{DIC} = 25$ pMC recharged under closed system conditions in a karst aquifer with a dilution factor, $q = 0.5$, would have an apparent age without correction for dilution (or minor fractionation):

Using $q = 1$

$$a_o^{14}C_{DIC} = 100\,\text{pMC},$$

$$t_{app} = -8267 \times \ln\left(\frac{a_t^{14}C_{DIC}}{a_o^{14}C_{DIC}}\right)$$

$$= -8267 \times \ln\left(\frac{25}{100}\right)$$

$$= 11,460\,\text{y BP}$$

and a corrected age using $q = 0.5$

$$a_o^{14}C_{DIC} = q \times a_o^{14}C_{atm} = 0.5 \times 100 = 50\,\text{pMC},$$

$$t_{corr} = -8267 \times \ln\left(\frac{a_t^{14}C_{DIC}}{a_o^{14}C_{DIC}}\right)$$

$$= -8267 \times \ln\left(\frac{25}{50}\right)$$

$$= 5730\,\text{y BP}$$

Considering that at that time, $a_o^{14}C_{atm}$ was close to 108, this age can be refined with that number to yield an age of 6370 years before present.

Many approaches and models have been developed over the past decades to calculate q for various groundwaters and reactive losses of ^{14}C (Pearson 1965; Pearson and Hanshaw 1970; Fontes and Garnier 1979; Plummer et al. 1994). Most are based on the use of geochemical parameters and ^{13}C of DIC and require some understanding of the conditions that prevailed at the time of recharge. If the climate has not changed, then analyzing the DIC of young (tritium-bearing) groundwaters in the recharge area may be acceptable to estimate $a_o^{14}C_{DIC}$ (e.g., Bajjali et al. 1997). However, additional dilutions to $^{14}C_{DIC}$ can occur beyond the recharge area along the flow path that require correction. Perhaps the most versatile model is the evolution of $\delta^{13}C$.

Dilution Correction with $\delta^{13}C_{DIC}$

Beyond the recharge environment, continued dilution of the initial radiocarbon content of DIC occurs not only through closed system dissolution of carbonates (no further exchange with soil CO_2) but also by *matrix exchange*. Groundwater that is essentially at equilibrium with calcite will exchange DIC with aquifer carbonate, where Ca^{2+} and CO_3^{2-} are continually exchanging with $CaCO_3$ (Gonfiantini and Zuppi, 2003), resulting in a dilution of $^{14}C_{DIC}$ loss according to

$$CaCO_3 + H^{14}CO_3^- \rightarrow Ca^{14}CO_3 + HCO_3^-$$

Similarly, ^{14}C losses to the matrix can occur due to the incongruent dissolution of dolomite.

$$Ca^{2+} + H^{14}CO_3^- + CaMg(CO_3)_2 \rightarrow 2Ca^{14}CO_3 + Mg^{2+} + HCO_3^-$$

In both cases, the addition of ^{14}C-free DIC from the aquifer matrix can be traced with stable ^{13}C, which has a characteristic value for both DIC from recharge ($\delta^{13}C_{rech}$ with ^{14}C-bearing DIC) and DIC from the rock ($\delta^{13}C_{rock}$ with ^{14}C-free DIC). The greater the degree of contribution from the rock through exchange or dolomite alteration, the more the groundwater $\delta^{13}C_{DIC}$ will evolve toward $\delta^{13}C_{rock}$. Pearson (1965), with development by Pearson and Hanshaw (1970), were the earliest to demonstrate the elegance of this tracer of dilution, calculated as

$$q = \frac{\delta^{13}C_{DIC} - \delta^{13}C_{rock}}{\delta^{13}C_{rech} - \delta^{13}C_{rock}}$$

where

$\delta^{13}C_{DIC}$ = measured ^{13}C of DIC in groundwater sample
$\delta^{13}C_{rech}$ = δ^{13}C of the DIC following open system weathering (≈ –24 to –14‰)
$\delta^{13}C_{rock}$ = δ^{13}C of the carbonate from the aquifer being dissolved or exchanged.

This is a versatile algorithm because it provides a simple and linear correction for the dilution of $^{14}C_{DIC}$ by closed system carbonate weathering in the recharge environment and by matrix exchange. There are geochemical and isotopic refinements that can be made for the calculation of this dilution factor (Mook 1976; Fontes and Garnier 1979), reviewed by Han and Plummer (2013). These include fractionation of ^{13}C and ^{14}C during the uptake of soil CO_2 by soil water ($\varepsilon^{13}C_{HCO_3\text{-}CO_2} \approx 9‰$; $\varepsilon^{14}C_{HCO_3\text{-}CO_2} \approx 18‰$ or 1.8%) which imparts a minor change to the initial ^{14}C activity. A second is the fractionation between DIC and rock matrix during the matrix exchange with carbonate minerals which under equilibrium exchange would be close to $\varepsilon^{13}C_{DIC-rock} \approx -3‰$. For the case where the aquifer matrix is considered to be 0‰, so then $\delta^{13}C_{rock} = -3‰$. However, if it is considered that a degree of closed system weathering has occurred during recharge, then this is unwarranted as there is no isotope exchange during carbonate dissolution, and so $\delta^{13}C_{rock} = 0‰$.

An interesting application for this correction to radiocarbon dates is for groundwater sampled at discrete depths from boreholes in the upper 200 m of Paleozoic bedrock on the eastern margin of the Michigan Basin in Southern Ontario (Intera Engineering Ltd. 2011). The bedrock comprises dolostones with karst horizons. Measured radiocarbon activities range from 16 to 22 pMC, but $\delta^{13}C_{DIC}$ values are quite enriched, with values up to –7.2‰.

The uncorrected radiocarbon ages for these samples suggest that they were recharged during the late Pleistocene when the region was still glaciated (Figure 7.12). However, the enrichments of ^{13}C in these groundwaters indicate significant exchange with the dolostones, which have $\delta^{13}C_{rock}$ near 2‰. Dilution factors, q, were calculated using a value for $\delta^{13}C_{rech} = -18‰$ for the DIC at the time of recharge and $\delta^{13}C_{rock} = 2‰$

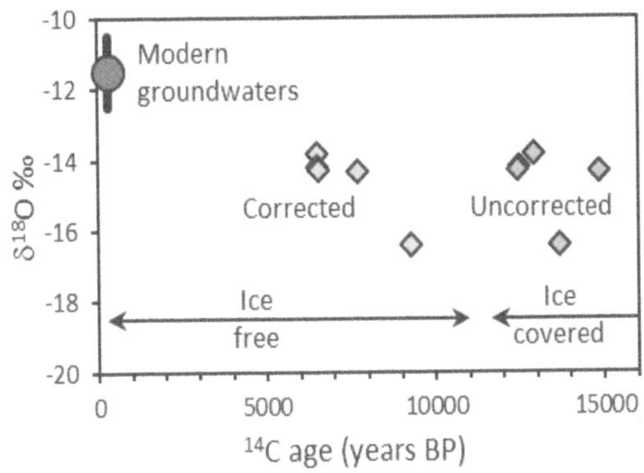

FIGURE 7.12 Corrected and uncorrected ^{14}C ages with δ^{18}O, showing the corrected ages indicating recharge dominantly in the Holocene, but with a component of recharge from Pleistocene meltwater. Modified from Intera Engineering Ltd. 2011.

for the dolostones. For the shallowest sample ($\delta^{13}C_{DIC} = -7.2‰$; $a^{14}C_{DIC} = 21.0\,\text{pMC}$), the age correction is calculated as

$$q = \frac{\delta^{13}C_{DIC} - \delta^{13}C_{rock}}{\delta^{13}C_{rech} - \delta^{13}C_{rock}} = \frac{-7.2-2}{-18-2} = 0.46$$

Corrected age:

$$t = -8267 \times \ln\left(\frac{21.0}{0.46 \times 100}\right)$$

$$= 6482\,\text{y BP}$$

Plotting corrected and uncorrected ages with δ^{18}O for these waters (Figure 7.12) shows they are a mixture of late Pleistocene glacial meltwater (δ^{18}O –20 to –30‰) and Holocene meteoric waters. In regions that were ice-covered at the last glacial maximum, high hydrostatic pressures at the base of the melting ice front generated circulation hundreds of meters deep. Today it is common to have meltwater in aquifers that are not actively flushed with modern recharge (Siegel and Mandle 1984; Clark et al. 2000; Figure 4.10).

Radiocarbon in silicate terrains may also be affected by $^{14}C_{DIC}$ dilution through dissolution and exchange with carbonate fracture minerals. However, if carbonate minerals are not present in fractures, the evolution of DIC during recharge under closed and open system conditions has little impact on $a_o^{14}C_{DIC}$ which can simplify age calculations. The evolution of both $\delta^{13}C_{DIC}$ and $a^{14}C_{DIC}$ are discussed as tracers of weathering in silicate terrains in Chapter 5.

Additional Dilution Processes Affecting $a_o^{14}C_{DIC}$

Sources of geogenic DIC that require consideration for radiocarbon dating of groundwater can be associated with redox evolution along the groundwater flow path where DIC contributions occur from sulfate reduction and methanogenesis of buried organic substrates in the aquifer. Volcanogenic and metamorphic CO_2 from regional tectonics can also contribute ^{14}C-free geogenic DIC that must be accounted for in correction calculations.

Geogenic CO_2. In a study of groundwater in granite in Japan, Takahashi et al. (2013) use the $\delta^{13}C$ correction algorithm from above to account for admixed geogenic DIC in freshwaters found in drillholes to over 200 m (Figure 7.13). In this case, the geogenic DIC was determined to be contributing ^{14}C-free DIC to the groundwater from a mixture of geogenic CO_2 and fracture minerals. A decrease in pH and an increase in DIC are indicators of subsurface contributions of carbonic acid, in this case geogenic CO_2. They determined a value for $\delta^{13}C_{geo}$ by back extrapolating a plot of DIC vs. $\delta^{13}C_{DIC}$ to infinite dilution with geogenic DIC (inset plot of 1/DIC vs. $\delta^{13}C_{DIC}$). Using this value for $\delta^{13}C_{diss/exch}$ in the algorithm given above, they produce dilution factors and corrected ages. The result shows a decrease in age in the upper 150 m of this granite, to a late Pleistocene age of 10,000 years to 15,000 years before present.

Deeply circulating groundwaters along plate margins or in volcanic settings may often have elevated P_{CO_2} values due to the incorporation of geogenic CO_2. This can be mantle-derived CO_2, which has a $\delta^{13}C$ value of about −6‰. An example of this input is the Mount Meager geothermal waters (Clark et al. 1982). Here, 45°C water discharges at the base of this Quaternary volcano with a pH of 6.85 and 150 ppm DIC with $\delta^{13}C_{DIC} = -6.3$. The spring waters have tritium and so have a recently recharged component. However, the radiocarbon activity, $a^{14}C_{DIC}$, is only 0.2 pMC and yields no reliable age estimate.

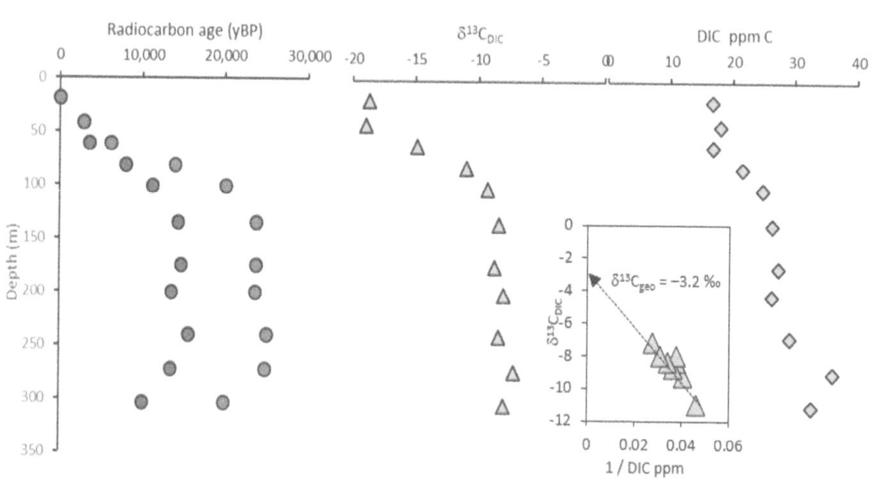

FIGURE 7.13 Radiocarbon ages in fresh groundwater in a granite pluton, Japan. The downward increase in both dissolved inorganic carbon (DIC) and $\delta^{13}C_{DIC}$ indicates a geogenic source of DIC in this fractured silicate bedrock. A value for $\delta^{13}C_{geo}$ required in the correction model was determined by extrapolation of 1/DIC vs. $\delta^{13}C_{DIC}$ to infinite DIC (i.e., infinite geogenic DIC).

Another geogenic source of CO_2 is from the thermal metamorphism of limestone by magma. In the presence of silica, the decarbonation of $CaCO_3$ occurs as low as 300°C, with the formation of wollastonite:

$$CaCO_3 + SiO_2 \rightarrow CaSiO_3 + CO_2$$

Such CO_2 is similar to or enriched above the limestone precursor. This is the case for the Mount Edziza geothermal waters (recall from Chapter 5) that have a $\delta^{13}C_{DIC}$ value of −1.2‰ and no measurable ^{14}C (Clark et al. 1989). In both these cases, any radiocarbon incorporated from recharge through soils has been highly diluted by geogenic CO_2, precluding any potential to be used for dating.

Buried organic carbon. Aquifers may contain organic carbon both as buried plant debris, such as peat horizons in unconsolidated sand and gravel aquifers, coal beds in sandstone and shale horizons, and residual bitumen and kerogen in marine rocks, such as limestone and dolostone. Under reducing conditions, bacteria will oxidize these organic substrates, typically using sulfate or ferrihydrite as an electron acceptor. In the absence of such electron donors, DIC can be generated through fermentative and methanogenic reactions. The ^{14}C activity of DIC generated from such sources must then be taken into account in any correction model (Aravena et al. 1995).

The contribution of buried organic carbon to groundwater DIC is generally evident from $\delta^{13}C_{DIC}$, which is generally around −27‰ for terrestrial organics and closer to −30‰ for marine organics. However, with the enrichment of ^{13}C that accompanies carbonate reactions (above), this tool is not always quantitative, and therefore a second parameter must be used.

Sulfate reduction. The production of DIC during sulfate reduction depends on the type of organic involved. Oxidation of buried terrestrial plant debris (e.g., peat) in which the carbon has an overall redox state near 0 (fixed carbon) would follow a reaction such as the following:

$$2CH_2O + SO_4^{2-} \rightarrow H_2S + 2HCO_3^-,$$

where 1 mole of H_2S would signify the contribution of 2 moles of DIC. The dilution factor, q_{H_2S}, can be calculated as

$$q_{H_2S} = \frac{m_{DIC}}{m_{DIC} + 2m_{H_2S}}$$

If a reduced carbon substrate is used (C redox state of −IV for methane), the reaction will be

$$CH_4 + SO_4^{2-} + H^+ \rightarrow H_2S + HCO_3^- + H_2O$$

where one mole of H_2S follows from the production of one mole of DIC. In this case, the dilution factor for the process of sulfate reduction alone can be expressed as

$$q_{H_2S} = \frac{m_{DIC}}{m_{DIC} + m_{H_2S}}$$

Sulfate reduction affects not only $^{14}C_{DIC}$ but also $\delta^{13}C_{DIC}$ due to the addition of ^{13}C-depleted DIC from the organics, for which $\delta^{13}C_{org}$ is generally about –25‰ to –30‰, which complicates corrections for carbonate weathering (Clark et al. 1995).

Methanogenesis. DIC from buried or sedimentary organic carbon in the aquifer can also contribute to groundwater DIC by microbial methanogenesis. The series of fermentative reactions (anoxic degradation of higher-weight organic compounds) that precede the production of CH_4 produce considerable amounts of DIC (see Chapter 6):

$$C_6H_{12}O_6 + 2H_2O \xrightarrow{\textit{fermentation}} 2CH_3COOH + 2CO_2 + 4H_2 \xrightarrow{\textit{methanogenesis}} 3CH_4 + 3CO_2 + 2H_2O$$

The DIC resulting from this complex series of biological reactions dilutes the ^{14}C-active DIC that accompanied groundwater recharge. Correction is complicated and not simply traced with $\delta^{13}C$ as was done for carbonate reactions. This is due to the fractionation of ^{13}C in this overall reaction. From the precursor carbohydrate with $\delta^{13}C$ of around –27‰, the ^{13}C is selectively partitioned into the DIC while the methane is enriched in ^{12}C. The overall fractionation in these reactions, $\varepsilon^{13}C_{CO_2-CH_4}$, is on the order of 50‰ to 70‰, leaving the DIC in such waters with $\delta\ ^{13}C_{DIC}$ values much greater than the –10‰ to –15‰ for soil-derived DIC, and often up to +10‰ or higher (see Figure 6.19).

Developing an algorithm based on methane concentrations to correct for these effects requires the assumption that CH_4 is retained in the groundwater sampled and that DIC has not been affected by carbonate exchange unless otherwise corrected. If the source of organic carbon is ^{14}C-free, then a correction factor calculated as follows may approximate the ^{14}C dilution from methanogenesis:

$$q_{CH_4} = \frac{m_{DIC} - 2m_{CH_4}}{m_{DIC}}$$

A more comprehensive approach to correcting for the various carbon reactions in methanogenic groundwater involves geochemical programs to calculate their effect on DIC as well as on $\delta^{13}C_{DIC}$ and $a^{14}C_{DIC}$. Aravena et al. (1995) undertake radiocarbon dating for methanogenic groundwater in the Alliston aquifer, a regional aquifer of glaciofluvial outwash sediments in southern Ontario, Canada, confined above by late Pleistocene clay till. Elevated DOC concentrations between 5 and 20 ppm C are generating on average about 3 mmol/kg methane concentrations (Table 7.4). This represents over 2 atm partial pressure ($K_{CH_4} = 0.0014 = m_{CH_4}/P_{CH_4}$, and so $P_{CH_4} = 0.003/0.0014 = 2.1\,\text{atm}$) held in solution by the hydrostatic head in the aquifer (>20 m or 2 atm pressure). When pumped by local groundwater users, such methane degases in the water distribution system and presents an explosion hazard.

TABLE 7.4
Concentrations, $\delta^{13}C$ and $a^{14}C$ of Methanogenic Groundwater in the Alliston Aquifer, Ontario

Well	DOC ppm C	DIC ppm C	$\delta^{13}C_{DIC}$ ‰	CH_4 mmol/kg	$\delta^{13}C_{CH_4}$ ‰	$a^{14}C_{DIC}$ pMC	Apparent age ky BP	q	Model age ky BP
OV7	7.4	67.2	2.0	2.08	−84	3.6	28.4	0.23	15
OV15	18	85.2	1.3	3.55	−77	4.7	23.4	0.15	9
OV23	15	85.6	−2.7	2.75	−73	2.0	31.2	0.17	18
OV34	7.1	46.8	−8	0.94	−86	2.6	31.1	0.19	16

On the basis of this ^{13}C enrichment observed for the DIC and the $\delta^{13}C_{CH_4}$ and δD_{CH_4} signature (Table 7.4), the source of the methane was identified as in situ biogenic production and not thermocatalytic methane from the bedrock (see methane in groundwater, Chapter 6). Uncorrected radiocarbon activities gave unrealistic apparent ages between 23,000 and 31,000 years before present, considering that this region was covered in Laurentide glacier ice at that time and that the aquifer itself is the glacial outwash sediment deposited after 21,000 years before present. Corrections for contributions to the DIC pool by methanogenesis were attempted using the USGS geochemical mass balance program NETPATH (Plummer et al. 1994). The results of NETPATH modeling show that methanogenesis has diluted the ^{14}C-active DIC derived from the soil by at least five times, giving a dilution factor of $q = 0.2$. Using this factor calculates corrected ages of 9,000–18,000 years, which place recharge at some point following deglaciation, when the region was largely ice-free.

^{14}C Dating with Dissolved Organic Carbon

Radiocarbon dating with soil-derived DOC holds some advantage over DIC. It is not affected by carbonate dissolution or exchange as is DIC, and losses by oxidation along the flow path will not significantly affect the radiocarbon content of DOC, just the overall concentration of DOC. There are, however, serious limitations to its application. As seen for the Alliston aquifer above, subsurface contributions to DOC in groundwater exist and would dilute the soil-derived component used for dating. Although corrections for subsurface DOC can be made (Aravena and Wassenaar 1993), this requires knowledge of the $a^{14}C$ of the buried organic carbon. Typically only the fulvic acid component can be reliably traced to an origin in the soil, and so detailed characterization and column separation (with XAD-8® resin) of the DOC fractions are required (Wassenaar et al. 1989).

Many of these complications regarding the origin of DOC in clastic aquifers with the potential for buried organic material do not encumber the DOC method in most bedrock aquifers. Thomas et al. (2021) use direct oxidation of low DOC groundwater in two adjacent geological settings in Nevada comprising volcanic rocks (no carbonate) and limestone (carbonate). In this case, the low DOC gives confidence that it is soil-derived, with no evidence for geogenic DOC, and so $^{14}C_{DOC}$ ages require no

TABLE 7.5
Carbon Isotopes for DIC and DOC in a Selection of Groundwater from Volcanic and Limestone Bedrock (Thomas et al. 2021).

δD (‰)	$\delta^{18}O$ (‰)	$\delta^{13}C_{DIC}$ (‰)	$^{14}C_{DIC}$ (pMC)	Age (DIC) (corrected)	$\delta^{13}C_{DOC}$ (‰)	$^{14}C_{DOC}$ (pMC)	Age (DOC) (uncorrected)	DOC (mg/L)
Volcanic	terrain							
−99.4	−13.3	−15.9	107.4	0	−23.3	110.4	0	1.91
−102.9	−13.2	−11.9	64.2	1315	−23.1	76.8	2182	0.36
−115.7	−15.2	−4.6	12.5	7579	−25.2	47.1	6860	0.28
−113.3	−14.5	−2.7	7.3	7621	−24.6	43.9	7442	0.05
−113.7	−15.0	−4.6	13.4	7022	−27.2	20.0	13941	0.30
Limestone	terrain							
−101.5	−13.7	−11.2	95.8	0	−24.4	101.8	0	2.30
−100.3	−13.7	−10.0	68.2	2198	−23.7	88.2	1038	0.80
−103	−14.0	−8.6	25.0	10,035	−20.2	71.6	2762	0.28
−103.3	−13.7	−5.8	1.3	uncorrected	−25.7	68.6	3116	0.39
−102.6	−13.6	−6.7	7.3	19,492	−24.4	56.8	5000	0.36

DIC corrections made here using $\delta^{13}C_{rock} = 0‰$ and $\delta^{13}C_{rech} = -11.24$ (Limestone) and −15.89 (Volcanic) and with adjusted values for $a_o^{14}C_{atm}$ using first order determination from Figure 7.11

correction. In contrast, both do require correction for $^{14}C_{DIC}$ ages due to geogenic DIC (Table 7.5). The modern groundwater with $a^{14}C$ near 100 pMC was sampled from both settings, which provide values for $\delta^{13}C_{rech}$ that can be used for correction (for volcanic terrain, $\delta^{13}C_{rech} = -15.9‰$; for carbonate terrain, $\delta^{13}C_{rech} = -11.2‰$) (Figure 7.14). Comparing the results allow the following conclusions:

- Radiocarbon DOC ages are considered reliable, some with modern $a^{14}C_{DOC} > 100$ pMC, with no evidence of geogenic DOC or significant sorption or matrix diffusion.
- Radiocarbon DIC ages in the limestone terrain are all older than the DOC ages, suggesting insufficient correction with the $\delta^{13}C_{DIC}$ model.
- Radiocarbon DIC ages in the volcanic terrain are largely similar to equivalent DOC ages.
- Groundwater dated to older than 5000 years before present have a negative shift in δD signifying a change to warmer climate conditions in the mid-Holocene.

PALEOGROUNDWATER

Reconstruction of groundwater ages and the identification of paleogroundwater allows characterization of the paleoclimate under which they recharged. Complementary tools that aid in the characterization of paleogroundwater include stable isotopes, which, as shown in Chapter 4, are sensitive to climate parameters of temperature,

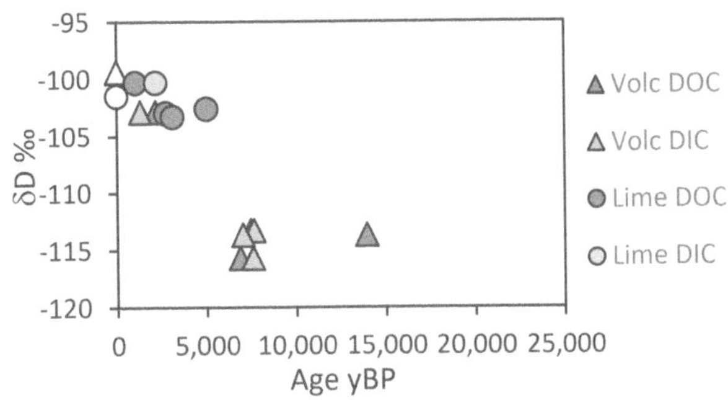

FIGURE 7.14 Plot of δD with radiocarbon age (dissolved organic carbon [DOC] and DIC) for groundwater in the volcanic terrain and limestone terrain (Thomas et al. 2021). DOC ages have no correction, while DIC ages are corrected with $\delta^{13}C_{DIC}$ for limestone dissolution and exchange. The shift in δD to lower values for groundwater older than about 6000 years before present indicates a climate shift in the mid-Holocene. Clear symbols for limestone and volcanic are the modern ($t = 0$) samples used for initial recharge $\delta^{13}C_{DIC}$ in the $\delta^{13}C$ correction calculation.

humidity, and precipitation. Noble gas concentrations record temperature at the time of recharge and so are an important tool for characterizing paleogroundwater.

Stable Isotopes and Paleogroundwater

The $\delta^{18}O$ and δD signals in climate records, such as ice cores and speleothem, are used to reconstruct past temperatures. Similarly, groundwater recharged under a different climate regime in the past will then preserve its isotopic signature and can be compared with modern groundwater to provide an indication of the climate shift. For groundwater recharged during the late Pleistocene to Holocene, this is related to climate shifts from the last glacial maximum at around 21 ka to the current interglacial climate from the early Holocene to present. This is manifested in the stable isotope contents of groundwater in two ways. One is the incorporation of meltwater from the massive continental ice sheets, typically observed in higher-latitude aquifers outcropping along the margin of the ablating ice as it retreated over the landscape. The other is from the recharge of cooler Pleistocene precipitation, which is typically observed in groundwater from nonglaciated regions at lower latitudes.

Groundwater recharged during continental deglaciation is noted by a depletion in ^{18}O and D (e.g., Clark et al. 2000; Grasby and Chen 2005). One of the earliest recognitions of such recharge comes from Siegel and Mandle (1984), showing groundwater recharged by glacial meltwater in a sandstone aquifer, which outcrops west of Lake Michigan (Figure 7.15). The associated isotope plot for these groundwaters shows a marked depletion in both ^{18}O and D relative to modern, tritium-bearing groundwaters in the region.

Groundwater recharge under past climate conditions is recorded by the stable isotope records in speleothem from many nonglaciated regions, where the ^{18}O in calcite [$CaCO_3$] records the $\delta^{18}O$ of the groundwater that formed the deposit. Such a speleothem from Devil's Hole, Nevada, holds a remarkable 500,000-year record of glacial

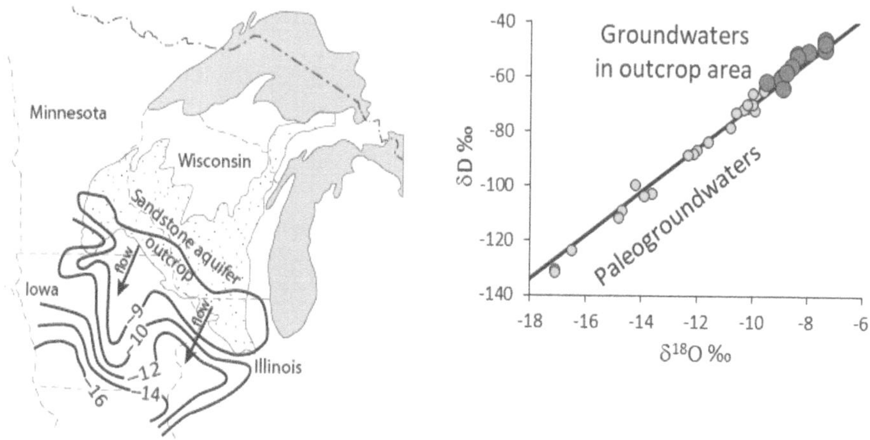

FIGURE 7.15 Glacial meltwater recharged into the Cambrian–Ordovician sandstone aquifer where it outcrops in the northern United States (after Siegel and Mandle 1984). Groundwaters are a mixture of modern precipitation ($\delta^{18}O > -9‰$) and glacial meltwater ($\delta^{18}O$ to $-16‰$).

and interglacial climate (Winograd et al. 1992; Figure 7.16). Groundwater discharging at Devil's Hole was recharged some 100 km to the east in southern Nevada in the Sheep Range mountains. It has been dated at 30,000 years before present, and so the climate record in the speleothem ends at that time, prior to the last glacial maximum (Figure 7.16). The speleothem record, dated with U-Th, corresponds strikingly well with the precipitation record from the Vostok ice core in Antarctica.

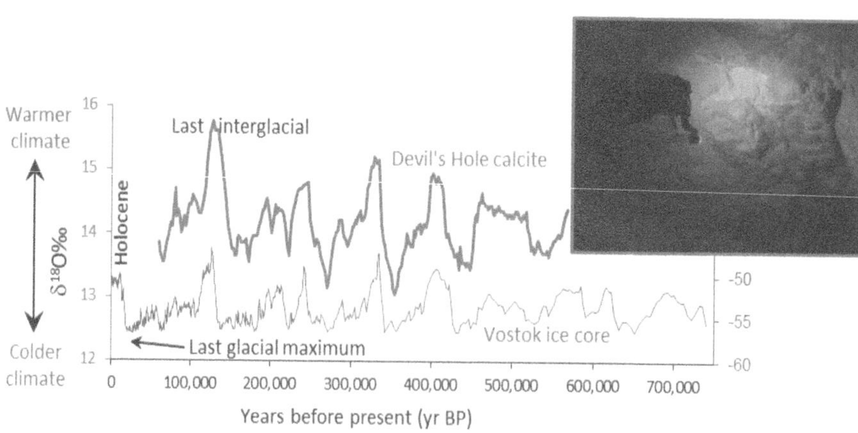

FIGURE 7.16 Change in climate temperature as recorded by calcite precipitated from regional groundwaters at the Devil's Hole springs, Nevada (data from Winograd et al. 1992). The photo shows the speleothem-lined cave where groundwater discharges. Peaks of enriched ^{18}O values documenting the warmer climates of the interglacial periods and depletions during cold climate periods correspond with the $\delta^{18}O$ climate record in glacier ice from Vostok Station, Antarctica (data from EPICA 2004).

FIGURE 7.17 Flowing artesian paleogroundwater from the Umm er Radhuma aquifer in southern Oman. Now used for center-pivot irrigation (inset), this fossil groundwater was recharged in the coastal mountains during a pluvial period from 6 to 12 ka before present.

Paleogroundwaters sampled in non-glaciated regions retain an isotopic signature of the climate that dominated at the time of their recharge. Differences in comparison with modern recharge from such regions reflect the change in climate. Shifts can arise due to changes in temperature, which affect both $\delta^{18}O$ and δD. Changes in humidity in the vapor source region can be manifested by changes in deuterium excess, *d*, and a differing meteoric water line. The Sultanate of Oman on the Arabian Peninsula, is currently in a hyper-arid phase. Groundwaters in the southern district discharge from a confined limestone aquifer that outcrops in the coastal mountains (Figure 7.17) and are dated to a pluvial period between 6000 and 12,000 years before present (Clark et al. 1995). These paleogroundwaters are isotopically depleted from modern rainfall in the country and fall on a different MWL with a lower deuterium excess (Figure 7.18, left chart).

Groundwaters from an aquifer underlying Chesapeake Bay in Maryland, USA, provide an example from a temperate region (see location in Figure 7.29, below). Here, paleogroundwaters were recharged during the late Pleistocene glacial period when global sea level was some 120 m lower than today. They are similarly depleted in $\delta^{18}O$ but with a larger deuterium excess than modern groundwater in this area (Figure 7.18, right chart; Plummer et al. 2012).

Noble Gas Paleo-recharge Temperatures

Recharge temperatures for paleogroundwaters are recorded by noble gas concentrations based on their increase in solubility with decreasing temperature. Kr and Xe reliably record recharge temperatures as both have significant change in concentration

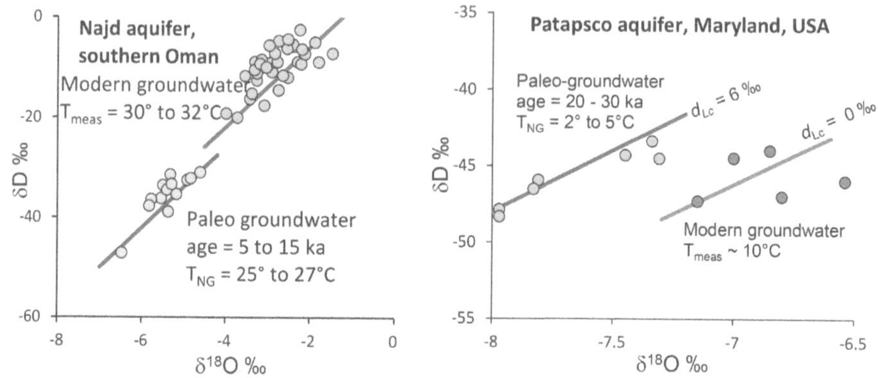

FIGURE 7.18 Stable isotope signature of paleogroundwater recharged under past climates. Left: The Najd paleogroundwaters with modern groundwaters from the interior (after Clark et al. 1987). Right: Paleorecharge during past glacial maxima (low sea level) in the northern flow path of the upper Patapsco aquifer flowing under Chesapeake Bay, Maryland. A drier and cooler climate (noble gas temperatures of 2 to 5°C) during ice-age conditions likely contributed to the higher deuterium excess of d_{Lc} = 6‰ and the ^{18}O depletion. Modified from Plummer et al. 2012.

with temperature. Helium and argon concentrations in groundwater can have geogenic contributions through radioactive decay (^4He from alpha decay in the U and Th decay series and ^{40}Ar from beta decay of ^{40}K) and so can gain concentrations along the flow path. Ne is the least sensitive to temperature of the three strictly atmospheric noble gases, and is used to correct for excess air gained during recharge.

The concentrations of the noble gases in air-saturated fresh water (i.e., in equilibrium with the atmosphere) are a product of their temperature-dependent solubility (Bunsen coefficient in units of $cc_{STP}/cc_{H_2O}/atm$, see Table 2.8), which are shown in Figure 7.19 and their atmospheric partial pressure, as shown in Figure 7.20. The

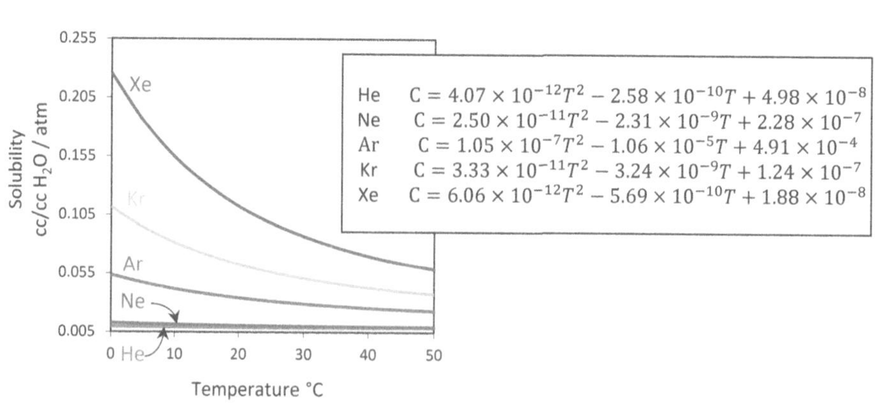

FIGURE 7.19 Solubility equations (cc_{STP}/cc_{H_2O}) for the noble gases from 0°C to 50°C for 1 atm partial pressure.

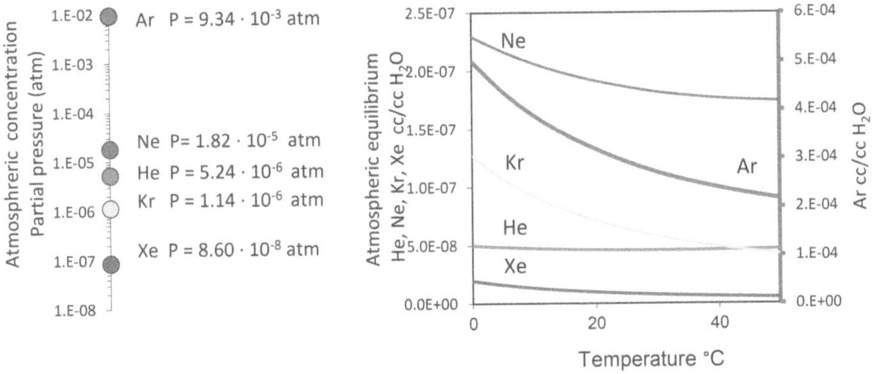

FIGURE 7.20 Noble gas partial pressures in the atmosphere (left) and their concentrations in air-saturated water (right).

concentrations of atmospheric noble gas in groundwater are controlled not only by their atmospheric abundance and temperature. Recharge at higher elevations lowers their concentration through reduced atmospheric pressure, as does high salinity (above a few g/L TDS [total dissolved solids]) (Figure 7.21).

A correction must also be made for additions to the dissolved noble gas concentrations gained by the incorporation of excess air as microbubbles that can be entrained during a recharge event. A correction for this excess air-derived noble gas component is made using the neon concentration, which is the atmospheric noble gas that is the least sensitive to temperature and so the best estimate of excess air. More recently, Aeschbach-Hertig et al. (2000) improved the model for incorporation of excess air during recharge to account for partial dissolution of entrained air bubbles. This excess air component together with salinity and altitude of recharge are incorporated into Matlab programs (Aeschbach-Hertig et al. 1999) that provide estimates of recharge temperature for each noble gas. The average of these estimates is then used (generally excluding He and Ar if there are geogenic inputs for these gases).

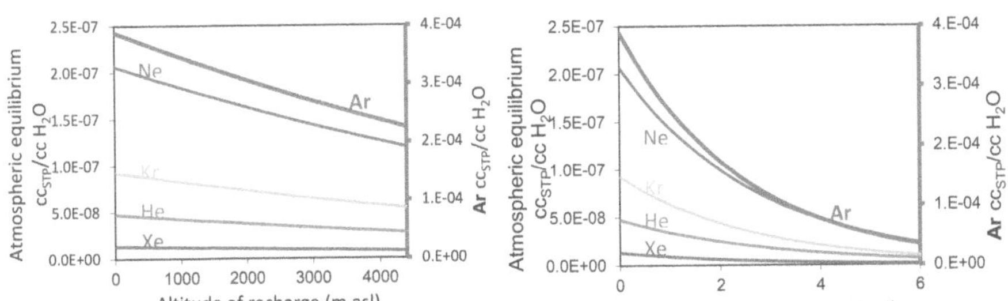

FIGURE 7.21 Noble gas concentrations for air-saturated water at 10°C with (left) altitude and (right) salinity.

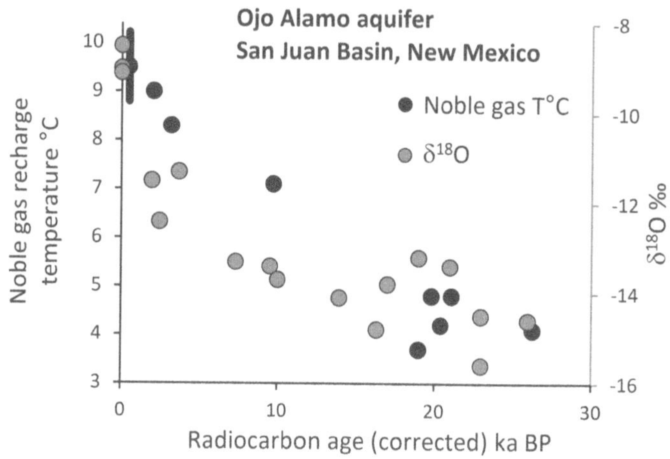

FIGURE 7.22 Noble gas recharge temperatures for Holocene and Pleistocene groundwaters dated with radiocarbon from the San Juan Basin, New Mexico, together with the $\delta^{18}O$ values for these waters, showing an isotope depletion in the older groundwaters, which have lower noble gas recharge temperatures. Range of temperatures for modern groundwaters shown with a black vertical line. Data from Phillips et al. 1986; Stute et al. 1995.

An example of noble gases recording a paleorecharge temperature is that of old groundwater from New Mexico (Figure 7.22). The lower recharge temperatures calculated from noble gases correlate with depletions in ^{18}O, showing the older groundwater to have recharged under cooler climate conditions at the last glacial maximum some 20,000 years ago.

DATING VERY OLD GROUNDWATER

Some important aquifers host groundwaters that are older than about 50,000 years and have radiocarbon activities that are below detection. The age of old groundwater can be made by the ingrowth of radiogenic isotopes of the noble gases, including ^{4}He, ^{21}Ne, ^{40}Ar, and ^{136}Xe, providing their production rates can be determined. Quantitative measurements can be made by the decay of atmospherically derived radioisotopes, providing the input function in the recharge environment can be established at the time (and climate) of recharge. Two such radioisotopes with appropriately long half-lives are ^{36}Cl ($T_{½}$ = 301,000 years) and ^{81}Kr ($T_{½}$ = 229,000 years). Both are conservative solutes and so have less complicated corrections and interpretations than radiocarbon, although their very low atmospheric production rates are an analytical challenge.

CHLORINE-36

The decay of atmospheric ^{36}Cl gained by groundwater during recharge can provide age estimates up to about five half-lives and so greater than one million years. It decays to abundant ^{36}Ar, and so age is calculated by the loss-of-parent decay equation, like radiocarbon. It is measured in environmental samples as a concentration relative to stable chloride, typically with $^{36}Cl/Cl$ ratios on the order of 10^{-15} to 10^{-13} or higher. These very low ratios of ^{36}Cl require analysis by AMS using total Cl extracted from water or mineral samples, giving an atom ratio of ^{36}Cl to atoms of stable Cl:

$$R^{36}\mathrm{Cl} = \frac{^{36}\mathrm{Cl}}{\mathrm{Cl}}$$

The concentration of ^{36}Cl as atoms per kilogram or per liter of water ($A^{36}Cl$) is independent of the Cl^- content and an important way of expressing ^{36}Cl data. It is calculated as

$$A^{36}\mathrm{Cl} = \frac{\mathrm{atoms}\,^{36}\mathrm{Cl}}{\mathrm{L}} = R^{36}\mathrm{Cl} \times m_{\mathrm{Cl}^-} \times 6.022 \times 10^{23}$$

where m_{Cl^-} = moles Cl^-/kg or moles/L.

Two key constraints are determining an initial atmospheric concentration of ^{36}Cl and accounting for ingrowth of geogenic ^{36}Cl along the flow path. Both $R^{36}Cl$ and $A^{36}Cl$ can help. The initial atmospheric concentration of $R^{36}Cl$ will not change during concentration by evaporation in the recharge area, whereas $A^{36}Cl$ will increase. Conversely, $A^{36}Cl$ will not necessarily change with increases in Cl^- along the flow path by leaching or evaporite dissolution (^{36}Cl-free additions of Cl^-), whereas $R^{36}Cl$ will decrease. Both will change through incorporation of ^{36}Cl-active chloride from subsurface sources where ^{36}Cl production is important, such as in granitic rocks.

The atmospheric production of ^{36}Cl is a function of the geomagnetic latitude, with a maximum in wet and dry fallout of about 20 atoms/m^2/s at about 45° north or south but falls off to less than 5 atoms/m^2/s near the equator and the poles (Andrews and Fontes 1992). Ratios for $^{36}Cl/Cl$ are typically up to 700×10^{-15} in precipitation but drop off at lower and higher latitudes and with proximity to the coast. Bentley et al. (1986) indicate that $^{36}Cl/Cl$ ratios drop off to values as low as 20×10^{-15} along the U.S. continental margins where ^{36}Cl-free marine aerosols dilute cosmogenic ^{36}Cl.

The atmospheric neutron flux impinging on evaporative salts in soils at the surface can activate stable ^{35}Cl to ^{36}Cl, which, together with minor production from spallation of ^{39}K (n-α) and ^{40}Ca (n-p, α), contribute *epigenic* ^{36}Cl to recharge waters. This can provide a variable and enriched input function (Andrews and Fontes 1992; Phillips 2000). Accordingly, $R^{36}Cl$ in recharge waters in temperate latitudes is typically in the range of 300 to 600×10^{-15} and in low latitudes from 100 to 200×10^{-15} (Bentley et al. 1986; Patterson et al. 2005).

Atmospheric and epigenic ^{36}Cl enters the subsurface with groundwater recharge. If the system remains closed to other sources of ^{36}Cl, losses should be due to radioactive decay to ^{36}Ar and will reflect the subsurface residence time of the groundwater. However, additions from subsurface or *geogenic* production must be taken into account. Alpha decay of U and Th irradiates light elements (O, Na, Mg, Al, Si,

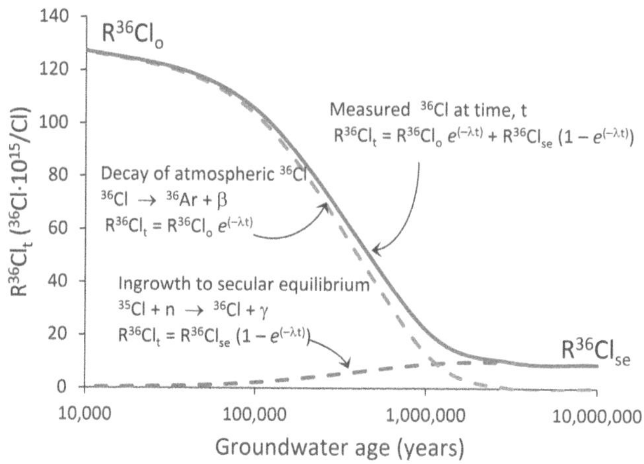

FIGURE 7.23 Decay of atmospheric ^{36}Cl combined with ingrowth toward secular equilibrium, giving total ^{36}Cl in groundwater with age.

and K) through alpha decay to produce an in situ neutron flux which can activate ^{35}Cl to produce in situ ^{36}Cl (Figure 3.11) (Andrews et al. 1989). This production follows an ingrowth function that stabilizes at some value $R^{36}Cl_{se}$ that represents a secular equilibrium between in situ production and decay (Figure 7.23). After about five half-lives, the $^{36}Cl/Cl$ ratio reaches a steady value of secular equilibrium, where cosmogenic ^{36}Cl has decayed away and in situ production is balanced by decay (Table 7.6). At this point, dating is not possible. Typical values for $R^{36}Cl_{se}$ have been suggested by Bentley et al. (1986) (Table 7.6).

The use of ^{36}Cl is also complicated by the potential for the gain of Cl^- during flow by dissolution of salt, diffusion from adjacent aquitards, or by mixing with other groundwater. Such gains of Cl^- will include ^{36}Cl accompanying these sources of Cl^-. Corrections often assume a value for $R^{36}Cl_{se}$ in the aquitard. Given a fixed initial $R^{36}Cl_0$ at recharge and an established value for secular equilibrium, $R^{36}Cl_{se}$,

TABLE 7.6
$^{36}Cl/Cl$ Ratios at Secular Equilibrium for Various Rock Types (Bentley et al. 1986)

Rock	$R^{36}Cl_{se} \times 10^{-15}$
Granite	30.1
Sandstone	4.68
Shale	12.5
Limestone	10.9

groundwater age is derived using the ^{36}Cl decay constant $\lambda^{36}Cl = 2.303 \times 10^{-6}\,yr^{-1}$ (Bentley et al. 1986):

$$t = -\frac{1}{\lambda^{36}Cl} \times \ln \frac{R_t^{36}Cl - R_{se}^{36}Cl}{R_o^{36}Cl - R_{se}^{36}Cl}$$

$$= -434{,}251 \times \ln \frac{R_t^{36}Cl - R_{se}^{36}Cl}{R_o^{36}Cl - R_{se}^{36}Cl}$$

Krypton-81

With its similarly long half-life of 229,000 years, ^{81}Kr is also very suitable for dating radiocarbon-free groundwaters that were recharged in the Pleistocene. It is produce almost exclusively in the atmosphere through cosmic ray induced neutron activation of ^{80}Kr (Loosli and Oeschger 1969). Unlike ^{36}Cl, it has no complications from subsurface production or dilution with subsurface Kr sources. However, a very low abundance (892 atoms/L in air-saturated water; Jiang et al. 2012; Aeschbach-Hertig 2014) makes analysis a challenge. Until recently, sampling required extraction from cubic meters of groundwater. ATTA (Jiang et al. 2012) currently requires extraction of Kr from only 100 to 200 L of water, with lower volumes anticipated in the near future.

Dating with ^{81}Kr is based simply on loss by decay of an initial atmospheric ratio, $R_{air}^{81}Kr = 5.2 \times 10^{-13}$, for krypton in the air. While atmospheric production may have had minor fluctuations over the Pleistocene due to changes in the cosmic ray flux, the atmospheric concentration is believed to have been essentially constant in the past. Given the constant atmospheric $^{81}Kr/Kr$ ratio, measured groundwater ratios for some time t can be normalized to the atmospheric ratio ($R_t^{81}Kr/R_{air}^{81}Kr$) and expressed as a percentage of this initial value. Dating is then based on the simple exponential decay equation

$$R_t^{81}Kr = R_{air}^{81}Kr \times e^{-\lambda t}$$

$$t = -1/\lambda^{81}Kr \times \ln \left[\frac{R_t^{81}Kr}{R_{air}^{81}Kr} \right]$$

$$= -330{,}377 \times \ln \left[\frac{R_t^{81}Kr}{5.2 \times 10^{-13}} \right]$$

^{36}Cl and ^{81}Kr Ages for Groundwater in the Western Desert of Egypt

The Nubian sandstone underlying much of the northeastern Sahara Desert is one of the world's great aquifers, hosting tens of thousands of cubic kilometers of ancient, low-salinity groundwater. The sandstone is draped over the Precambrian shield, deepening northward from outcrop areas to the southwest. The groundwater has been determined to be radiocarbon-free and an excellent case study for the comparative use of ^{36}Cl and ^{81}Kr (Sturchio et al. 2004; Patterson et al. 2005). The aquifer and sampling transect connecting some of the oases are shown in Figure 7.24.

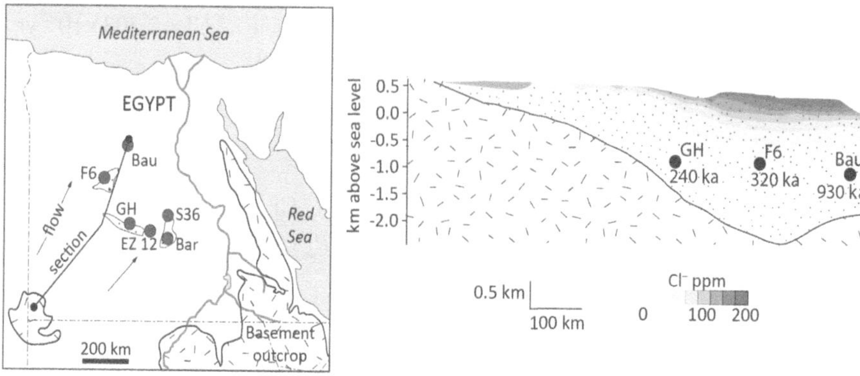

FIGURE 7.24 The Nubian sandstone aquifer overlying the crystalline basement in the Dahkla Basin of the Egyptian western desert, showing sample locations and salinity gradient. Modified from Patterson et al. 2005.

Measurements of ^{36}Cl and ^{81}Kr ratios and groundwater ages for six wells studied in the Dahkla Basin are given in Table 7.7.

The ^{36}Cl ages were calculated with the above equation using a value of 5×10^{-15} for $R^{36}_{se}Cl$ and 131×10^{-15} for the initial $R^{36}_{0}Cl$. The initial value was measured for a groundwater sample in the recharge area to the southwest. The conceptual model for Cl^- and ^{36}Cl measured in the deep groundwater samples shown in Figure 7.24 is that each has a given concentration of Cl^- dissolved during recharge. The variability in concentration reflects the degree of evaporation on surface, but as $R^{36}Cl$ is conservative during evaporation, all start with the same initial value of 131×10^{-15} for $R^{36}_{0}Cl$. This model for recharge and decay of ^{36}Cl for these samples is presented in Figure 7.25. Decay follows the dashed lines showing the decrease in both $R^{36}Cl$ and $A^{36}Cl$ over time from initial values for Cl^- and ^{36}Cl in the recharge area.

TABLE 7.7

^{36}Cl and ^{81}Kr Ratios and Ages Wells in the Nubian Sandstone Aquifer, Western Egypt (Patterson et al. 2005)

Well	Depth (m)	Cl^- (mg/L)	$R^{36}Cl$ ($\times10^{-15}$)	$A^{36}Cl$ (10^6 at/L)	^{36}Cl Age (ky BP)	$^{81}Kr/Kr$ ($\times10^{-12}$)	^{81}Kr Age (ky BP)	$\delta^{18}O$ (‰)	δD (‰)
Recharge area	surface	112	131	249	–			−8.1	−62.6
GH	1200	20	76.5	26.1	240	5.73	210	−11.1	−81.2
EZ 12	720	59	72.7	72.7	270	3.34	390	−10.9	−81.7
F6	800	24	65.0	26.6	320	4.00	330	−10.3	−78.9
Bar	600	92	45.6	71.1	490	2.49	490	−10.8	−81.4
Bau	1200	52	20.2	17.7	930	5.23	1000	−10.8	−81.3
S36	750	95	12.2	19.6	1200	1.40	680	−11.0	−81.6

^{36}Cl measurements made by AMS at PRIME Lab, Purdue, ^{81}Kr analyzed by ATTA at Argonne NL

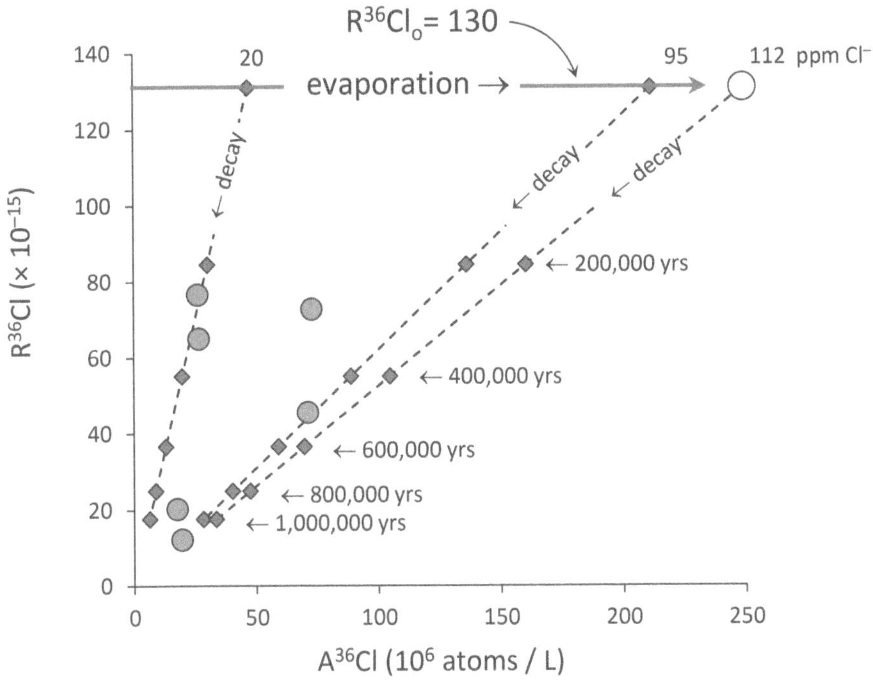

FIGURE 7.25 R_t^{36}Cl versus A_t^{36}Cl showing Pleistocene recharge for groundwater in the Egyptian Dahkla Basin of the Nubian sandstone aquifer (after Patterson et al. 2005). Dashed lines show decay over 1,000,000 years from varying initial Cl^- concentrations, marking 200,000-year intervals.

The ^{36}Cl ages are compared in Figure 7.26 with ages for ^{81}Kr as calculated directly with the above equation, cover a remarkable range from 240,000 years to over 1,000,000 years before present, and fit well with hydrodynamic modeling of the groundwater flow over these distances (Patterson et al. 2005). These ages also fit with paleoclimatic models. Lower and variable Cl^- concentrations in the oldest groundwater (20 to 95 ppm) in recharge water as compared with the recharge area today (112 ppm) are clear from the decay lines in Figure 7.25, suggesting wetter climate in the past. This is supported by the stable isotope data for these groundwaters (Figure 7.26 right plot), showing that modern recharge is enriched in ^{18}O and D relative to the mid Pleistocene groundwater. The consistency in ages between these two methods is impressive, with only one sample in this transect that deviates to a much higher ^{36}Cl age. This may reflect a source of geogenic Cl^- in that part of the aquifer, whereas Kr has no subsurface sources.

A study to determine the age of deep groundwater in the Paradox Basin of the Colorado Plateau found the groundwater geochemistry to be greatly complicated by mixing with remnant Paleozoic seawater and dissolution of evaporites (Kim et al. 2022). Using ^{81}Kr, which is unimpacted by salinity, provided a reliable tracer of

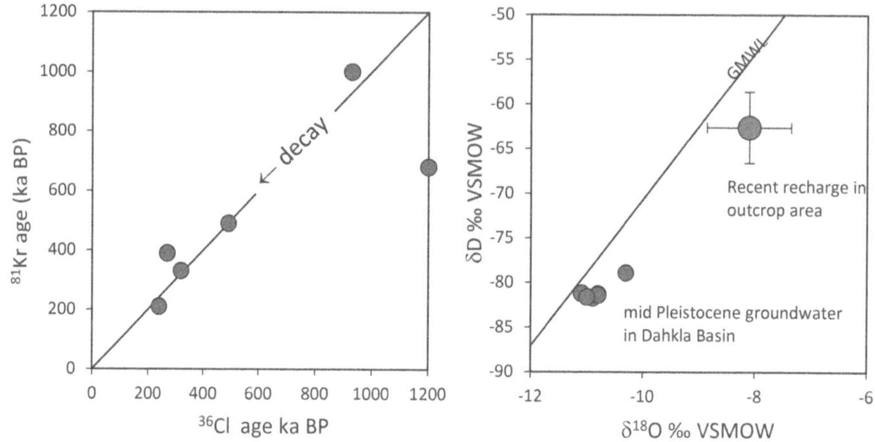

FIGURE 7.26 Left: Comparison of ^{81}Kr groundwater ages with ^{36}Cl groundwater ages for the same wells. Right: These mid-Pleistocene ages are consistent with a cooler wetter climate as shown by the stable water isotopes (data from Sturchio et al. 2004; Patterson et al. 2005).

topographically driven meteoric recharge down to a depth of 3000 m over the interval of 400,000 to 1.1 million years.

Geogenic Noble Gases

The stable noble gases, He, Ne, Ar, Kr, and Xe, have very well-constrained atmospheric concentrations (Table 2.8). Each of these gases comprises a range of stable isotopes with well-documented atmospheric molar fractions, shown in Figure 7.27.

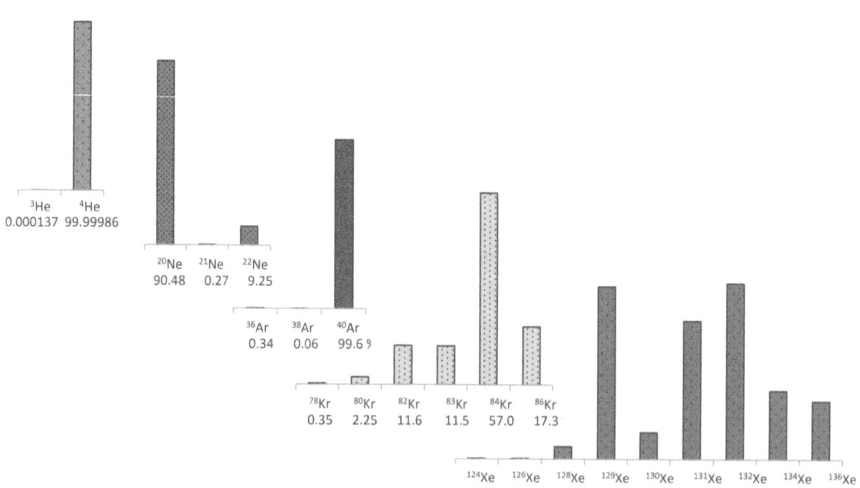

FIGURE 7.27 Stable isotope abundances (percent) of the five stable atmospheric noble gases.

While most were produced during nucleosynthesis at the formation of our solar system, others continue to be produced by nuclear reactions in the subsurface, including radioactive decay, spontaneous uranium fission, and neutron capture. Measuring the increase in abundance of these geogenic noble gases in deep groundwater provides a measure of time since the waters were isolated from the atmosphere.

Alpha-decay of the actinides along the U and Th decay series produces ^{4}He, which can accumulate within the environment of production:

$$^{238}U \rightarrow {}^{206}Pb + 8\,{}^{4}He$$
$$^{235}U \rightarrow {}^{207}Pb + 7\,{}^{4}He$$
$$^{232}Th \rightarrow {}^{208}Pb + 6\,{}^{4}He$$

Similarly, beta decay of 40K produces 40Ar, the abundant stable isotope of Ar, which also can accumulate in the subsurface. Providing the release of these radiogenic nuclides into groundwater from mineral grains in the aquifer is relatively efficient, their concentration can be used as an estimate of age. ^{4}He and ^{40}Ar are the abundant isotopes of atmospheric helium and argon, so these atmospheric contributions in the groundwater must be known and accounted for, according to the temperature–solubility and excess-air considerations discussed earlier.

Geogenic ^{4}He is mainly produced in the ^{238}U decay chain, which releases eight helium nuclides by the time it decays to stable lead (Figure 7.28). Lesser contributions come from ^{232}Th, which is more abundant but with a much longer half-life. ^{235}U is the most active of these three primordial actinides but has an abundance of only 0.73% of total uranium.

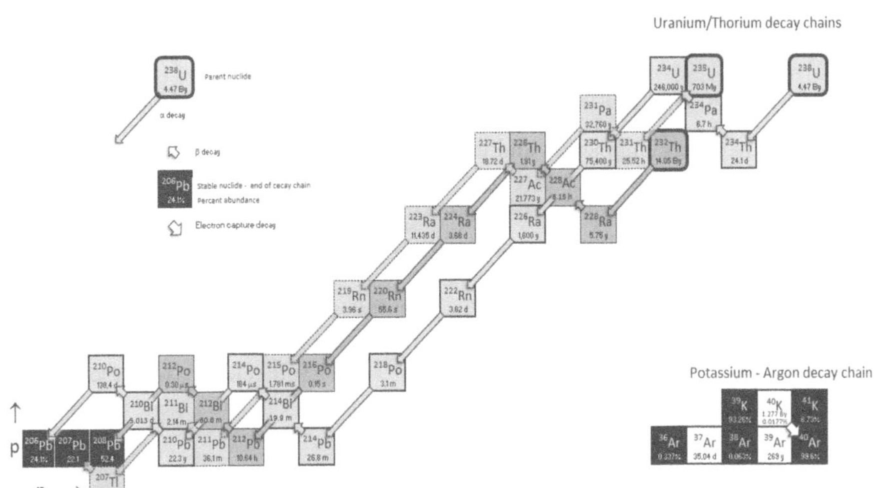

FIGURE 7.28 The U and Th decay series showing alpha and beta decay steps and intermediary radioisotopes. Black shading for stable isotopes. Decay in the ^{238}U series produces 8 ^{4}He nuclides, 7 ^{4}He in the ^{235}U series, and 6 ^{4}He in the ^{232}Th–decay series. Shown also is the ^{40}K decay to ^{40}Ar by electron capture.

Assuming that there are no gains or losses along the flow path, the ^{4}He age of groundwater is calculated from the ratio of the radiogenic helium concentration, $^4He_{radiogenic}$ (essentially all measured ^{4}He with correction for atmospheric He incorporated during recharge), to the helium accumulation rate, $^4He_{acc\text{-}rate}$:

$$t = \frac{^4He_{radiogenic}\left(cc_{STP}/cc_{H_2O}\right)}{^4He_{acc\text{-}rate}\left(cc_{STP}/cc_{H_2O}/y\right)}$$

Calculation of the helium accumulation rate, $^4He_{acc\ rate}$, in groundwater is based on the concentrations of U and Th in the aquifer matrix, the matrix density and porosity, and the decay factors for these radionuclides (Andrews et al. 1989):

$$^4He_{acc\ rate} = RF_{He}\frac{\rho_r}{\rho_w}\frac{1-n}{n_{eff}}\left(1.19\times10^{-13}\left[U\right]+2.88\times10^{-14}\left[Th\right]\right)cc_{STP}/cc_{H_2O}/y$$

RF_{He} is the helium release factor for diffusion of ^{4}He from the mineral lattice where the α particles are produced from U and Th decay. For long time periods this is generally taken to be 1 (full release of produced He from minerals to groundwater).

ρ_r is the density of the rock, in g cm^{-3}
ρ_w is the density of the water, in g cm^{-3}
n is the total fractional porosity of the rock
n_{eff} is the effective porosity of the rock (connected porosity contributing to flow) and [U] and [Th] are the uranium and thorium contents of the rock matrix in ppm

Helium Dating of Groundwater in Patapsco Sand Aquifer, Maryland

The stable isotope composition and noble gas temperatures of the Patapsco aquifer paleogroundwater in Maryland (Figure 7.29) were presented above in Figure 7.18. The low sea level during glacial time allows enhanced recharge and flow to the east under Chesapeake Bay (Plummer et al. 2012). The $^{14}C_{DIC}$ activity becomes less than detection within 40 km along the flow path.

The ^{4}He accumulation rate ($^4He_{acc\text{-}rate}$) for this aquifer was calculated from the concentrations of U and Th in aquifer minerals, according to the equation presented above, and using a helium release factor, RF, of 1 (full release from the mineral matrix to the groundwater).

$$\begin{aligned} ^4He_{acc\ rate} &= RF_{He}\frac{\rho_r}{\rho_w}\frac{1-n}{n_{eff}}\left(1.19\times10^{-13}\left[U\right]+2.88\times10^{-14}\left[Th\right]\right) \\ &= 1\times\frac{2.6}{1}\times\frac{1-0.39}{0.25}\left(1.19\times10^{-13}\left[1.9\right]+2.88\times10^{-14}\left[7.5\right]\right) \\ &= 2.80\times10^{-12}\ cc_{STP}/cc_{H_2O}/y \end{aligned}$$

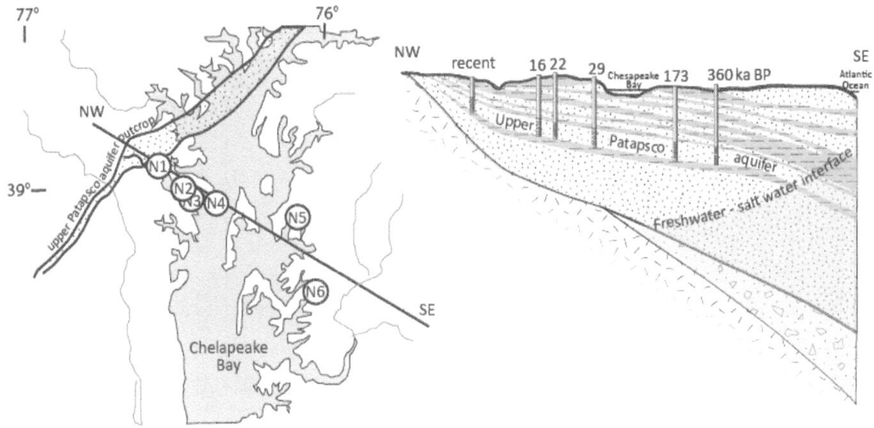

FIGURE 7.29 Location and cross section of the Patapsco aquifer, Maryland, showing the outcrop and subsurface geology. Wells sampled along the northern flow path are identified on the plan view and in section with ^{4}He ages in thousands of years before present (ka BP). Modified from Plummer et al. 2012.

Refinement of the accumulation rate was made by using ^{14}C ages in the zone overlapping with measureable ^{4}He (15–25 km) to calibrate the production of radiogenic ^{4}He (Plummer et al. 2012). The accumulated radiogenic helium, $^4He_{radiogenic}$, was calculated from the measured concentration of $^4He_{meas}$ with corrections for dissolved air ($^4He_{dissolved}$) and excess air ($^4He_{excess}$) gained during recharge (Table 7.8).

TABLE 7.8
Helium and Neon Data for Wells along the Northern Flow Path, Patapsco Aquifer, Maryland

Well	T_{NG}°C	Ne_{meas}	$Ne_{dissolved}$	$Ne_{excess\ air}$	He_{meas}	$He_{excess\ air}$	$He_{dissolved}$	$He_{recharge}$	$^4He_{radiogenic}$	^{4}He Age Years
N-1	10	2.5E–7	2.1E-07	4.6E-08	5.8E-08	1.3E-08	4.8E-08	6.1E-08	LD	
N-2	3.9	2.8e–7	2.2E-07	6.2E-08	1.13E-07	1.8E-08	4.9E-08	6.7E-08	4.6E-08	16,000
N-3	3.3	2.6E–7	2.2E-07	4.0E-08	1.23E-07	1.2E-08	4.9E-08	6.1E-08	6.2E-08	22,000
N-4	1.5	2.7E–7	2.2E-07	4.6E-08	1.43E-07	1.3E-08	4.9E-08	6.3E-08	8.0E-08	29,000
N-5	5.4	2.6E–7	2.2E-07	4.3E-08	5.47E-07	1.2E-08	4.9E-08	6.1E-08	4.9E-07	175,000
N-6	4.3	2.6E–7	2.2E-07	4.0E-08	1.10E-06	1.2E-08	4.9E-08	6.0E-08	1.0E-06	360,000

Source: Plummer et al. 2012.
Notes:
1 Helium from excess air incorporated during recharge is calculated from measured neon minus the neon dissolved from air at the recharge temperature, according to the equations given in Figure 7.19.
2 The ^{4}He age in years is calculated from the radiogenic ^{4}He and the ^{4}He production rate, ^{4}He age = $He_{radiogenic} / {^4He}_{acc\ rate}$.

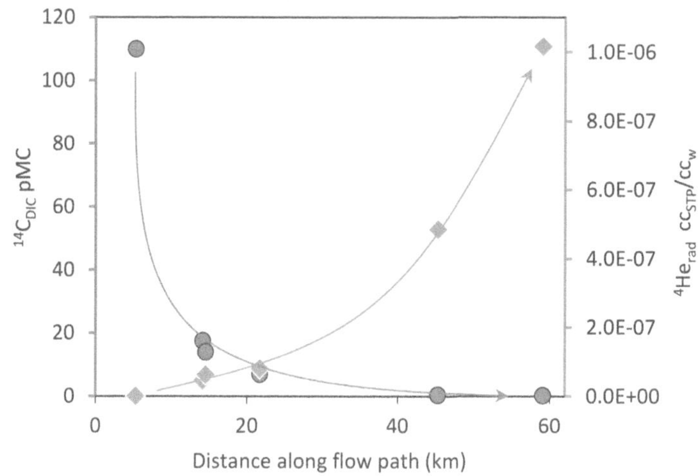

FIGURE 7.30 Increase in ^{4}He with decrease in ^{14}C along northern groundwater flow path in the Patapsco aquifer, Maryland. High ^{4}He concentrations are found beyond 40 km, where ^{14}C activities are near detection. The high radiogenic ^{4}He measurements beyond 40 km provide age estimates up to 360,000 years before present. Plummer et al. 2012.

The corrected $^4He_{radiogenic}$ normalized to the accumulation rate (2.80×10^{-12} cc_{STP}/cc_{H_2O}) according to the dating equation above, provide these age estimates.

The age estimates increase from modern groundwater (no measureable radiogenic ^{4}He) in the recharge area to over 360,000 years before present (Figure 7.30). These very old estimates are interpreted to represent recharge during past Pleistocene glaciations when the low sea level stand increased the hydraulic gradient and increased recharge and flow. Refinements of these age estimates were made by using ^{14}C ages in the zone overlapping with measureable ^{4}He (15–25 km) to calibrate the production of radiogenic ^{4}He (Plummer et al. 2012.)

Like ^{4}He, the accumulation of ^{40}Ar (the dominant isotope of argon) from decay of ^{40}K provides a geochronometer for groundwater and is similarly calculated from its concentration normalized to its accumulation rate:

$$t = \frac{^{40}Ar_{radiogenic}\left(cc_{STP}/cc_{H_2O}\right)}{^{40}Ar_{acc\text{-}rate}\left(cc_{STP}/cc_{H_2O}/y\right)}$$

$$^{40}Ar_{acc\text{-}rate} = RF_{Ar}\frac{\rho_r}{\rho_w}\frac{1-n}{n_{eff}}3.887\times10^{-14}\left[K\right]cc_{STP}/cc_{H_2O}/y$$

Here, [K] is the weight percent of potassium in the rock. This formula takes into account that the abundance of radioactive ^{40}K is only 0.0177% of stable potassium (Figure 7.28). The half-life for ^{40}K is long (1.277 By), but only one daughter ^{40}Ar is produced, while U and Th decay produces 8 and 6 ^{4}He nuclides, respectively, and so total accumulation of ^{40}Ar is considerably less. Further, unlike alpha decay, which damages the crystal lattice, beta decay is low impact. This limits diffusion of ^{40}Ar from the mineral into the groundwater, and so the release factor, RF, may be less than 1.

^4He and ^{40}Ar Ages in Crustal Brines

Brines encountered at depth in the Con Mine, Yellowknife, NWT, Canada, were studied as part of research on fluids deep within the crystalline terrain of the Canadian Shield (Greene et al. 2008). The mine follows a gold-bearing shear zone dipping almost 2 km through faulted metavolcanics within the granitic terrain of the Canadian Shield. Shallow groundwaters from the Holocene to the modern age are underlain by glacial meltwaters that recharged at the end of the Pleistocene along the steep gradients below the melting Laurentide ice sheet (Douglas et al. 2000). The deepest mine levels have inflows of high-salinity $Ca^{2+}-Cl^{-}$ brine, with TDS up to 290 g/kg.

The deep brines are believed to have been recharged during Devonian time, when the region was covered by highly evaporated epicontinental seas. The highest salinity brine was analyzed for radiogenic ^4He and ^{40}Ar in an attempt to constrain their subsurface residence time (Figure 7.31).

The concentrations of both ^4He and ^{40}K are orders of magnitude higher than found in air-equilibrated water or in air-equilibrated brine so this correction is not necessary. The greater accumulation of ^4He than ^{40}Ar (six orders of magnitude, compared with two orders of magnitude) reflects the high production of ^4He from the alpha decay chains for ^{238}U (8 ^4He atoms), ^{235}U (7 ^4He atoms), and ^{232}Th (6 ^4He atoms) (Figure 7.28), despite their similar abundances in these rocks. The resulting ages of 221.4 and 294.7 million years for ^4He and ^{40}Ar, respectively, are younger than the Devonian Period when marine evaporites formed and when brine infiltration would have occurred. However, some diffusive loss of these gases, He in particular, would be expected over this period of time, and so a late Paleozoic age for the brine was concluded.

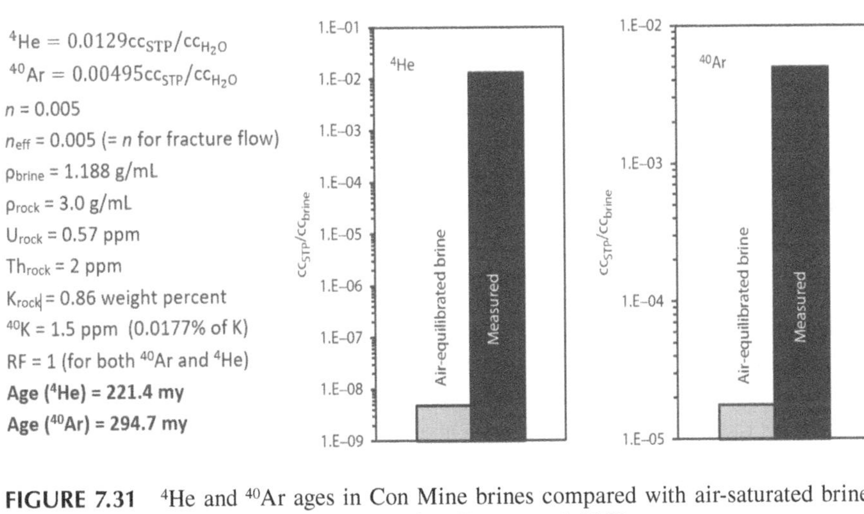

FIGURE 7.31 ^4He and ^{40}Ar ages in Con Mine brines compared with air-saturated brine, yielding ages of about 400 million years. After Greene et al. 2008.

^{129}I and ^{136}Xe from ^{238}U Spontaneous Fission

Long subsurface residence times can be estimated by ^{129}I, which has a half-life of 15.7 million years. As a halide, it is useful in studies of saline systems, such as marine sediments, and crustal brines. Iodine-129 has an atmospheric source through n-p reaction by cosmic radiation of atmospheric ^{129}Xe. Before anthropogenic inputs to environmental ^{129}I from the reprocessing of spent nuclear fuel, natural fallout provided some 200,000 to 400,000 atoms/L to surface waters. Now surface waters can have upward of 10 million atoms/L. Decay of atmospheric sources is seldom used as an approach to dating. Subsurface production by spontaneous fission of ^{238}U (Figure 3.12) generates greater concentrations of ^{129}I:

$$^{238}U_{SF} \rightarrow {}^{129}I + {}^{107}Pd + 2n$$

^{129}I decays by beta emission to stable ^{129}Xe,

$$^{129}I \rightarrow {}^{129}Xe + \beta^-$$

Spontaneous fission of ^{238}U is well known for the production of fission tracks in quartz grains and used for mineral dating. Decay of ^{238}U by spontaneous fission occurs with a half-life ($T_{½\,sf} = 8.15 \times 10^{15}$ y), far longer than that of ^{238}U decay by alpha emission ($T_{½\alpha} = 4.47 \times 10^{9}$ y). The fission yield of ^{129}I is only 0.03%, but it is sufficient for significant accumulation over periods of time up to about five half-lives, or 80 million years. In situ production of ^{129}I is then solely a function of the concentration of uranium in the host rock and will build up to the point of secular equilibrium, where the rate of production is equal to the rate of decay (for example, as shown for ^{36}Cl in Figure 7.23).

In another study of the deep crustal brines from the Con Mine, Bottomley et al. (2002) measured the ^{129}I concentration to be 340 million atoms/L. This was close to the value of 380 million atoms/L calculated for in situ secular equilibrium using a uranium concentration of 1 ppm. They concluded that the brines have a subsurface residence time in that environment for greater than five half-lives, or more than 80 million years, which is consistent with the 200 to 300 million-year ages found with ^{4}He and ^{40}Ar above.

Spontaneous fission of ^{238}U produces stable isotopes of xenon at higher yields than for ^{129}I, and with the greatest yield of 6.3% for ^{136}Xe:

$$^{238}U_{sf} \rightarrow {}^{136}Xe + {}^{100}Mo + 2n$$

Enrichment of ^{136}Xe above its atmospheric ratio (Figure 7.27) is a simple and conservative geochronometer. Zuo et al. (2022) used helium and xenon isotopes to date porewater in an Ordovician shale aquiclude under investigation as cap rock for isolation of radioactive waste from a potential deep geological repository (Figure 7.32). Measurements were made of noble gas isotopes in outgassed samples from drill cores; as an example, a sample from a depth of 567 mbgs had 11.5 million atoms of excess $^{136}Xe_{sf}$ per gram of rock produced by spontaneous fission from 8.22×10^{15}

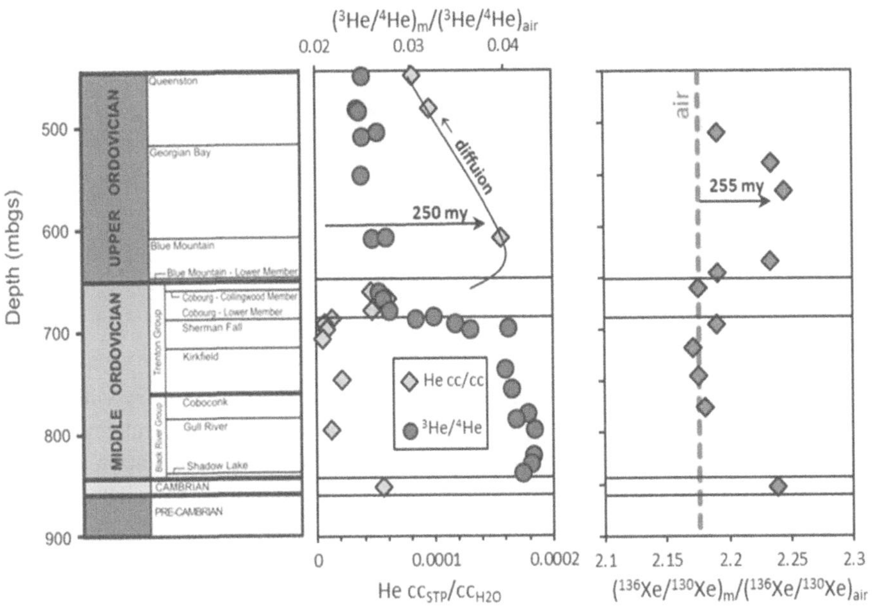

FIGURE 7.32 Paleozoic marine brines dated with ingrowth of ^{4}He and ^{136}Xe in porewaters from an Ordovician shale aquiclude, Bruce Nuclear site, Canada (Zuo et al. 2022). The ^{3}He/^{4}He ratio in the shales matches that calculated for authigenic ingrowth of ^{3}He from ^{6}Li fission and ^{4}He from α decay of U and Th with an age of 250 million years (Zuo 2022).

atoms of ^{238}U per gram (3 ppm U). Using the decay constants for U decay by α-decay ($\lambda_\alpha^{238}\text{U} = 1.55\times10^{-10}$) spontaneous fission ($\lambda_{sf}^{238}\text{U} = 8.51\times10^{-17}$) and fission yield ($Y_{^{136}\text{Xe}} = 0.063$) in the following equation produces a $^{136}\text{Xe}_{sf}$ age of 255 million years.

$$t_{(\text{Xe}_{sf})} = \frac{1}{\lambda_\alpha^{238}\text{U} + \lambda_{sf}^{238}\text{U}} \times \ln\left[\frac{^{136}\text{Xe}_{sf} \times \left(\lambda_\alpha^{238}\text{U} + \lambda_{sf}^{238}\text{U}\right)}{^{238}\text{U} \times Y_{sf} \times \lambda_{sf}^{238}\text{U}}\right] + 1$$

Here, the ^{136}Xe age closely approximates that of ^{4}He, giving confidence in this aquiclude for isolation of radioactive waste.

8 Contaminant Geochemistry and Isotopes

INTRODUCTION

The contamination of groundwater resources has generated an industry with hydrogeologists, engineers, and environmental scientists involved in research and studies from environmental assessments to design and installation of remediation schemes. Geochemistry and isotopes play an important role at all stages. Geochemistry and isotopes are used to trace both the origin and transformation of contaminants in groundwater, as well as predict their fate and impacts on surface waters and water supplies. In this chapter, the major areas of application are examined, using the background on geochemistry and isotopes developed in the earlier chapters.

SORPTION AND RETARDATION

The exchange of solutes between reactive surfaces and the bulk solution is an important process for the retardation of contaminant migration during advective groundwater flow. Aquifer minerals and organics are reactive surfaces that can attract and retain contaminants. Like cation exchange processes, the amount exchanged or *sorbed* onto the solid surface is proportional to the amount in solution and can often be characterized as an equilibrium exchange process, like cation exchange. However, sorption reactions are considered for minor and trace solutes and are modeled with a distribution coefficient rather than an exchange selectivity coefficient.

DISTRIBUTION COEFFICIENT, K_D, AND THE SORPTION ISOTHERM

The simplest model for the movement of a reactive solute through an aquifer is that of reversible sorption. Here, the amount of the solute sorbed onto the surface is proportional to its concentration in solution; the higher its concentration, the greater the mass sorbed onto the surface. This is expressed by a distribution coefficient or K_d, which relates the mass of the solute sorbed onto the surface to that in solution:

$$K_d = \frac{\text{mass sorbed}}{\text{mass dissolved}} = \frac{\mu g/g}{\mu g/mL} = \frac{S}{C_{eq}}(mL/g)$$

giving units of volume per mass, or mL/g in this case. Distribution coefficients depend not only on the chemical characteristics of the solute or contaminant but

DOI: 10.1201/9781003353836-8

also on the aquifer medium. In particular, the cation exchange capacity (CEC), solid organic fraction, types of minerals, presence of iron oxyhydroxides, and so forth all affect the distribution coefficient that controls sorption.

Given the range of controlling factors, the K_d value for a given groundwater–aquifer combination is empirically measured. This is done by batch tests, whereby a water sample with given concentration of the solute is mixed with a given mass of the medium. The difference between the solute concentration before mixing ($C_{initial}$ in, say, mg/L) and after equilibrium is established (C_{eq}) yields the mass sorbed, S, per mass of sediment.

$$S = \frac{\left(C_{\text{initial}} - C_{\text{eq}}\right) \times \left(\text{L}_{\text{solution}}\right)}{\text{g}_{\text{sediment}}} = \frac{\text{mg}}{\text{g}}$$

Comparison of the sorbed mass, S, with the equilibrium concentration (C_{final}) gives the distribution coefficient for that concentration at the batch reaction temperature for the test sediment.

$$S = K_d C_{eq}$$

Conducting batch tests for a range of concentrations yields the sorption isotherm for that solution–sediment system (Figure 8.1). For these linear isotherms, K_d is the slope of the S versus C relationship. If a solute is conservative, such as Cl^-, then there is no sorption onto the solid surface and the sorption isotherm sits on the abscissa. For increasingly reactive solutes, the isotherm becomes increasingly steep. In the example for Sr^{2+} and NH_4^+ in Figure 8.1, strontium is more reactive than ammonium.

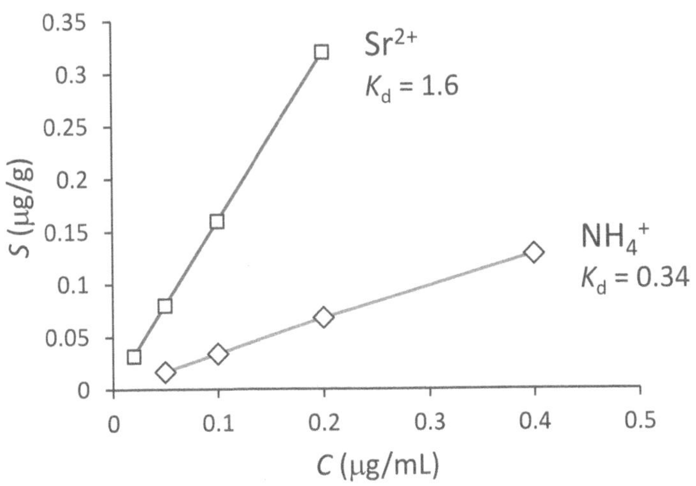

FIGURE 8.1 Sorption isotherms for two reactive solutes. Sr^{2+} sorption onto a medium- to fine-grained quartz sand aquifer was investigated by Lee et al. (1998) to remediate a ^{90}Sr plume. Böhlke et al. (2006) investigated NH_4^+ sorption onto glacial outwash sand aquifer.

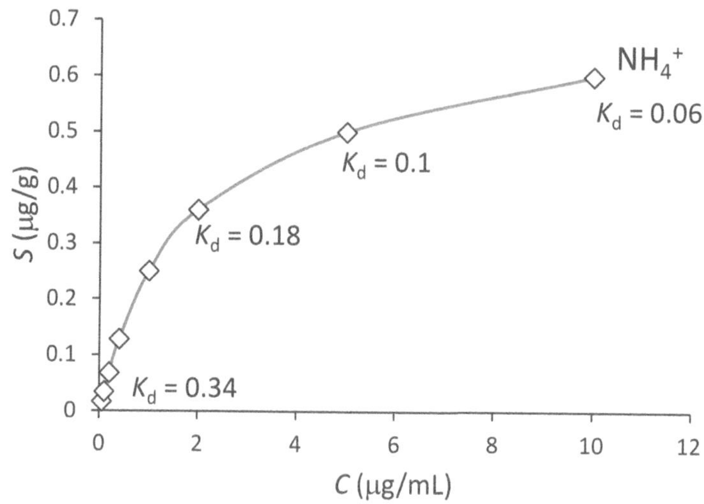

FIGURE 8.2 Sorption isotherm for ammonium at higher concentrations. Increased saturation of exchange sites leads to a reduction in the distribution coefficient, K_d, leading to reduced retardation and more conservative transport.

For higher concentrations of reactive solutes and for highly reactive solutes, the linear sorption isotherm becomes inadequate, and nonlinear representations are required for modeling reactive transport. In the case of ammonium, higher concentrations would lead to a saturation of the available exchange sites on the aquifer medium, and the K_d decreases (Figure 8.2). At this point, the retardation factor decreases and the solute behaves in a more conservative fashion, with a transport velocity approaching that of the groundwater.

Retardation

The effect of sorption on solute transport in advective groundwater flow is to retard its movement along the flow path. Figure 8.3 shows the retardation of ammonium

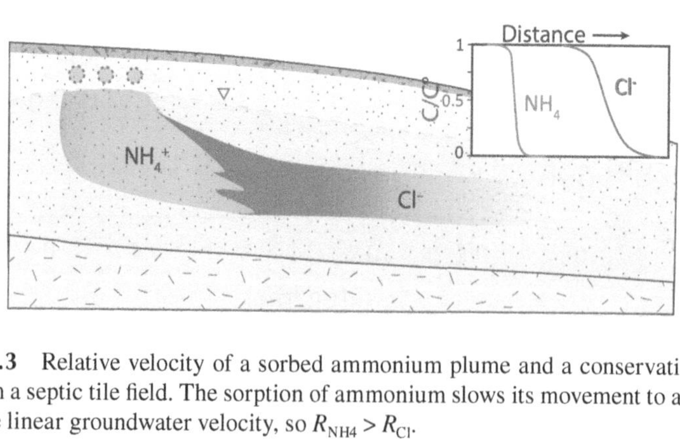

FIGURE 8.3 Relative velocity of a sorbed ammonium plume and a conservative chloride plume from a septic tile field. The sorption of ammonium slows its movement to a fraction of the average linear groundwater velocity, so $R_{NH4} > R_{Cl^-}$.

in groundwater by sorption onto the aquifer medium, where the NH_4^+ plume moves slowly relative to the movement of a non-sorbing solute, Cl^-. The degree of retardation is affected by the available surface area for sorption, which is a function of both porosity, θ, and sediment bulk density, ρ(g/cm³). The relative velocity of a nonconservative solute or contaminant that is sorbed by equilibrium exchange with the medium is characterized by its retardation factor, *R* (unitless).

$$R = 1 + \frac{\rho}{\theta} \times K_d$$

$$\text{and so } R = \frac{v_{gw}}{v_c}$$

where v_{gw} is the average linear groundwater velocity, and v_c is the retarded solute velocity. Conservative solutes then have a retardation factor of $R = 1$, while the reactive solute with $R > 1$ moves more slowly that the groundwater.

Sorption and Solid Organic Carbon

Nonpolar compounds such as hydrocarbons and chlorinated solvents do not sorb onto charged surfaces. However, they can sorb onto organic substrates in an aquifer, which can retard their transport. Nonpolar compounds are soluble in organic substrates and so partition onto these phases as a plume migrates through the aquifer. Further, the availability of negatively charged sites on complex organic compounds fosters chelation of multivalent cations, such as the transition metals. Chelating compounds have multiple charge sites to complex metals and remove them from solution. In the case of waters with high dissolved organic carbon (DOC), chelation with such compounds can increase the transport of metals.

Characterizing the sorption capacity of a soil for a given organic compound or metal then requires some assessment of its organic content and its affinity for sorption. This is represented for most compounds of interest as the partition coefficient for that compound in water relative to a fatty alcohol, octanol ($CH_3(CH_2)_7OH$). This is the octanol–water partition coefficient, K_{O-W}. Depending on the degree of polarity, an organic contaminant will preferably partition into the organic liquid, octanol, or into water. This partitioning function, in conjunction with the fraction of organic carbon in the aquifer, can define how the organic contaminant will sorb onto the aquifer. The coefficient, K_d, defining the distribution of that compound between the soil or aquifer with a given organic carbon fraction, f_{OC}, and groundwater is defined as follows:

$$K_d = K_{O-W} \times f_{OC}$$

From this relationship, sorption and retardation increase for organic molecules with higher K_{O-W} in aquifers with more organic carbon.

NITROGEN SPECIES AND GROUNDWATER CONTAMINATION

Nitrate is perhaps the most pervasive and significant contaminant in groundwater today (Shukla and Saxena 2019). Nitrogen is critical for agriculture, and its use has increased the concentrations of nitrogen species in groundwater, surface water, and the atmosphere globally. Tracing and mitigating sources of nitrogen species requires an understanding of the geochemistry and transformation of this important redox element.

NITROGEN AND CIVILIZATION

Nitrogen is an essential nutrient for life, with a key structural role in protein and plant tissue. Algae, for example, have a chemical composition close to $C_{106}H_{263}O_{110}N_{16}P(S)$. The largest reservoir of nitrogen is atmospheric N_2, but requires transformation to mineralized or *fixed* forms for assimilation by life. The triple covalent bond of N_2 makes this molecule remarkably difficult to break such that it can be incorporated as nucleotides in DNA and as amino groups in protein. Nitrogenase enzymes manage this by transferring electrons to N_2 and producing reduced nitrogen as NH_3. Prokaryotes accomplished this reaction billions of years ago, but these nitrogen-fixers remain restricted to certain cyanobacteria in marine environments and root nodules in legumes, such as clover, soybeans, and alfalfa. Cycling from NH_3 to organic N and NO_3^- ultimately returns nitrogen back to the atmosphere as N_2. Another pathway for nitrogen fixation includes lightning in the atmosphere which can oxidize several kilograms of N_2 per strike, producing NO and NO_2.

Nitrogen is central to the development of civilization over the Holocene as the importance of N-rich manure for agriculture became recognized. By early in the last millennium, saltpeter (KNO_3) for agriculture and for black powder dominated international trade, with supplies mainly from urea purified from soil in the cattle pens of India. By the late 19th century this source was eclipsed by the rich deposits of guano from the arid islands off the coast of Peru and the even richer accumulations of nitrate from Chile's hyperarid Atacama Desert. These natural deposits supplied NO_3^- not only for agriculture but also for the production of dynamite, Alfred Nobel's stabilized form of nitroglycerin, which fueled the huge construction projects of the late 1800s, such as the Suez Canal, train tunnels through the Rocky Mountains, and trinitrophenol for high-explosive artillery in World War I. Competition for these coveted supplies spawned efforts to synthesize nitrate from atmospheric N_2, mainly by arcing, requiring hydroelectric generation and with limited yields. However, in 1909, Fritz Haber managed to reduce N_2 to ammonia at high pressure using H_2 as an electron donor and a hot ferric iron catalyst:

$$N_2 + 3H_2 \rightarrow 2NH_3 \quad \Delta G_r^\circ = -16.5\,\text{kJ/mol} \quad \text{Synthesis of ammonia}$$

Brought to full industrial scale by Carl Bosch, the method is responsible for the agricultural revolution of the 20th century that supported a threefold increase in the human population. Today, nitrogen from the Haber–Bosch process is found in over 90% of the world food supply. However, the widespread use of fixed nitrogen compounds has led to unprecedented contamination of surface water and groundwater, making nitrogen perhaps the most pervasive anthropogenic contaminant in the environment today.

Nitrogen Species and Water Quality

Once nitrogen is fixed from atmospheric N_2, it can be transformed between a variety of redox states and species in a complex cycle involving inorganic and bacterially mediated reactions under both aerobic and anaerobic conditions. The principal species and their redox state include the following:

Nitrate	NO_3^-	N^{+V}	Stable oxide of N, highly soluble as an anion
Nitrogen dioxide	NO_2	N^{+IV}	Produced by lightning and combustion
Nitrite	NO_2^-	N^{+III}	Intermediate in conversions between NO_3^- and NH_4^+
Nitrogen oxide	NO	N^{+II}	Produced by lightning and combustion, $NO_2 + NO = NO_x$
Nitrous oxide	N_2O	N^{+I}	Bacterially produced gas, in atmosphere at < 1 ppm
Nitrogen	N_2	N	Elemental nitrogen gas
Hydroxylamine	NH_2OH	N^{-I}	Intermediate species during oxidation of NH_4^+
Ammonia	NH_3	N^{-III}	Unionized ammonia gas
Ammonium	NH_4^+	N^{-III}	Ionized ammonia (dominates at pH below 9.23)
Organic N	$-NH_2$	N^{-III}	Reduced N in organic compounds including protein.

Of these nitrogen species, nitrogen (N_2), ammonium (NH_4^+), nitrate (NO_3^-), and nitrous oxide (N_2O) represent the largest reservoirs in groundwater. The natural cycle of nitrogen from fixation of atmospheric nitrogen to organic nitrogen in plants and its transformation back to atmospheric nitrogen is shown in Figure 8.4.

There are, however, two predominant water quality issues with nitrogen species related to manure and nitrogen fertilizers in agriculture, wastewater from septic systems, and industrial releases. These are elevated NO_3^- (and NO_2^-) in drinking water and elevated NH_3 in surface waters. There is no maximum acceptable concentration (MAC) for ammonia in drinking water, as it is naturally part of our metabolic system. However, unionized ammonia does pose a problem to aquatic life at concentrations exceeding a few micrograms per liter. The concentration of unionized ammonia is calculated from total ammonia for a given temperature and pH. Most jurisdictions place the limit for unionized at some 20 μg/L. At pH 8 and 25°C, this corresponds to a total ammonia concentration of only 0.35 mg/L.

Nitrate affects human health when it is converted to nitrite in the stomach. Nitrite interferes with the blood's capacity to carry oxygen. This leads to methemoglobinemia or *blue baby* syndrome in infants. The MAC for nitrate in drinking water is universally set at 10 mg/L as N, and for NO_2^- at 1 mg N/L. In surface waters, nitrate generally has no limit, although high concentrations promote excessive algae and weed growth.

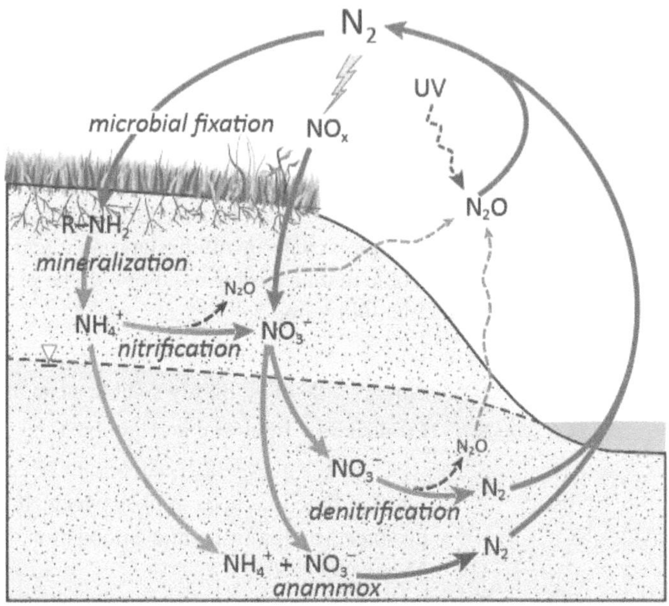

FIGURE 8.4 The natural nitrogen cycle, showing the principal reservoirs and reactions from the fixation of atmospheric nitrogen by microbial activity to the mineralization of organic carbon and transformation back to atmospheric N_2. N_2O produced during bacterial nitrification and denitrification is photo-dissociated in the atmosphere to N_2. Oxidation of N_2 to atmospheric nitrous oxide by lightning also contributes to NO_3^- in soils through wet and dry fallout.

The Nitrogen Cycle

Oxidation of N_2O and NO_2 with O_2 in the atmosphere produces nitric acid, HNO_3, contributing to soil nitrate via precipitation or dry fallout, but represents a minor contribution to the inventory of fixed nitrogen species in soils and soil water. The cycling of nitrogen in soils and groundwater involves a number of possible transformations, which occur under a range of redox conditions.

Microbial fixation of N_2:

$$N_2 + 8H^+ + 16Mo_{ATP} + 8e^- \rightarrow 2NH_3 + H_2 + 16Mo_{ADP} + 16P$$

The most common N-fixation reaction is mediated by the Fe-Mo adenosine triphosphate nitrogenase ATP enzyme (Kim and Rees 1994; Seefeldt et al. 2009) common to cyanobacteria and operating for over 3 billion years. It produces ammonia and hydrogen, which are then synthesized into proteins as R–NH_2.

Mineralization of organic N:

$$-NH_2 + 2H^+ \rightarrow NH_3$$

Organically bound nitrogen is a component of all proteins and plant biomass. Aerobic and anaerobic degradation of such carbon compounds releases this reduced nitrogen in the form of ammonia. Where this occurs in unsaturated materials such as soils or manure, the ammonia can volatilize, or dissolve into water as ammonium, NH_4^+.

Decomposition of urea:

$$CO(NH_2)_2 + H_2O \rightarrow 2NH_3 + CO_2$$

Urea is often applied as fertilizer in granulated form and breaks down by a bacterially mediated reaction (*urease* enzyme) to release ammonia for plants.

Ionization of ammonia:

$$NH_3 + H^+ \leftrightarrow NH_4^+ \quad K_{NH_3} = 10^{-9.23}$$

Ionization of ammonia to highly soluble ammonium occurs at neutral pH. The two species have equal concentrations at pH 9.23. Analytical reports generally provide a single value of total ammonia that includes both ionized and unionized ammonia.

Volatilization of ammonia:

$$NH_{3(aq)} \rightarrow NH_{3(g)} \quad K_{NH_{3\,g}} = 10^{1.76}$$

The relatively high Henry's law constant for ammonia allows an aqueous concentration of 246 mg/L at 25°C for a NH_3 partial pressure of 1 atm. As the partial pressure of NH_3 in the air is negligible, diffusive loss from manure piles or surface applications occurs. Volatilization from groundwater below the water table is greatly minimized by the slow rate of aqueous diffusion.

Sorption of ammonium:

$$Na - clay + NH_4^+ \rightarrow NH_4 - clay + Na^+$$

Ammonium is sorbed onto clay minerals in soils and aquifers with a selectivity coefficient close to that of K^+. Accordingly, NH_4^+ migration in aquifers is significantly retarded. Erosion of NH_4-bearing soils is one of the major sources of ammonia contamination in surface waters.

Aerobic nitrification of ammonium:

$$NH_4^+ + 2O_2 \rightarrow NO_3^- + H_2O + 2H^+ \qquad \Delta G_r^\circ = -266.5\,kJ/mol$$

Both ionized and unionized ammonia can be oxidized to NO_3^- in water by reaction with elemental oxygen (O_2). The significant energy yield makes this favorable for bacteria. It follows a two-step reaction of oxidation to nitrite and then to nitrate by a mixture of species, including *Nitrosomonas*, *Nitrobacter*, and *Nitrosospira*. Dissolved oxygen is required, limiting this reaction to manure piles, soils, and shallow groundwaters. Such losses from farm fields are costly and require slow-release fertilizers like urea to allow plants to assimilate it before it is oxidized.

Denitrification:

$$5CH_2O + 4NO_3^- + 4H^+ \rightarrow 2N_2 + 5CO_2 + 7H_2O \qquad \Delta G_r^\circ = -252.47\,\text{kJ/mol}$$

Where a low-p*e* electron donor exists, such as carbon or sulfide, nitrate can be used as an electron acceptor with nearly the same energy yield as O_2. *Pseudomonas denitrificans* reduces NO_3^- to N_2 using fixed carbon (biomass) as a source of energy. Chemotrophs such as *Thiobacillus denitrificans* use sulfide (H_2S or pyrite) as a substrate for denitrification. This reaction generally will not occur in oxygenated waters where aerobic bacteria outcompete denitrifiers for available carbon substrates. In anaerobic waters with low nitrate concentrations (NO_3^--limited), denitrification to N_2 gas may not be complete, resulting in the production of N_2O gas.

Anaerobic oxidation of ammonium – anammox:

$$NH_4^+ + NO_2^- \rightarrow N_2 + 2H_2O \qquad \Delta G_r^\circ = -362.8\,\text{kJ/mol}$$

This is a recently discovered yet thermodynamically favorable bacterial reaction (Jetten 1999). It is found to operate in anaerobic environments where both ammonium and nitrate species are present, such as in wastewater streams, anoxic marine waters, and contaminated groundwaters (Clark et al. 2008; Moore et al. 2011). The anammox bacteria use NH_4^+ as an electron donor and NO_2^- generated through partial reduction of NO_3^- as an electron acceptor in a reaction that converts both species to elemental nitrogen, N_2. It is apparently the only known biologically mediated reaction for the conversion of ammonium to N_2 in natural systems. The nitrite for this reaction can be produced by partial denitrification ofNO_3^-, giving the overall reaction $3NO_3^- + 5NH_4^+ \rightarrow 4N_2 + 9H_2O + 2H^+$ (Mulder et al. 1995) or through partial oxidation of NH_4^+ under low redox conditions.

^{15}N and ^{18}O in the Nitrogen Cycle

The isotope ^{15}N is present in the nitrogen reservoir on the earth at a concentration of about 0.4% of the common nitrogen isotope, ^{14}N, and can be used to trace the origin and the transformations of nitrogen species. As nitrogen transformations are dominantly biologically mediated redox reactions, fractionation tends to be significant, and with Rayleigh distillation as the reactant reservoir is consumed, leads to readily observed enrichment that signifies reaction.

The ^{15}N content of nitrate or ammonium is determined by the ^{15}N of its nitrogen source plus the associated fractionation. Most industrial nitrogen compounds, including ammonium and nitrate, are synthesized from atmospheric N_2, and as the reaction is *quantitative* (i.e., complete conversion of the reactant), these compounds have $\delta^{15}N$-values close to 0‰. The $\delta^{18}O$ of nitrate is established during oxidation or *nitrification* from ammonia. During low-temperature nitrification, the $\delta^{18}O$ of the ambient water plays an important role. In the production of KNO_3 chemical fertilizers, the $\delta^{18}O$ of atmospheric O_2 dominates. The trends in isotope values during the geochemical transformations presented above are as follows:

Ammonium volatilization, NH_4^+ to NH_3 $\varepsilon^{15}N_{NH_4-NH_{3(g)}} = 20.5‰$ (Wickens 2003; Ti et al. 2021)This reaction involves dissociation of NH_4^+ to $NH_{3(aq)}$ and diffusion of $NH_{3(aq)}$ to $NH_{3(g)}$, both of which favor the light isotope, ^{14}N, with enrichment of ^{15}N in the residual NH_4^+. NH_3 volatilization is minimal under circumneutral pH conditions where the unionized ammonia fraction of the total ammonium pool is less than 1% ($K_{NH_4-NH_3} = 10^{-9.23}$and so the NH_3 fraction is 50% at pH 9.23, 5% at pH 8.23, but only 0.5% at pH 7.23). Rayleigh enrichment of $\delta^{15}N_{NH_4^+}$ is typically less than 2‰ to 3‰ in the dissolved NH_4^+ pool unless the soils are particularly alkaline (Figure 8.5).

Nitrification of ammonium, NH_4^+ to NO_3^- $\varepsilon^{15}N_{NH_4^+-NO_3} \approx 30‰$ (Kendall and Aravena 2000). The oxidation of ammonium to nitrate favors ^{14}N, resulting in ^{15}N enrichment of the residual ammonium (Figure 8.5). The $\delta^{15}N$ value of the initially produced nitrate should be depleted by about 30‰ below $\delta^{15}N_{NH_4^+}$, but both $\delta^{15}N_{NO_3^-}$ and $\delta^{15}N_{NH_4^+}$ will increase as the NH_4^+ reservoir is converted toNO_3^-, with $\delta^{15}N_{NO_3^-}$ evolving toward the initial $\delta^{15}N_{NH_4^+}$ value (Wassenaar 1995). The nitrification of ammonium

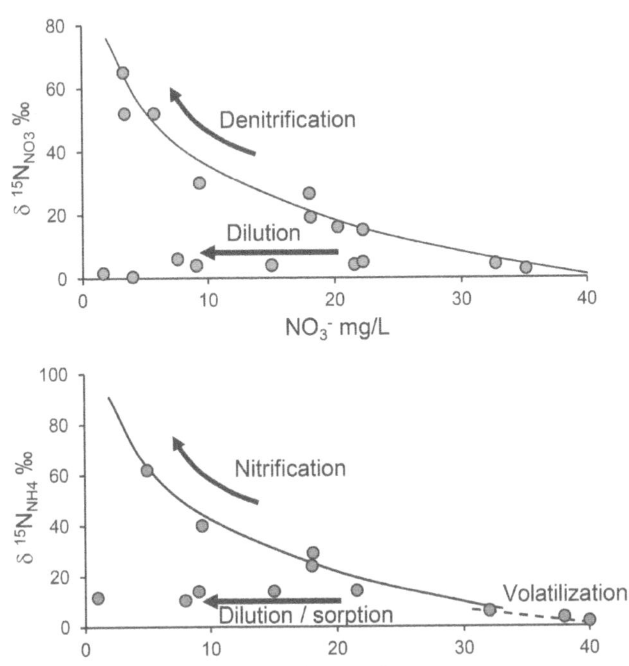

FIGURE 8.5 Trends for 15N during nitrate and ammonium attenuation by reaction and by dilution. Sorption of NH_4^+ can have a minor enrichment but is seldom observed above background variability.

can be traced with $\delta^{18}O$, which differentiates the resulting NO_3^- from synthetic KNO_3 fertilizer. Nitrification generally takes one oxygen from the air ($\delta^{18}O_{O_2}$ = 23‰) and two oxygen atoms from the ambient soil water (Kendall and Aravena 2000; Anderson and Hooper 1983). For example, water with $\delta^{18}O$ of −10 ±2% would yield a value for $\delta^{18}O_{NO_3^-}$ between −1% and 2%. This contrasts with synthetic KNO_3, which is produced with only atmospheric O_2 and so has $\delta^{18}O_{KNO_3}$ between approximately 20‰ and 25‰ (Figure 8.5). As nitrate does not readily exchange ^{18}O with water, this signal can only be modified by denitrification or anammox reaction.

Denitrification, NO_3^- to N_2, $\varepsilon^{15}N_{NO_3^- - N_2} \approx$ 25‰ to 30‰, and $\varepsilon^{18}O_{NO_3^- - N_2} \approx$ 10‰ to 15‰ (Kendall and Aravena 2000). Denitrification is a strongly fractionating process that results in an increase in $\delta^{15}N$ and $\delta^{18}O$ in the residual (unreacted) nitrate in the groundwater (Figure 8.5). This strong enrichment is characteristic of this reaction, and is used to unambiguously distinguish denitrification (reactive loss of NO_3^-) from simple dilution by dispersion in groundwater, which is nonfractionating.

Anammox $NH_4^+ + NO_3^-$ to N_2, $\varepsilon^{15}N_{NH_4^+ - N_2} \approx$ 4‰ (Clark et al. 2008). Few studies have looked at isotope fractionation during the anammox reaction. As in nitrification and denitrification, the residual ammonium and nitrate in the groundwater will have more positive $\delta^{15}N$ values. Robertson et al. (2012) observed an isotopic enrichment factor from 4.3‰ to 7.7‰ in a field septic system study. The low fractionation in the field studies likely relates to the greater reservoir of ammonium sorbed on the aquifer that buffers the enrichment of $\delta^{15}N_{NH_4^+}$ in dissolved NH_4^+.

Nitrogen in Agricultural Settings

Rudolph et al. (1998) show in a province-wide study that over 25% of groundwater in agricultural areas had NO_3^- levels above drinking water standards, largely attributable to contributions from field applications of manure. Rural regions are also dependent on septic tanks and tile fields for waste disposal. In both settings, concerns include not only NO_3^- and NH_4^+ sources and transformations but also the potential for the transport of pathogenic bacteria into aquifers. A surface-derived source is evident from the general decrease with depth from denitrification. Rudolph et al. (1998) show NO_3^- concentrations up to three and four times the drinking water limit of 10 mg N/L in the upper 2 m below the water table, decreasing to less than 20 mg N/L by 8 m depth. Ammonium concentrations are typically much less due to nitrification in agricultural soils and to sorption onto the soil and aquifer matrix.

Potassium nitrate fertilizer is derived from NH_3 produced by the Haber–Bosch process using atmospheric N_2 ($\delta^{15}N$ = 0‰) and oxidized by atmospheric oxygen ($\delta^{18}O$ = 23‰) (Figure 8.6). Urea, $CO(NH_2)_2$, is also produced from atmospheric N_2, but mineralization involves oxygen from the soil water, and so the resulting NO_3^- has a more depleted $\delta^{18}O$ value. Subsequent attenuation by dilution during recharge will not affect these signatures, whereas losses through denitrification or anammox will. The associated fractionation and Rayleigh distillation enrich the residual NO_3^- in both isotopes (Figure 8.6).

The Abbotsford aquifer is a transboundary aquifer on the Fraser River in British Columbia and Washington state that has chronic nitrate contamination, with levels in

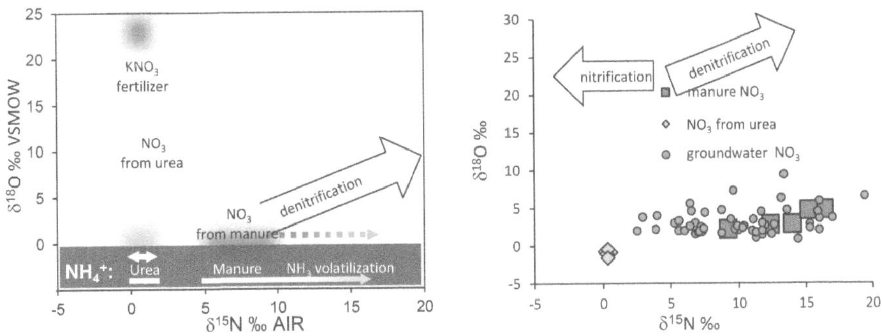

FIGURE 8.6 Left: Sources and approximate isotopic composition of NO_3^- ($\delta^{15}N$ and $\delta^{18}O$ in the upper part of the chart, in white) and NH_4^+ ($\delta^{15}N$ alone in the lower part of the chart, in black) in agricultural settings. Right: Case study—Abbotsford aquifer, British Columbia. (After Wassenaar 1995.)

water wells routinely exceeding drinking water guidelines. The source was linked by Wassenaar (1995) to poultry manure stockpiles on the highly permeable gravel sediments. In Figure 8.6 (right-hand chart), the distribution of $\delta^{15}N$ and $\delta^{18}O$ data for NO_3^- shows a clear source in the manure, with a trend of ^{15}N depletion due to nitrification. No contribution from nitrification of urea, which is used on many farms, is observed. Further, the absence of any enrichment trend for $\delta^{15}N$ and $\delta^{18}O$ shows that there is little to no natural attenuation by denitrification, which was anticipated due to the paucity of organic carbon and highly oxidizing redox conditions (Wassenaar 1995). This conclusion led to new best management practices adopted for the handling and storage of manure.

Septic Effluent Infiltration

Infiltration from septic tile beds in rural areas represents a significant input of NO_3^- to groundwater. Unlike the spatially extensive source area of agricultural fields, these are point sources. Loadings of nitrogen species can be significant, with concentrations well in excess of drinking water standards, such as observed at the seasonal use campground at Long Point, Ontario (Robertson et al. 2012). Here, multilevel groundwater samplers were used to monitor concentrations (Figure 8.7) and ^{15}N (Figure 8.8) for NH_4^+ and NO_3^- during the infiltration of a plume from the May 24 opening weekend. Concentrations together with ^{15}N data were used to demonstrate three zones of reaction:

Top of plume—nitrification of the NH_4^+ to NO_3^-
Core of plume—anammox oxidation of NH_4^+ using NO_3^- to produce N_2
Below plume—denitrification of NO_3^- to N_2

DNA evidence was then used to identify the strains of anammox bacteria involved in the attenuation of these nitrogen species in the core of the plume.

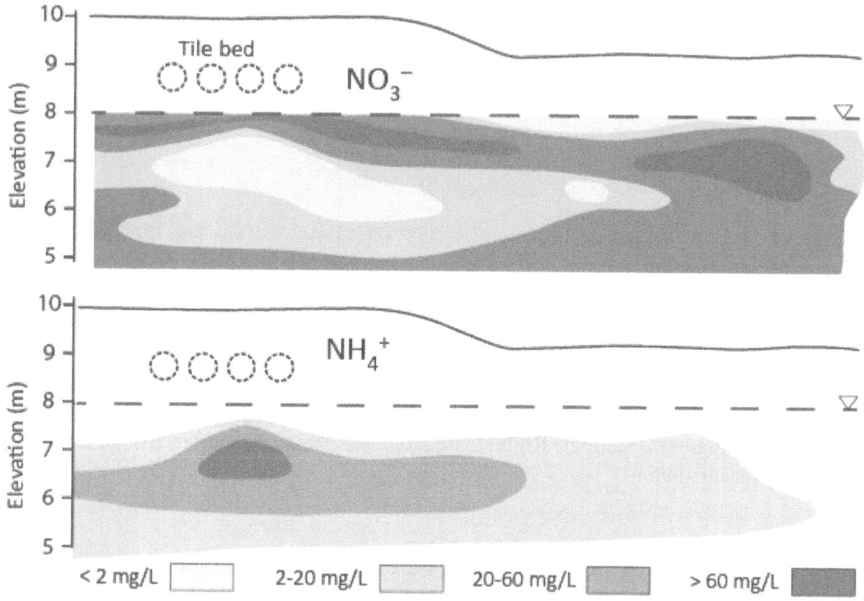

FIGURE 8.7 NO_3^- and NH_4^+ plumes beneath a septic effluent tile bed. (Modified from Robertson et al. 2012.) The length of the section is 30 m.

Permeable Reactive Barriers for NO_3^-

Attenuation of nitrate through denitrification and simple dilution are not effective strategies for this pervasive contaminant in groundwater, particularly where impacts on local drinking water wells and surface waters are at risk. Robertson and Cherry (1995) were the first to install permeable barriers of reactive media to passively treat groundwater contaminants. In this case, they used sawdust as an electron donor to stimulate denitrification.

Robertson et al. (2000) describe installations at three sites, where waste cellulose from wood chips, sawdust, and leaf litter comprised the carbon source for denitrification in groundwaters flowing through the porous material. Losses of NO_3^- were greatest for the highest initial concentrations, with denitrification of up to 90% for nitrate starting at 60 ppm N and over 50% for initial NO_3^- levels of less than 5 ppm N.

These permeable contaminant barriers were installed as horizontal beds in the vadose zone beneath septic tank infiltration and as vertical walls or trenches intercepting horizontally-flowing groundwater with nitrate plumes. Since this time, permeable reactive barriers (PRBs) have been installed to treat a wide range of contaminants in groundwater including fuels, chlorinated solvents and acid mine drainage (AMD; see later sections).

Boron as a Tracer of Nitrogen from Wastewater

In addition to its potential impact on water quality, boron can also serve as a tracer for sources of wastewater contamination. Natural boron concentrations in

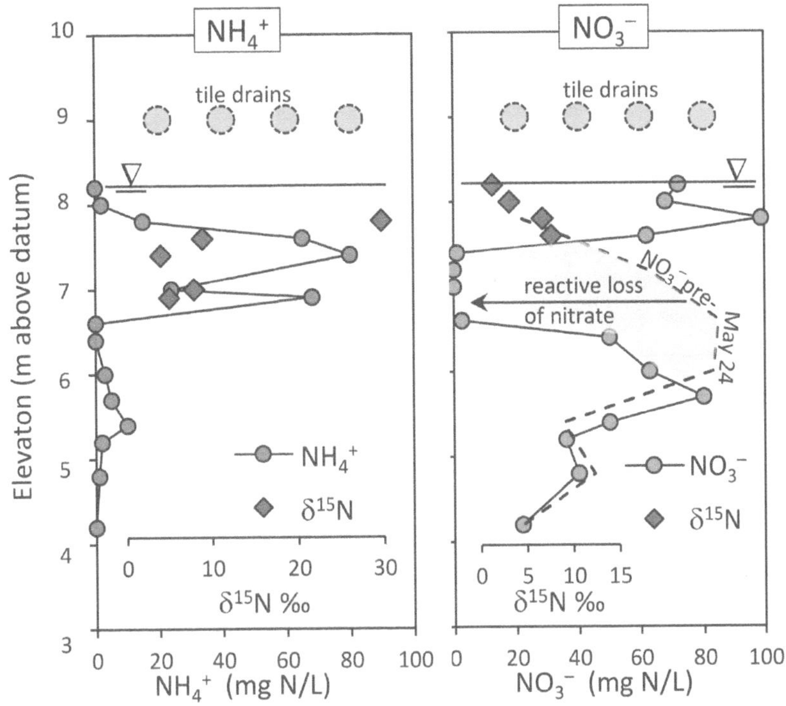

FIGURE 8.8 Concentrations and $\delta^{15}N$ for NO_3^- and NH_4^+ vertically through the plume at a location beneath the tile bed, showing downward movement of the May 24 NH_4^+ plume. (Ater Robertson et al. 2012.) The concentration shows the NH_4^+ plume at the 7–8 m elevation and with nitrate occurring at the top of this plume but dropping off with depth. The $\delta^{15}N_{NO_3^-}$ is depleted at the top of the plume, while $\delta^{15}N_{NH_4^+}$ is enriched, signifying nitrification. Deeper within the NH_4^+ plume there is reactive loss of nitrate and lower NH_4^+, attributed to anammox reaction. Below the plume, only NO_3^- remains, which is attenuated by denitrification. Monitoring of this plume over time shows nitrogen loss of over 80% by these reactions.

groundwater are typically only a few ppb but are significantly enriched in liquid manure and septic effluents. Natural B in groundwater can be derived from mixing with seawater (B = 4.5 ppm) or from the weathering of sandstones and igneous rocks. Boron is also found naturally in certain evaporites as borax ($Na_2B_4O_5[OH]_4{\cdot}8H_2O$), which is mined for use in detergents due to its strong complexing ability as well as other industrial uses. In water, B^{III} dominates as neutral trigonal borate $B(OH)_3^{\circ}$ but hydroxylates to tetrahedral $B(OH)_4^-$ above pH 8 with a dissociation constant of $10^{-9.24}$. Transformations between these two species serve to fractionate its two stable isotopes, ^{10}B and ^{11}B, and provide a tool for tracing B and its associated compounds.

Figure 8.9 shows the ranges for the principal natural and contaminant sources of B. Boron from cleaning agents in wastewater systems has $\delta^{11}B$ values close to the boron standard NBS951 with values between 10‰ and −10‰, whereas fertilizers from animal manure are typically enriched with values from 10‰ to as high

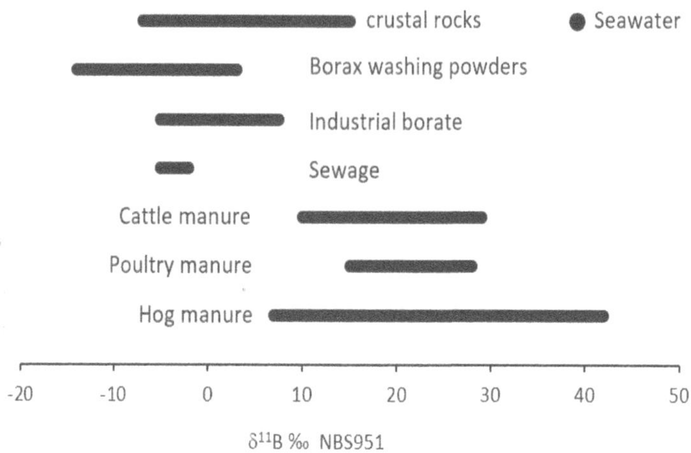

FIGURE 8.9 Ranges for $\delta^{11}B$ in sources of B. (Data sources: Widory et al. 2004, 2013; Kloppmann et al. 2009.)

as 40‰, which is the value for modern marine borate. Widory et al. (2004, 2012) use $\delta^{11}B$ in conjunction with isotopes of nitrate to distinguish nitrate contamination by manure applications from sewage inputs (Figure 8.9). Sorption along groundwater flow paths can enrich the ^{11}B when the pH is above pH 8, when the anion $B(OH)_4^-$ becomes important. In this case, ^{10}B is preferentially sorbed, which imparts an enrichment of ^{11}B in solution. However, Kloppmann et al. (2009) show that B transport is essentially nonfractionating at neutral pH, where $B(OH)_3^{\circ}$ dominates and is therefore a reliable tracer of source.

Industrial Sources of Ammonium

Sources and degradation pathways of industrial contamination by nitrogen species in the Elmira aquifer in southern Ontario were traced with $\delta^{15}N$ (Clark et al. 2008). At this site, remediation of the aquifer for chlorobenzene by pump and treat was complicated by ammonium contamination from chemical wastewater infiltration ponds and infiltration of both NH_4^+ and NO_3^- from adjacent fertilizer-blending activities (Figure 8.10).

Apportioning the sources of ammonium based on $\delta^{15}N_{NH_4^+}$ was complicated by reaction in the subsurface, which modified the source $\delta^{15}N$ values. This is shown in Figure 8.10, where NH_4^+ samples from the chemical wastewater ponds and from the fertilizer blending operations are distinctly different. However, NH_4^+ sampled along the plume from the fertilizer-blending site became progressively enriched in ^{15}N, plotting outside the mixing envelope, and indicating reactive loss of ammonium. Sampling in the subsurface showed that both NH_4^+ and NO_3^- decreased together along the groundwater plume from the fertilizer-blending site, while $\delta^{15}N_{NH_4^+}$, $\delta^{15}N_{NO_3^-}$, and

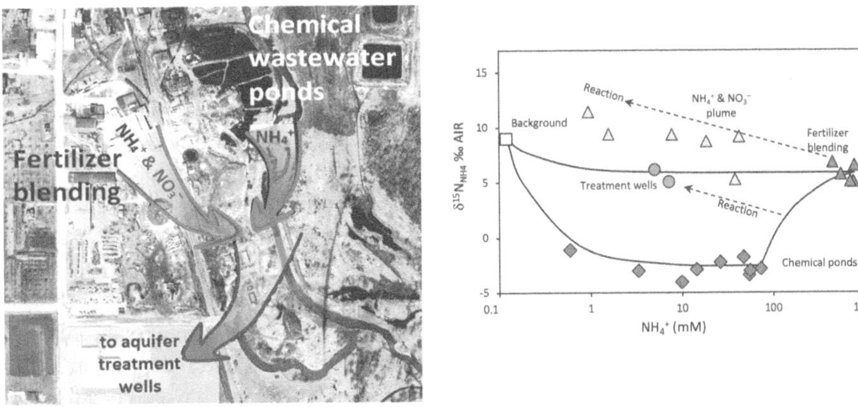

FIGURE 8.10 Industrial site showing sources of ammonium that have impacted the municipal aquifer at Elmira, Ontario. Solid arrows show flow through the surficial aquifer. Lower arrow shows recharge to the deep aquifer and the pathway for nitrogen species. Three-component mixing diagram for NH_4^+ and $\delta^{15}N_{NH_4^+}$ in groundwaters in the source areas and in the deep aquifer. (Modified from Clark et al. 2008.)

N_2 partial pressure increased, as shown in Figure 8.11. Anammox was concluded, a reaction that was previously unknown in groundwater.

Molecular techniques to amplify and clone DNA in samples from this site identified several anammox strains, including *Candidatus Brocadia* and *Candidatus Scalindua* (Moore et al. 2011). Apportionment calculations concluded that the treatment wells were impacted by NH_4^+ from both sources and that the $\delta^{15}N_{NH_4^+}$ measured in the treatment wells had been enriched from an intermediate value by anammox activity.

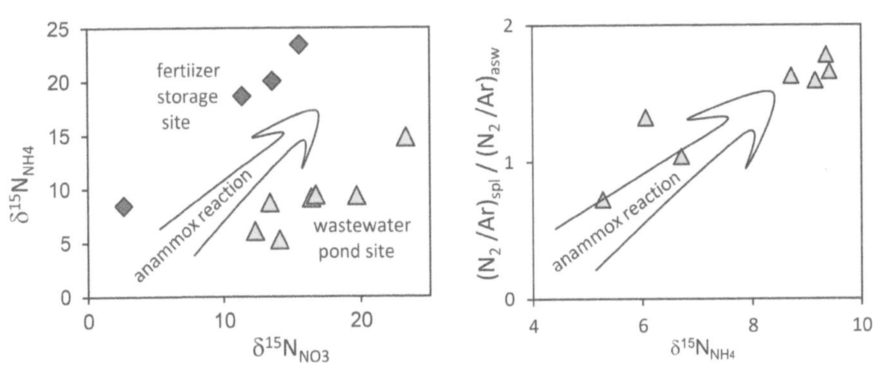

FIGURE 8.11 Left: Enrichment in both $\delta^{15}N_{NH_4^+}$ and $\delta^{15}N_{NO_3^-}$ during the anammox reaction of NH_4^+ and NO_3^-. Right: N_2 overpressuring due to anammox activity, measured as N_2/Ar normalized to the N_2/Ar ratio in air-saturated water.

Atmospheric Sources of NO_3^- and ClO_4^-

The brine basins, or *salars*, of the Atacama Desert in northern Chile host rich nitrate deposits that were exploited for decades in the 19th century for agriculture and for explosives. Similar accumulations are found in Death Valley in the Mojave Desert in California. Although their accumulation as salts in these hyperarid closed basins was not surprising, the origin of the nitrogen remained a mystery. More recently, these deposits were found to contain significant concentrations of another strong electron acceptor—perchlorate, ClO_4^-. In perchlorate the chlorine atom is oxidized to Cl^{7+}, making it a highly reactive oxidant that is manufactured for solid rocket boosters and fireworks. It is also toxic to humans where it disrupts the uptake of iodine by the thyroid. The extensive use of Atacama nitrate in the past and industrial perchlorate sources today contribute ClO_4^- to groundwater, which can have concentrations well above regulatory limits of 1–18 ppb (Sturchio et al. 2009). Desert nitrates were believed to originate through leaching of soils and accumulation in arid basins by evaporation until analysis of their stable isotope composition demonstrated an atmospheric source (Böhlke et al. 1997).

The oxidation of atmospheric N_2 by lightning produces NO and NO_2, which oxidize to nitrate with atmospheric ozone according to the following reactions,

$$N_2 + O_2 \rightarrow 2NO$$
$$NO + O_3 \rightarrow NO_2 + O_2$$
$$2NO_2 + O_3 + H_2O \rightarrow 2HNO_3 + O_2$$

Nitrate then accumulates on the surface by dry fallout of nitric acid aerosols or is washed out by precipitation. Price et al. (1997) determined that lightning over the continents contributes some 12 million tons of nitrate to the surface annually, which represents almost 10% of the natural nitrogen fixation that takes place through nitrogenase enzyme activity in the roots of certain plants, such as clover and legumes. In contrast, the Haber–Bosch process produces some 160 million tons of nitrogen fertilizers annually (FAO 2008). Nonetheless, fallout of atmospheric nitrate can accumulate as surface salts in arid regions with closed basins.

Böhlke et al. (1997) use both $\delta^{15}N$ and $\delta^{18}O$ of nitrate to show the involvement of atmospheric N_2 and O_2, showing $\delta^{15}N$ values for the Atacama NO_3^- clustering around 0‰ and $\delta^{18}O$ even greater than the atmospheric value of 23‰. This is diagnostic, considering that bacterially mediated nitrate involves oxygen from local groundwaters and so has $\delta^{18}O$ near −10‰. However, the involvement of O_3 allows another isotope to be used—$\delta^{17}O$. Oxygen-17, present as only 0.038% of all oxygen, fractionates the same as ^{18}O, but with about half of the mass effect. As it brings no additional information, it has not been of interest in hydrological studies. However, recently it has been shown that production of atmospheric ozone involves reaction steps where the input of atmospheric oxygen occurs with little to no mass fractionation, resulting in an enrichment of ^{17}O in O_3. This mass-independent fractionation is observed in atmospheric nitrate, which involves O_3 during transformation of NO_x to HNO_3, during the reactions above (Figure 8.12).

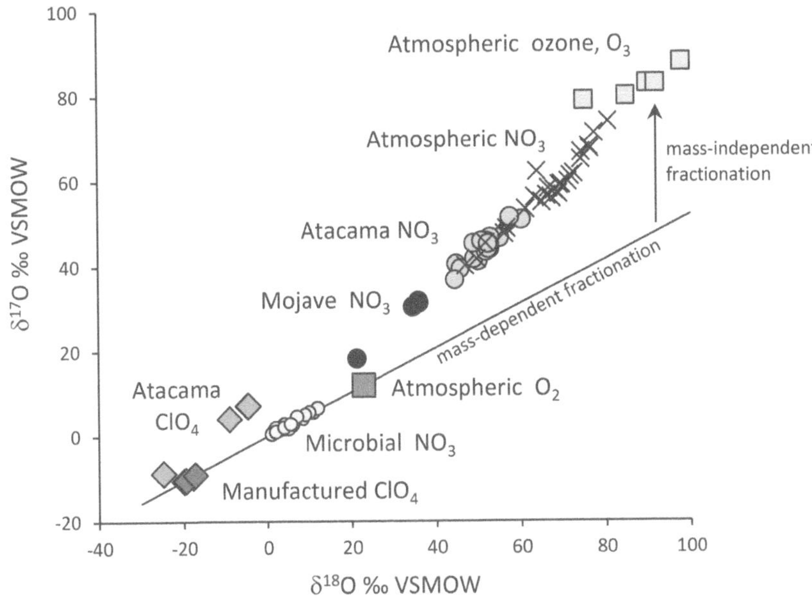

FIGURE 8.12 The oxygen isotope contents of atmospheric nitrate and perchlorate, showing a $\delta^{17}O$ anomaly that indicates formation by atmospheric ozone. Atmospheric NO_3^- from Proemse et al. (2013); Ozone, Atacama, and Mojave NO_3^- from Michalski et al. 2004); Microbial NO_3^- from Kaiser et al. (2007); Atacama and manufactured ClO_4^- from Bao and Gu (2004).

The production and fallout of atmospheric perchlorate is believed to occur through the photo-oxidation of atmospheric chloride and chlorate ions during the polar sunrise in high latitudes, but in mid and low latitudes, the pathway is unclear. Bao and Gu (2004) use $\delta^{17}O$ together with $\delta^{18}O$ to show that the perchlorate associated with the atmospheric nitrate deposits of the Atacama Desert is derived in part from atmospheric ozone and not from industrial sources. The origin of perchlorate can be demonstrated by isotopes of Cl as well. Sturchio et al. (2009) use low concentrations of ^{36}Cl (recall from dating in Chapter 8) together with $\delta^{37}Cl$ close to 0‰ as evidence for ClO_4^- contamination in many groundwaters in the United States from manufactured perchlorate.

ORGANIC CARBON COMPOUNDS

Organic debris accumulating in marine sediments over Phanerozoic time, through burial and heating, has been transformed to hydrocarbon, with a high-energy yield (e.g., 817.78 kJ/mol for combustion of methane). Two-thirds of what we exploit of this resource goes to transportation as gasoline and diesel for internal combustion engines, and paraffins for jet engines with the balance being used for a range of industrial compounds, solvents, insulators, and polymers. Their inevitable spills and leaks in the environment have spawned research on their mobility and degradation

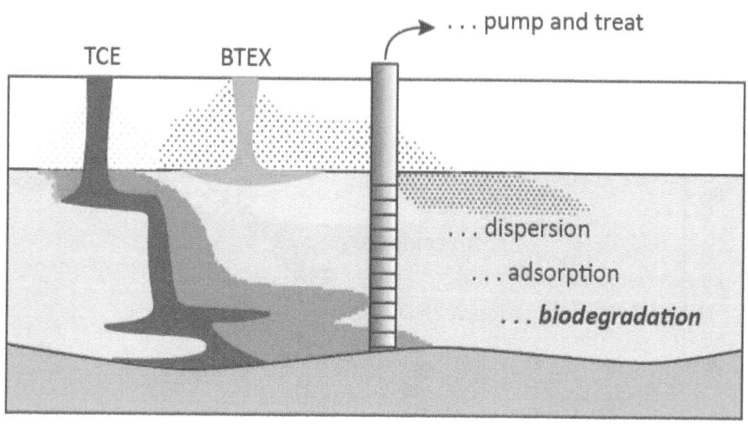

FIGURE 8.13 Pathway for hydrocarbons (e.g., gasoline composed of benzene, toluene, ethylbenzene and xylene [BTEX], with density less than that of water) and chlorinated organics (e.g., trichloroethylene [TCE], with density greater than water) into groundwaters and possible attenuation/mitigation processes.

(Figure 8.13). This section presents the use of geochemical and isotopic tools to trace contamination and biodegradation.

Crude oil contains a range of aliphatic (chain-structured hydrocarbon) and aromatic (benzene-ring structures) compounds, most with functional groups attached, including compounds such as methyl $–CH_3$, hydroxyl –OH, carbonyl –CO, carboxyl –COOH, amine $–NH_2$, and sulfanyl –SH. Refined petroleum products generally fall into two groups (Figure 8.14). BTEX is composed of about 11% benzene and the balance as *alkyl*benzene compounds—benzene with an alkane group such as a methyl group (CH_3)—making toluene and xylene, or an ethyl group, $–CH_2–CH_3$, making ethylbenzene. Carbon in the benzene ring is triple-bonded with its two neighbors, with redox state C^{-I}. In the methyl group, the carbon is C^{-III}, providing more energy

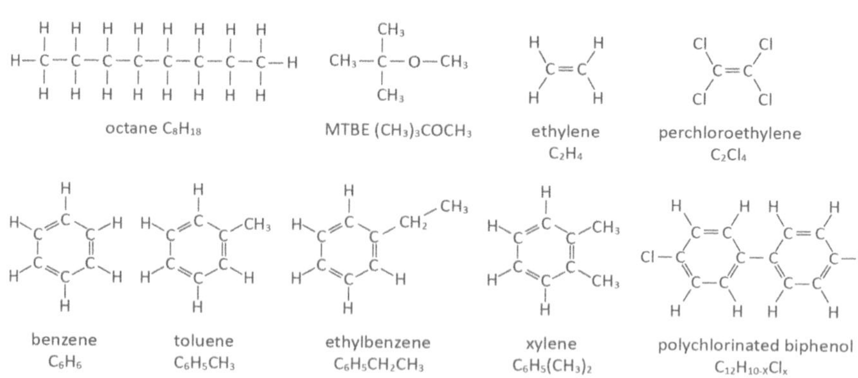

FIGURE 8.14 Aliphatic (upper row) and aromatic (lower row) hydrocarbon and chlorinated hydrocarbon compounds commonly found as contaminants in groundwater.

during combustion. Alkanes (or paraffins) are single-bonded carbon chains (double-bonded aliphatic HCs are alkenes, and triple-bonded are alkynes). They have the highest energy density because the linked carbons, $-CH_2-$, are in the C^{-II} state, with the methyl groups at the ends in the C^{-III} state. The simplest, of course, is methane, with C^{-IV} covalently bonded with its four H^+ atoms.

Of particular concern in hydrogeology are the organohalogens—chlorinated or brominated organic compounds. These dense hydrophobic organic liquids can sink through aquifers, leaving a residual phase that can sustain a chronic contamination of groundwater over many decades. A range of compounds exists, including polychlorinated biphenols used as electrical insulators and chlorinated alkenes such as perchloroethylene (PCE), used as solvents for dry cleaning and degreasers in industry (Figure 8.14). Organobromine compounds such as polybrominated diphenyl ether (PBDE) used as a fire retardant can also be of concern.

BTEX now contains up to 10% ethanol, C_2H_5OH, as an additive to reduce CO_2 emissions. The addition of ethanol to fuel reduces its energy density (lower mileage), and the roughly equal cost in fossil fuel for production and distribution makes it CO_2 neutral. As an oxygenate it offers an alternative to methyl tert-butyl ether (MTBE) to reduce CO emission, although it is found to increase other emissions and poses a greater hazard to water quality than MTBE (Williams et al. 2003).

Diesel fuel (natural petrodiesel) is largely alkane molecules with about 25% aromatics, including alkylbenzenes, which have a high ignition temperature achieved by high compression rather than spark ignition. Biodiesel is similar but uses vegetable oil with long-chain fatty acids (soluble aliphatic chains of single carbons with a carboxyl group, O–C–OH) like alkanes but transformed with methanol, C_2H_5OH, to a fatty acid methyl ester (FAME).

The solubility of these compounds in groundwater varies greatly from a few milligrams per liter for the aliphatic components to several grams per liter for the hydrophilic oxygenates (Table 8.1). The concentration of these fuel components and additives in groundwater depends not only on their pure phase solubility, S_{pure}, but also on their mass fraction of the product, f_m, that is leaking from a storage tank or a spill. This is the effective solubility, S_{eff}, and is established by Raoult's law:

$$S_{eff} = S_{pure} \times f_m$$

For example, most gasoline (BTEX) has about 11% benzene and 10% MTBE. If the tank beneath the filling station is leaking into the groundwater, then the concentration of these two components would be as follows:

$$\begin{aligned} S_{eff\,(benzene)} &= S_{benzene} \times f_{benzene} & S_{eff\,(MTBE)} &= S_{MTBE} \times f_{MTBE} \\ &= 1{,}780 \times 0.11 & &= 48{,}000 \times 0.10 \\ &= 196\,\text{mg/L} & &= 4{,}800\,\text{mg/L} \end{aligned}$$

Measurement of concentrations near these values in groundwater would be evidence that there is a nearby free-product source that is supporting such high concentrations.

TABLE 8.1
Solubilities (U.S. Environmental Protection Agency [EPA]) and Henry's Law Constants[a] (Sander 1999) for Major Fuel Components and Solvents

Compound	K_H (M/atm)	Solubility in Water (mg/L) at 25°C
Benzene	0.18	1780
Toluene	0.15	535
Xylene	0.12–0.19	130–196
Diesel[b]	0.0002–0.0008	40
MTBE	1.6	48,000
Catechol	4600	430,000
Ethanol[c]	200	miscible
PCE	0.0054	200
TCE	0.097	1100
cis-DCE	0.24	3500
Trichloromethane	0.25	4000

[a] Note that Henry's law constants for volatile compounds are often expressed as the volatility constant, $K_{H\text{-inv}}$, which is the inverse of the more familiar gas solubility constant, K_H (mol/L/atm, such that $K_{H\text{-inv}} = 1/K_H$ Units of $K_{H\text{-inv}}$ are atm/mol/L or atm × L/mol, but are often seen as atm × m³/mol, where K_H (mol/L/atm) = $1/(K_{H\text{-inv}} \cdot 1000)$ (atm × m³/mol).

[b] Grade 1-D for cars, trucks, and boats containing kerosenes (alkanes—paraffins and cycloparaffins).

[c] Ethanol is the dominant water-soluble component of biofuels, including biodiesel. Other components of biodiesel include long-chain FAMEs that are of very low solubility in water.

Biodegradation of Hydrocarbons

Hydrocarbon solubility is unique to each compound. Because of their high volatility, they have a Henry's law constant that defines the solubility of the vapor phase (Table 8.1). Accordingly, hydrocarbons can be dissolved in groundwater from a vapor phase present within the unsaturated zone or from a liquid phase pooled on the water table or residually coating grains in the unsaturated zone. Remediation options range from excavation or pumping to in situ degradation that may involve injection of oxidants or passive biodegradation. Biodegradation of hydrocarbons can follow a number of pathways that are typically mediated by aerobic, denitrifying, and sulfate-reducing bacteria or methanogenic archaea. These involve multistep enzymatic reactions, resulting ultimately in the production of CO_2 accompanied under anaerobic conditions by N_2, H_2S, and CH_4.

Benzene and alkylbenzene components of BTEX are sparingly soluble in groundwater (Table 8.1). Under oxidizing conditions, BTEX follows a multistep oxidation, as described by Yu et al. (2001), that increases both their solubility and their bioavailability for further degradation. Benzene, for example, is partially hydrolyzed through an initial step where carbon atoms are oxidized from a C^{-I} state to a C^{+I} state, producing highly soluble dihydroxybenzene or catechol compounds (Table 8.1). This increases the oxygen content of the product, which becomes available for further oxidation in groundwaters.

$$C_6H_6 + O_2 \rightarrow C_6H_6O_2 \text{ (hydroxylbenzene or catechol)}$$

Further oxidation ultimately produces CO_2:

$$C_6H_6O_2 + 6\tfrac{1}{3}O_2 \rightarrow 6CO_2 + 3H_2O$$

With an overall reaction and energy yield of

$$C_6H_6 + 7\tfrac{1}{2}O_2 \rightarrow 6CO_2 + 3H_2O \quad \Delta G^\circ = -3202.32\,\text{kJ/mol}$$

However, groundwaters with high organic carbon loadings can quickly become anoxic, making redox reactions operating lower on the electromotive potential scale more competitive. Unlike the formation of catechols during aerobic oxidation of BTEX, there is no evidence for significant generation of extracellular intermediates under anaerobic conditions. Burland and Edwards (1999) show the degradation of ^{14}C-labeled benzene by nitrate reduction in microcosm studies using soils from a contaminated aquifer. The stoichiometry of the reaction (six moles NO_3^- per mole of C_6H_6) is seen in the experimental measurements showing the loss of benzene tracking the loss of nitrate (Figure 8.15).

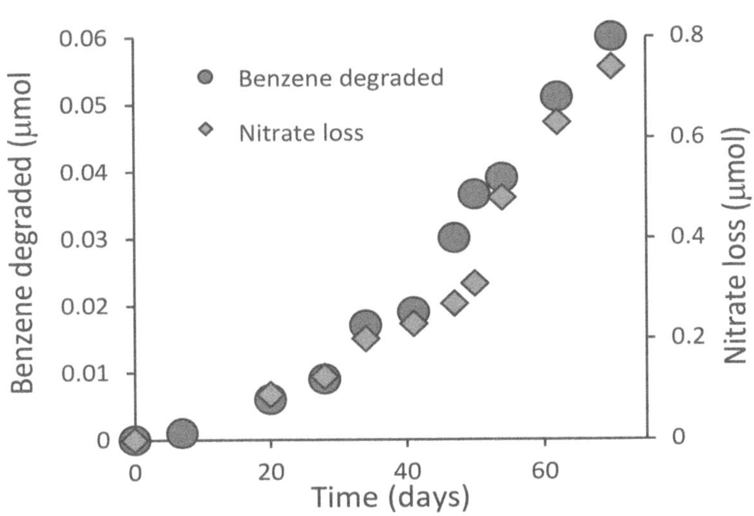

FIGURE 8.15 Biodegradation of benzene by denitrification showing loss of both benzene and nitrate over time in microcosm studies. ^{14}C-labeled benzene was used, and the CO_2 reaction product was sampled. Up to 95% of the ^{14}C activity was recovered as CO_2, with the balance attributed to cell growth. (Modified from Burland and Edwards 1999.)

$$C_6H_6 + 6NO_3^- \rightarrow 6HCO_3^- + 3N_2 \quad \Delta G^\circ = -2992.86\,\text{kJ/mol}$$

As observed above for nitrate and ammonium reactions, the loss of organic carbon by degradation can be distinguished from simple dilution in the field using isotopes. Recall in Chapter 4 that isotopes undergo a Rayleigh-type enrichment or depletion as a reservoir of the reactant is depleted. This Rayleigh enrichment was observed for the biodegradation of ethylbenzene in microcosm studies using nitrate as the electron acceptor (Meckenstock et al. 2004). The stoichiometry of the oxidation of ethylbenzene is not as simple as that of benzene (above), but also yields bicarbonate and nitrogen gas:

$$C_6H_5CH_2CH_3 + 8\tfrac{2}{5}NO_3^- + \tfrac{2}{5}H^+ \rightarrow 8HCO_3^- + 4\tfrac{1}{5}N_2 + 1\tfrac{1}{5}H_2O$$

Nonetheless, the reaction produced ^{13}C enrichments of up to 8‰ in the residual ethylbenzene after 50 days (Figure 8.16).

Further down the electromotive potential scale, benzene degradation can take place by ferric iron reduction, which has an energy yield still in the same range as for aerobic and denitrifying degradation, and by sulfate reduction, which has a significantly lower energy yield per mole of benzene:

$$C_6H_6 + 30Fe(OH)_{3\,fer} + 54H^+ \rightarrow 6HCO_3^- + 30Fe^{2+} + 72H_2O$$
$$\Delta G^\circ = -2443.68\,\text{kJ/mol}$$
$$C_6H_6 + 3\tfrac{3}{4}SO_4^{2-} + 3H_2O + 1\tfrac{1}{2}H^+ \rightarrow 6HCO_3^- + 3\tfrac{3}{4}H_2S$$
$$\Delta G^\circ = -248.24\,\text{kJ/mol}$$

Ferrihydrite as an electron donor is required in high quantities, as it involves only a single electron transfer, making this a poor remediation option. Sulfate reduction,

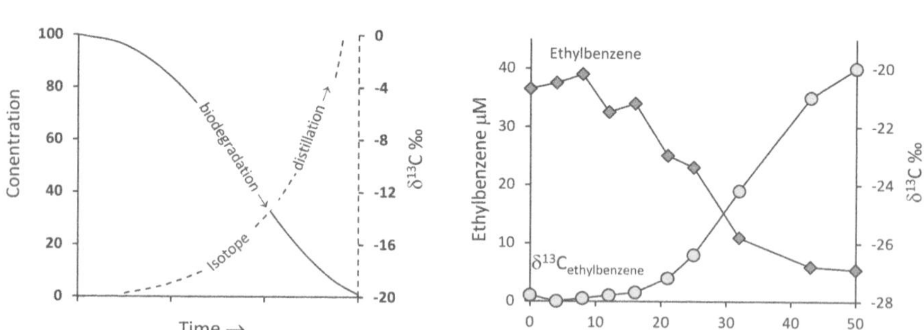

FIGURE 8.16 Isotope enrichment by Rayleigh distillation during biodegradation, theoretical (left) and observed (right) for ^{13}C in ethylbenzene during biodegradation by denitrification. (After Meckenstock et al. 2004.)

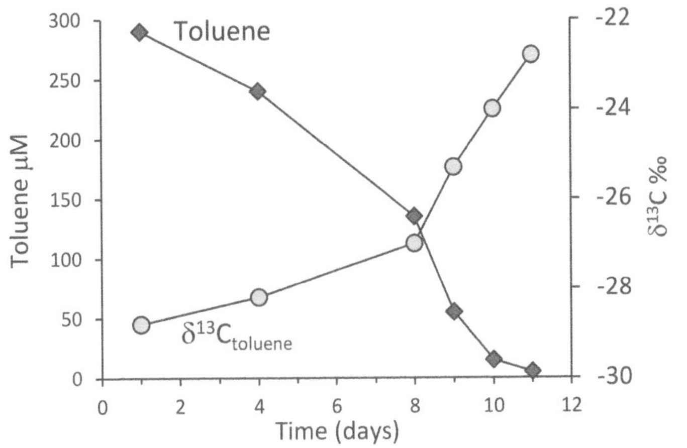

FIGURE 8.17 Biodegradation of toluene by sulfate reduction in a microcosm, with $\delta^{13}C$ enrichment in the residual toluene. (After Meckenstock et al. 2004.)

however, involves a six-electron transfer and, as sulfate solutions, can be injected into aquifers; this represents a viable in situ treatment option. Meckenstock et al. (2004) monitored H_2S and toluene together with analysis of $\delta^{13}C$ in residual toluene (compound-specific isotope analysis [CSIA]) to demonstrate biodegradation by sulfate reduction (Figure 8.17). By the end of the 11-day experiment, 300 µmol of toluene had been consumed and 1100 µmol of sulfide produced. This gives a stoichiometric ratio of just under 4 mol of sulfate per mole of toluene, which approximates the reaction above for benzene.

Neither of these reactions is very favorable from an environmental perspective, given the high production of ferrous iron or hydrogen sulfide. Further, these electron acceptors need to be plentiful in the aquifer for any significant amount of BTEX degradation to take place. In their absence, there is a rapid jump to methanogenic conditions for the degradation of these compounds. Although the energy yield is low, autoclastic methanogenesis is one of the most common pathways for the degradation of aromatics and aliphatic hydrocarbons:

$$C_6H_6 + 6\tfrac{3}{4}H_2O \rightarrow 3\tfrac{3}{4}CH_4 + 2\tfrac{1}{4}HCO_3^- + 2\tfrac{1}{4}H^+ \quad \Delta G^{\circ} = -34.23\,\text{kJ/mol}$$

If a substrate of hydrogen gas is available from other fermentative reactions in the aquifer, then it can act as an electron donor for benzene reduction to methane:

$$C_6H_6 + 9H_2 \rightarrow 6CH_4 \quad \Delta G^{\circ} = -428.7\,\text{kJ/mol}$$

In a series of microcosm experiments under anaerobic conditions, Mancini et al. (2003) demonstrate that very strong Rayleigh enrichments are associated with this low-electromotive potential pathway for benzene biodegradation. Figure 8.18 is

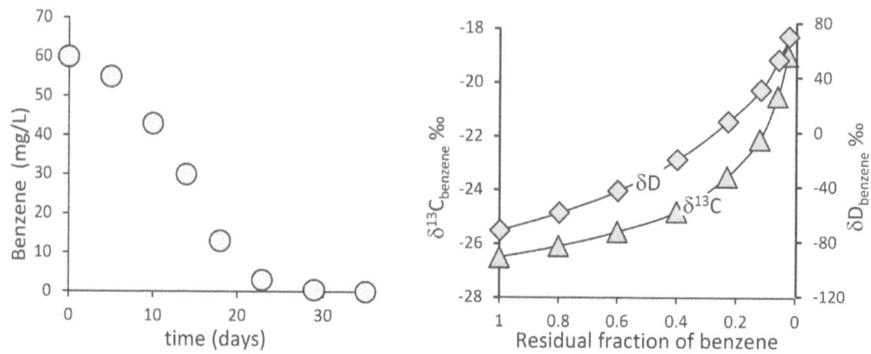

FIGURE 8.18 Compound-specific isotope analysis of benzene during methanogenic biodegradation, showing a decline in benzene to less than detection over a 20-day period that is associated with enrichments in both ^{13}C and D of the residual benzene. (After Mancini et al. 2003.)

adapted from their full data sets to show these enrichments. From these isotope values, the enrichment factor for the reaction can be calculated. Recall from Chapter 4 the formulation of a Rayleigh equation in permil units.

$$\delta_{\text{reactant}} = \delta_{\text{initial reactant}} + \varepsilon_{\text{product-reactant}} \times \ln f_{\text{residual reactant}}$$

Rearranging this equation and using the values given in Table 8.2 for the isotope enrichment trends in Figure 8.18 allows calculation of the enrichment factor for the biodegradation of benzene to CH_4 and dissolved inorganic carbon (DIC):

$$\varepsilon^{13}C_{\text{CH}_4\text{/DIC-benzene}} = \frac{\left(\delta^{13}C_{\text{benzene-initial}} - \delta^{13}C_{\text{benzene-residual}}\right)}{\ln f_{\text{residual benzene}}}$$

TABLE 8.2
Isotope Enrichment Factors, $\varepsilon^{13}C$ and εD, for Methanogenic Biodegradation of Benzene

$f_{\text{residual benzene}}$	$\delta^{13}C_{\text{benzene}}$	$\varepsilon^{13}C_{\text{benzene-CH}_4\text{/DIC}}$	$\delta D_{\text{benzene}}$	$\varepsilon D_{\text{benzene-CH}_4\text{/DIC}}$
1	−26.5		−72	
0.6	−25.6	1.9	−43	57
0.4	−24.8	1.9	−20	57
0.12	−22.1	1.9	31	49
0.025	−9.0	1.9	70	38

Source: Mancini et al. 2003.

Degradation of Fuel Oxygenates: MTBE and Ethanol

The addition of oxygenates to fuel is designed to reduce emissions of CO and ground-level ozone. However, their high solubility (Table 8.1) gives these minor components of BTEX a disproportionately high concentration in groundwaters impacted by leaking storage tanks and spills. Ethanol can be readily attenuated by both aerobic and anaerobic biodegradation, while MTBE is often found to be more recalcitrant (Church et al. 1997). Both are mineralized under oxidizing conditions:

$$C_2H_5OH_{ethanol} + 3O_2 \rightarrow 2CO_2 + 3H_2O \qquad \Delta G^\circ = -1325.32\,\text{kJ/mol}$$
$$(CH_3)_3\,COCH_{3MTBE} + 7\tfrac{1}{2}O_2 \rightarrow 5CO_2 + 6H_2O \quad \Delta G^\circ = -3158.5\,\text{kJ/mol}$$

Under anaerobic conditions, biodegradation follows the electrochemical chain of electron acceptors from nitrate to sulfate (Finneran and Lovley 2001):

$$(CH_3)_3\,COCH_{3MTBE} + 6NO_3^- + H^+ \rightarrow 5HCO_3^- + N_2 + 4H_2O$$
$$(CH_3)_3\,COCH_{3MTBE} + 30Fe(OH)_3 + 55H^+ \rightarrow 5HCO_3^- + 30Fe^{2+} + 76H_2O$$
$$(CH_3)_3\,COCH_{3MTBE} + 3\tfrac{3}{4}SO_4^{2-} + 2\tfrac{1}{2}H^+ \rightarrow 5HCO_3^- + 3\tfrac{3}{4}H_2S + H_2O$$

Under methanogenic conditions, evidence suggests that the degradation includes the formation of the intermediary compound *tert*-butyl alcohol (TBA).

$$(CH_3)_3COCH_{3\,MTBE} + H_2 \rightarrow (CH_3)_3COH + CH_4$$
$$(CH_3)_3COH + H_2O \rightarrow 3CH_4 + CO_2$$

Giving an overall reaction as

$$(CH_3)_3COCH_{3MTBE} + 2\tfrac{3}{4}H_2O \rightarrow 3\tfrac{3}{4}CH_4 + 1\tfrac{1}{4}HCO_3^- + 1\tfrac{1}{4}H^+$$

Microbial degradation of MTBE in groundwaters has been recognized as a favorable treatment technology, but recognizing biodegradation from that of dispersive attenuation has been difficult at the field scale. Thornton et al. (2011) used both ^{14}C-labeled MTBE and natural abundance ^{13}C to trace its degradation under both aerobic and anaerobic conditions. Their radiolabeled MTBE was recovered as $^{14}CO_2$ along the aquifer, demonstrating active biodegradation. However, although a 7‰ enrichment was observed in highly aerobic microcosm experiments, their field study found no significant ^{13}C enrichment in MTBE along the groundwater flow path.

Under highly reducing conditions for methanogenic degradation of MTBE where the intermediary TBA is produced, ^{13}C does become a good tool to track the amount of degradation (Figure 8.19).

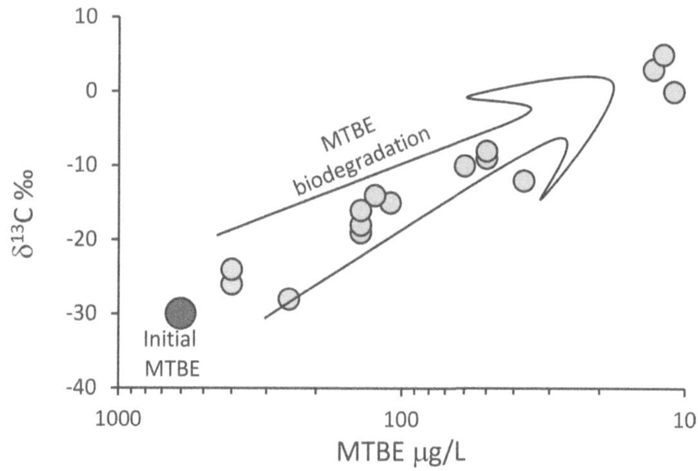

FIGURE 8.19 The Rayleigh enrichment of ^{13}C during the biodegradation of methyl tert-butyl ether (MTBE) to *tert*-butyl alcohol (TBA) and methane in well-stratified sand to clay sediments. Initial value for $\delta^{13}C_{MTBE}$ in gasoline is close to −30‰. (Modified from Kolhatkar et al. 2002.)

Biodegradation of Organohalogens

Highly chlorinated compounds are found to degrade slowly in aerobic groundwaters, whereas under anaerobic conditions, reductive dechlorination can reduce such compounds to more readily oxidized forms (Fennell et al. 1997; Hunkeler et al. 1999). In such reactions, hydrogen atoms are replaced by chlorine atoms carrying a +I valence state for coordination with carbon. PCE is a common example. Perc has been widely used as a solvent in dry cleaning, leading to widespread contamination arising from poor disposal practices. Under anaerobic conditions, it can be sequentially dechlorinated through a series of reductive reaction steps that produce vinyl chloride (VC) and then ethene.

$$C_2Cl_{4\,PCE} + 2e^- \rightarrow C_2HCl_{3\,TCE} + Cl^-$$

$$\hookrightarrow + 2e^- \rightarrow cis-C_2H_2Cl_{2\,DCE} + Cl^-$$

$$\hookrightarrow + 2e^- \rightarrow C_2H_3Cl_{VC} + Cl^-$$

$$\hookrightarrow + 2e^- \rightarrow C_2H_{4\,ethene} + Cl^-$$

Reductive dechlorination typically requires some organic substrate as the electron donor, which is often provided by labile organic compounds that may accompany such spills or occur naturally in the aquifer. Tracing these reactions in aquifers by measurement of these reaction products can demonstrate degradation of these compounds and the viability of in situ biodegradation as a remediation technique. However, when the source of such chlorinated organics is poorly characterized, evidence for active degradation may be uncertain. As seen above for BTEX degradation, CSIA provides unambiguous evidence for reactive degradation.

Hunkeler et al. (1999) used compound specific isotope analysis of TCE and its degradation compounds to demonstrate microbial degradation in both microcosms and contaminated groundwaters. Noteworthy in the microcosm experiments were the strong isotope enrichments for *cis*-DCE (dichloroethylene), VC, and ethane (Figure 8.20). These enrichments are due to a combination of two processes. First, the precursors PCE and TCE experience strong enrichments due to Rayleigh distillation during reaction to *cis*-DCE. These enrichments are passed onto the *cis*-DCE and VC as they accumulate. Second, as the *cis*-DCE are dechlorinated, both experience a Rayleigh enrichment, resulting in very substantial enrichments. Note that the accumulation of *cis*-DCE and its degradation products, VC and ethane, have a higher aqueous concentration than the initial PCE. This is due to desorption of the added PCE from the aquifer material as PCE reaction begins. About half of the PCE added to the microcosm at the outset was sorbed onto the solid material.

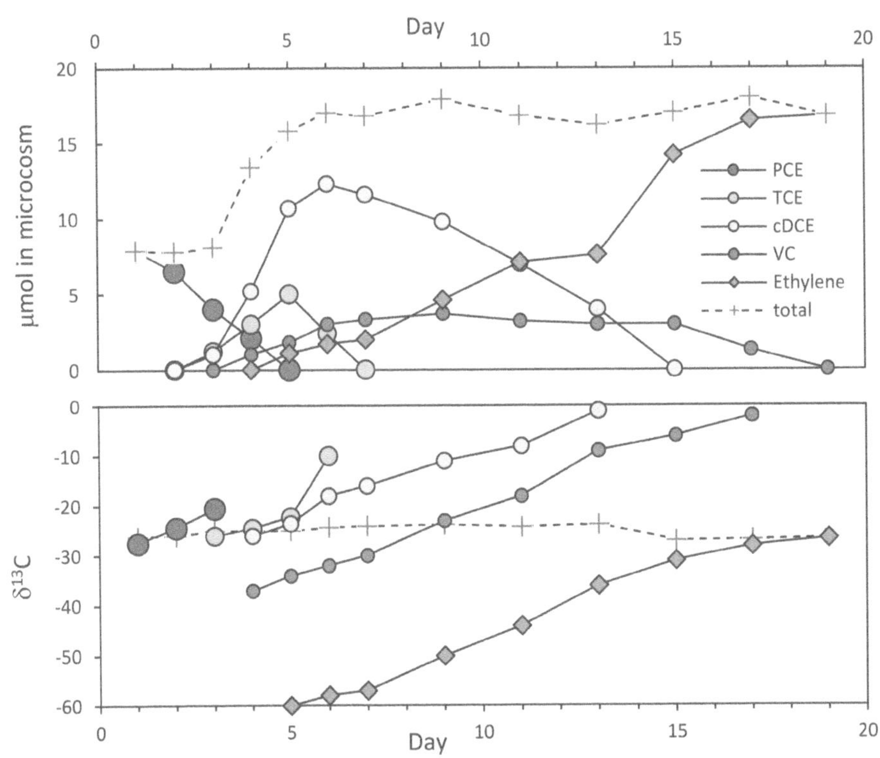

FIGURE 8.20 Concentration and $\delta^{13}C$ of perchloroethylene (PCE) and its degradation products over a 20-day microcosm experiment, and the $\delta^{13}C$ of PCE and degradation products in anaerobic groundwaters from a sandy aquifer. (After Hunkeler et al. 1999.)

The diagnostic capability of stable isotopes for such reactions was then applied in a sandy silt aquifer that had been contaminated in the past by PCE. Here, this study finds the same enrichments in the product compounds, providing unequivocal evidence for bioremediation of this groundwater.

Abiotic Degradation of Organochlorine Compounds in Permeable Reactive Barriers

Management of groundwaters contaminated by chlorinated organic carbon compounds by pump-and-treat methods, usually involving sorption on reactive carbon, is recognized as expensive remediation in perpetuity (Mackay and Cherry 1989). In situ degradation of contaminants, often by injecting nutrients and bacterial strains to stimulate biological activity, faces problems of aquifer heterogeneity. Research over the past two decades has focused on the inorganic degradation of chlorinated carbon compounds such as PCE and tetrachloromethane using zero-valence (metallic) iron installed as a geochemically reactive barrier to intercept the contaminant plume in the aquifer (Gillham and O'Hannesin 1992, 1994; Gillham et al. 1993). In such reductive dechlorination reactions, the covalently bonded Cl^+ is reduced to Cl^- using Fe^0 as the electron donor. Multiple dechlorination steps take place simultaneously or in rapid succession, with intermediate degradation products present at only low residual concentrations. Dissociation of water accompanies reductive dechlorination, providing H^+ to replace the Cl^+ and forming ethylene or methane.

$$C_2Cl_{4\,PCE} + 4Fe^0 + 4H_2O \rightarrow C_2H_4 + 4Cl^- + 4OH^- + 4Fe^{2+}$$

$$CCl_{4\,CT} + 4Fe^0 + 4H_2O \rightarrow CH_4 + 4Cl^- + 4OH^- + 4Fe^{2+}$$

The hydrocarbon products of reductive dechlorination are considerably more benign than the chlorinated compounds and more susceptible to aerobic or anaerobic biodegradation in the aquifer.

Following the 1991 pilot-scale installation at Base Borden in southern Ontario (O'Hannesin and Gillham 1998), PRBs with zero-valence iron have been used worldwide. Henderson and Demond (2007) report that the few underperforming installations were more likely to be compromised by poor hydraulic characterization of the aquifer than by performance of the reactive barrier itself.

Fugitive Gases

The increased production of natural gas from shales by high-pressure fracturing, or fracking, has heightened concern for impacts on shallow groundwater resources. Dissolved methane in groundwater becomes an explosion hazard as it degases in domestic water distribution systems that are under lower pressure than in the aquifer. However, methane can be produced naturally by microbial methanogenesis in shallow groundwaters, making it difficult to associate methane found in water supplies with shale-gas activities. Aravena et al. (1995) provide a good example of the former, where groundwaters in the Alliston aquifer of Southern Ontario have significant methane accumulations that pose a hazard to rural residents. Osborn et al. (2011) provide a good example of the latter, where high methane in domestic wells is attributed to leaking gas wells. In both cases, the stable isotope composition of methane was used to identify the source of methane. However, in addition to the $\delta^{13}C$ and δD composition of methane, the $C_1/(C_2 + C_3)$ ratio, radiocarbon and He data can discriminate between sources.

Biogenic methane tends to be very depleted in ^{13}C due to preferential selection of ^{12}C during methanogenesis. Depending on the pathway, $\delta^{13}C_{CH_4}$ can range from −50‰ to less than −100‰. Values for δD_{CH_4} are even more depleted, with a range from about −150‰ to less than −350‰.

$$\begin{array}{ll} \textit{Acetatoclastic} & CH_3COOH \rightarrow CH_4 + CO_2 \\ \textit{CO}_2\textit{ reduction} & CO_2 + 4H_2 \rightarrow CH_4 + 2H_2O \\ \textit{Thermocatalyitic} & \text{Kerogen} \rightarrow CH_4 + C_2H_6 + C_3H_8 + \cdots + CO_2 \end{array}$$

Thermocatalytic methane forms during the high-temperature decomposition of kerogen in marine sediments. Kerogen is the high-molecular-weight humic material generated during burial of organics (carbohydrate, proteins, etc.) with up to 1000 atomic mass units (daltons). As burial depth increases, kerogen cracks to lighter-weight hydrocarbons, losing oxygen, nitrogen, and sulfur from its structure in the process. The oil window for such reactions is generally from about 60°C to 150°C. Rapid burial of kerogen to greater depths brings it into the gas window, above about 150°C, favoring production of the lighter alkane hydrocarbons found in natural gas—methane (CH_4), ethane (C_2H_6), and propane (C_3H_8) plus minor heavier alkanes or condensates. These are the C_1 to C_3 alkanes. The composition of natural gas from different fields can be characterized by the molar ratio of methane to the higher-weight alkanes, determined as

$$\frac{CH_4}{C_2H_6 + C_3H_8} \text{ or } \frac{C_1}{C_2 + C_3}$$

In thermogenic natural gas, the $C_1/(C_2+C_3)$ ratio is generally less than 100 (Prinzhofer and Huc 1995). In contrast, archaeal activity in anaerobic waters, such as groundwaters or bogs, generates almost exclusively methane, and so the $C_1/(C_2+C_3)$ ratio for such gases is typically greater than 1000 (Whiticar et al. 1986).

As discussed in Chapter 6 (Redox Evolution), methane produced by archaeal reduction of CO_2 with H_2 tends to be more depleted in ^{13}C but enriched in D due to the incorporation of hydrogen from water. By comparison, acetoclastic archaea are less discriminating, producing methane that is more enriched in ^{13}C but more depleted in D because the hydrogen comes from the methyl group in the acetate (Figure 8.21). In contrast to archaeal methanogenesis, thermogenic methane is produced at high temperature by inorganic reaction (cracking) of higher-weight carbon compounds and so experiences less isotope fractionation. The $\delta^{13}C$ ranges from the value for marine kerogen, about −30‰, up to about −50‰.

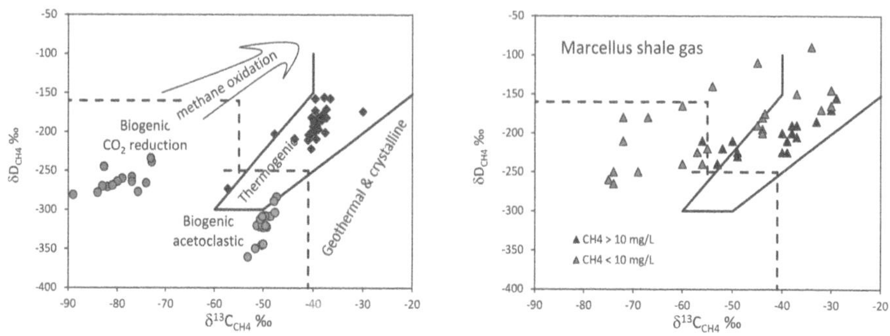

FIGURE 8.21 Ranges for stable isotope composition of methane from bacterial and thermogenic sources (based on Schoell 1980; Osborn et al. 2011). Left: Selection of biogenic (open discs) and thermogenic (closed diamonds) from Southern Ontario. (Data from Sherwood Lollar et al. 1994; Aravena et al. 1995; Clark et al. 2013.) Right: Methane from shallow groundwaters in the vicinity of the Marcellus shale gas field showing the high concentrations in water wells to be associated with production from the Marcellus shale. (Data from Osborn et al. 2011.)

The stable isotope values of methane in groundwater are indicative of its source. However, there can be some overlap due to mixing and methanotrophic activity such as aerobic methane oxidation or sulfate reduction:

$$CH_4 + 2O_2 \rightarrow HCO_3^- + H_2O + H^+ \qquad \Delta G^\circ = -773.24\,\text{kJ/mol}$$
$$CH_4 + SO_4^{2-} \rightarrow HCO_3^- + HS^- + H_2O \qquad \Delta G^\circ = -17.16\,\text{kJ/mol}$$

Both are bacterially mediated and so discriminate against ^{13}C and D, which imparts an enrichment on the residual methane. In the case of sulfate reduction, the residual SO_4^{2-} will also become enriched in ^{34}S. Van Stempvoort et al. (2005) observe depletion in $\delta^{13}C_{DIC}$ together with enrichment in both $\delta^{13}C_{CH_4}$ and $\delta^{34}S_{SO_4^{2-}}$ to demonstrate natural attenuation of fugitive methane from a leaking gas well in groundwaters from a site near Lloydminster, Alberta.

Radiocarbon is an additional tool that can often be used to discriminate between biogenic and thermogenic methane. Biogenic methane in shallow systems is typically generated by degradation of recently buried organics, which will contain ^{14}C. This is the case in the Alliston aquifer, which hosts buried peat horizons from the late Pleistocene and Holocene (Aravena et al. 1995), where measured ^{14}C activities of the methane in the overburden are up to 15 pMC (Figure 6.19). In contrast, thermogenic methane will have essentially no ^{14}C activity, as it is derived from old marine carbon.

Commercial natural gas sources are not exclusively derived from thermogenic sources. Coalbed methane is generated by bacterial activity within the formations, often stimulated by infiltration of fresh waters during the last deglaciation (McIntosh et al. 2002). Shale gas also has a high proportion generated by microbial activity (Curtis 2002; Martini et al. 2003), which precludes the use of $\delta^{13}C_{CH_4}$ and δD_{CH_4} as tools to distinguish natural shallow methane in groundwaters from fugitive gas from production wells. In such cases, radiocarbon cannot be used. As before, unlike shallow biogenic methane sources, which have modern radiocarbon activities (Nakagawa and Yoshida 2002; Chanton et al. 1995), methane from coalbeds and marine carbon substrates is radiocarbon-free. Even when the circulating waters contain ^{14}C-active DIC that can contribute to methane production by CO_2 reduction, this component is typically diluted below detection by the in situ degradation of the alkane precursors that produce both CH_4 and CO_2 (Jones et al. 2008).

Fugitive CO_2 from deep injection for carbon capture and storage (CCS) can be traced with radiocarbon. Although the $\delta^{13}C$ of such fossil-fuel-derived CO_2 should be highly depleted, near –30‰, reaction with carbonate in the subsurface can shift this value into the range commonly found for DIC in shallow groundwaters and soils (–20 to –15‰), and so $\delta^{13}C$ alone is not diagnostic. However, shallow DIC in young groundwaters will have measurable radiocarbon, while CO_2 from CCS will be radiocarbon-free. The $\delta^{13}C$ of CO_2 from CCS would also differ from that of ^{14}C-free geogenic CO_2 originating in metamorphic and volcanogenic terrains, which are generally close to 0‰.

Finally, radiogenic ^4He is a diagnostic tracer of fugitive gases of deep thermogenic origin. Geogenic CH_4 is accompanied by ^4He in a ratio that is characteristic of the source formation and will differ from shallow biogenic CH_4. Jackson et al. (2013) show that drinking water wells nearest the Marcellus shale-gas field (methane data in Figure 8.21) have $^4He/CH_4$ ratios distinctly lower than that of other potential methane sources.

LANDFILL LEACHATE

Modern landfill design and construction rely upon engineered barriers of clay and geopolymers at the base, together with internal systems to collect leachate and gases and often a cap to reduce infiltration and leachate production. Leachates can then be treated, often in municipal treatment plants, before discharge. Older landfills, by contrast, lack these infrastructures and so represent a legacy threat to groundwater resources that must be monitored and remediated. Understanding the biodegradation processes during leachate production and interaction with shallow groundwaters is fundamental to assessing impact and planning remediation.

Landfill Waste

The composition of municipal waste is dominated by organic refuse, categorized by the United States Environmental Protection Agency (EPA) according to waste forms both before and after diversion for recycling and composting (Table 8.3).

Considering that legacy landfills operated largely prior to the increasingly aggressive municipal waste diversion programs of recent decades, their inventory can be considered to be over 70% organic biomass, which can biodegrade. The steel components of the metal stream can oxidize as well, while most plastics and glass will remain relatively inert on the time scales considered for degradation. On top of this is placed the daily cover material, usually derived from local sources of surficial materials, including sand, soils, and till.

Under humid conditions, the degradation reactions in this biomass-dominated inventory quickly evolve to methanogenesis. Eleazer et al. (1997) have documented

TABLE 8.3
Composition of Municipal Solid Waste in United States in 2011

	% of 250 Million Tons before Recycling and Composting	% of 164 Million Tons after Recycling and Composting
Paper and cardboard	28	14.8
Food waste	14.5	21.3
Yard trimmings	13.5	8.8
Plastics	12.7	17.8
Metals	8.8	8.8
Rubber, leather, textiles	8.2	10.6
Wood	6.4	8.4
Glass	4.6	5.1

Source: EPA, *Municipal Solid Waste Generation, Recycling, and Disposal in the United States: Facts and Figures for 2011*, EPA, Washington, DC, 2013.

with microcosm experiments the methane production that can be achieved from the various organic components. CH_4 and CO_2 are the two most significant gases produced.

Biodegradation and Landfill Leachate

Following disposal and compaction, landfilled material typically begins to saturate with meteoric waters and oxygen, leading to a series of aerobic and then anaerobic degradation reactions.

Salt mobilization	$NaCl \rightarrow Na^{+} + Cl^{-}$
Humification	$CH_2O + O_2 \rightarrow HA + FA + CO_2 + H_2O$
Iron oxidation	$Fe^{\circ} + \frac{1}{2}O_2 + H_2O \rightarrow Fe^{2+} + 2OH^{-}$
$CaCO_3$ weathering	$CaCO_3 + CO_2 + H_2O \rightarrow Ca^{2+} + 2HCO_3^{-}$
Silicate weathering	$CaAl_2Si_2O_8 + 3H_2O + 2CO_2 \rightarrow Ca^{2+} + 2HCO_3^{-} + Al_2Si_2O_5(OH)_4$
N_{org} mineralization	$R - NH_2 + H_2O + H^{+} \rightarrow NH_4^{+} + R - OH$
Iron reduction	$4Fe(OH)_3 + CH_2O + 7H^{+} \rightarrow HCO_2^{-} + 4Fe^{2+} + 10H_2O$
Sulfate reduction	$CH_3COOH + SO_4^{2-} \rightarrow H_2S + 2HCO_3^{-}$
Iron oxidation	$Fe^{\circ} + 2H_2O \rightarrow Fe^{2+} + H_2 + 2OH^{-}$
Fermentation	$C_6H_{12}O_6 \rightarrow CH_3CH_2CH_2COOH + 2CO_2 + 2H_2$
Acetogenesis	$CH_3CH_2COOH + 2H_2O \rightarrow CH_3COOH + CO_2 + 3H_2$
Methanogenesis	$CH_3COOH \rightarrow CH_4 + CO_2$
CO_2 reduction	$CO_2 + 4H_2 \rightarrow CH_4 + 2H_2O$

This series of reactions shows an evolution from the oxidizing conditions at the surface of the landfill, where rainfall infiltrates with oxygen and begins the humification and mineral weathering processes, to anaerobic conditions, where fermentative and methanogenic reactions proceed. Plasterboard or Gyproc® waste is not permitted in municipal landfills, as the gypsum [$CaSO_4 \cdot 2H_2O$] supports sulfate reduction to toxic H_2S. In addition to these primary products of aerobic and anaerobic biodegradation of the organic material, Christensen et al. (2001) note the trace (<ppm) concentration of heavy metals, Cd, Cr, Cu, Pb, Ni, and Zn, likely released during reduction of ferrihydrite and other oxyhydroxides, and the presence of xenoorganics, such as BTEX components, pesticides, and solvents.

The leachate generated by these reactions is represented for a series of landfills evaluated by Christensen et al. (2001), together with a specific example from the Trail Road Landfill, near Ottawa, Canada (Figure 8.22), which has a significant mineral component (total dissolved solids [TDS] > 10 g/L) and a pH of 7.4. The very high HCO_3^- is derived from both the aerobic and anaerobic biodegradation reactions and acetoclastic methanogenesis. Carbonic acid from CO_2 produced in these reactions is consumed by mineral weathering, contributing to Ca^{2+} and Mg^{2+} concentrations. $Na^+$and Cl^- are common in leachate from food waste. The redox-sensitive species all demonstrate an evolution driven by the high DOC. The system has exhausted most electron acceptors, such as NO_3^- and SO_4^{2-}, which are negligible in leachate. The high concentrations of Mn and Fe exceed the solubility limits for their oxidized species (Figure 6.14), and so have been reduced to Mn^{2+} andFe^{2+}. Similarly, the high concentration of sulfide and the presence of CH_4 indicate that the redox evolution has dropped to methanogenesis, which is the final step in the redox evolution driven by organic carbon.

Tracing Leachate Biodegradation with ^{13}C

The biodegradation reactions given earlier can be traced through analysis of isotopes in the different compounds of leachate. Notably, ^{13}C is useful to trace the role of DIC, bulk DOC, and specific DOC compounds during leachate generation. The bulk DOC from the Trail Road Landfill, shown in Figure 8.22, was separated into

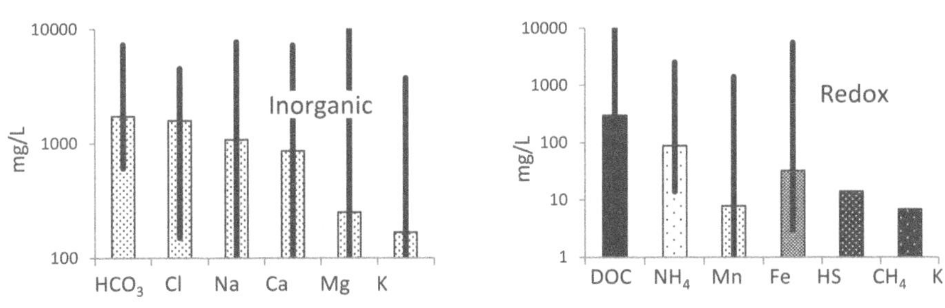

FIGURE 8.22 Geochemistry of municipal landfill leachate with an example (histograms) from the Trail Road Landfill, Ottawa, Canada (Mohammadzadeh and Clark 2008) and ranges (lines) for a series of landfills. (Data from Christensen et al. 2001.)

specific components for their concentration and $\delta^{13}C$ (Mohammadzadeh et al. 2005). Here the major compounds of the DOC show the pattern for ^{13}C partitioning during degradation and methanogenesis:

$$\begin{array}{ccccccccc} \text{Primary organics} & \rightarrow & \text{Humic acid} & \rightarrow & \text{Fulvic acid} & \rightarrow & \text{Acetate} & \rightarrow & CH_4 & + & CO_2 \\ -28.1‰ & & -28.1‰ & & -20.3‰ & & -12.0‰ & & -68.4‰ & & +10.7‰ \end{array}$$

Bacteria preferentially degrade the ^{12}C-fraction of a particular substrate (fulvic acid [FA], acetate) and discriminate against the ^{13}C-enriched fraction. The distribution of ^{13}C then reflects the rate of production and consumption of a given compound, according to the Rayleigh distillation function. As a particular substrate compound is consumed, the residual will become enriched in ^{13}C. The significant enrichments of ^{13}C in FA and acetate show that these components have been significantly consumed to produce methane and CO_2. The distribution of ^{13}C between these two terminal products of biodegradation in the leachate shows that a mass balance is preserved:

$$2CH_2O_{\text{primary organics}} \rightarrow CH_4 + CO_2$$
$$-28.1‰ \approx (-68.4 + 10.7)/2$$
$$\approx -28.8‰$$

Methanogenesis and δD

Another important isotope effect is observed for δD, which is enriched in leachate by about 15‰ (Figure 8.23) due to exchange during methanogenesis and through anaerobic oxidation of iron. In the following reactions, the production of H_2, and ultimately CH_4, is associated with strong fractionation for deuterium, which is

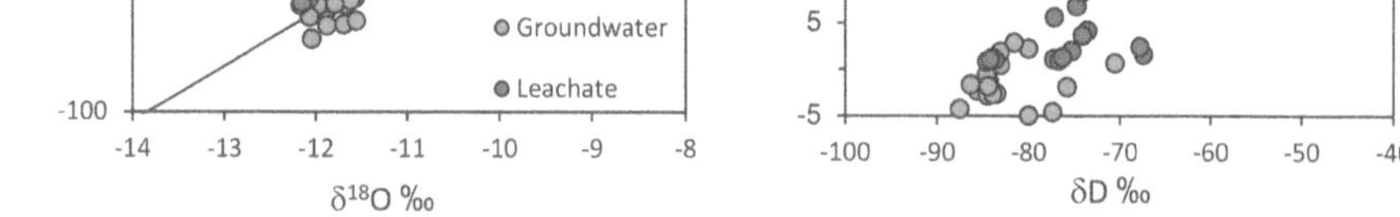

FIGURE 8.23 Deuterium enrichment in landfill leachate due to hydrogen exchange with water during H_2 generation for methanogenesis. The right-hand diagram shows the deuterium enrichment relative to the local meteoric water line, while the left shows the strong deuterium excess in the leachate waters (using Ottawa LMWL, $d_{Lc} = 7.5\ ^{18}O - 5.8‰$). Most groundwaters plot near the meteoric water line with d_{Lc} close to 0‰, whereas some exhibit minor deuterium enrichment due to leachate contamination.

partitioned into the water directly or through exchange with H^+ in the fatty acids (with deuterium, D noted):

$$C_6H_{11}DO_6 \rightarrow CH_3CH_2CH_2COO^- + D^+ + 2CO_2 + 2H_2 \quad \textit{Fermentation}$$

$$Fe^\circ + 2HDO \rightarrow Fe^{2+} + H_2 + 2OD^- \quad \textit{Iron reduction}$$

The fractionation factor $\alpha D_{H_2O-H_2} = 3.76$, and so hydrogen gas in equilibrium with water will be depleted in D by over 700‰ (recall from fractionation in Chapter 3). The fractionation factor for methane–water exchange, $\alpha D_{H_2O-CH_4} = 1.027$, is minor by comparison but would also act to partition D into the water. The deuterium depletion is transferred into methane, leaving a residual enrichment in the water. Recalling that Rayleigh distillation requires substantial reaction to have a measurable isotope shift, modification of δD_{H_2O} by the overall process of H_2 production and methanogenesis would then require production of several moles of methane per liter of leachate. Using an enrichment factor for deuterium fractionation between H_2 and H_2O of –700‰, the amount of methane produced per liter of leachate can be estimated from the deuterium excess in the leachate, according to methanogenesis from H_2 and CO_2:

$$d_{\text{Lc leachate}} - d_{\text{Lc precipitation}} = 10‰$$

Rayleigh distillation of D_{leachate}:

$$\delta D_f = \delta D_o + \varepsilon D_{H_2O-H_2} \times \ln f$$

$$10‰ = -700‰ \times \ln f$$

$$f = 0.986$$

Therefore, 0.014 L/L or 14 mL/L or 1.4% of the water has reacted. This amounts to 0.014 L × 55.6 mol H_2O/L = 0.8 mol H_2O. Considering the 2 H per H_2O and 4 H per CH_4, this would amount to a production of 0.4 moles or 9 L of $CH_{4(STP)}$ per liter of leachate.

These calculations are of course not well constrained but demonstrate that a tremendous production of gas is required to affect the isotopic composition of water and that the conditions within the landfill must remain unsaturated to have this volume of gas generated from this volume of water. It further shows that monitoring the D and ^{18}O contents of leachate could be used as a measure of methane production in the landfill.

ACID DRAINAGE

The exposure of pyrite and other base metal sulfides from crustal environments to the humid, aerobic exogene setting fosters a host of oxidation reactions. The energy released in converting sulfide to sulfate and ferrous iron to ferric iron is used by various bacteria to fix CO_2 for cell construction and for their fuel. In the process, metals and H^+ are released in abundance to the solution. The consequences for the immediate landscape as well as for local water bodies are tremendous, with dieback of vegetation and impacts on aquatic life (Figure 8.24).

FIGURE 8.24 Vegetation die back in a permafrost setting impacted by acid drainage from pyritic shales, Eagle Plains, Yukon. Inset shows an installation of a peeper-type water geochemistry sampler (Lacelle and Léveillé 2010).

Mitigation strategies are generally based on minimizing oxidation pathways. Engineered wetlands can buffer metal transport to runoff. PRBs can be designed to attenuate a particular geochemical parameter or solute. Finally, conversion of the source of AMD to anaerobic conditions by saturating them to exclude O_2 has become a part of remediation efforts in mine closure plans.

Generation of Acid Mine Drainage

Oxidation reactions generate acid. Oxidation of sulfide to sulfate involves the transfer of eight electrons and so is a particularly acidic reaction. These reactions have been presented in Chapter 7, on redox evolution. The chain of reactions begins as a near surface reaction by O_2 on sulfide minerals, usually ubiquitous pyrite found in abundant association with base metal sulfides. O_2 diffusion through water is slow and impedes the reaction, and so oxidation reactions are greatly favored by unsaturated conditions.

$$FeS_2 + 3\tfrac{1}{2}O_2 + H_2O \rightarrow Fe^{2+} + 2SO_4^{2-} + 2H^+$$

Thiobacillus thiooxidans

The ferrous iron released by sulfide oxidation is also oxidized by O_2, producing Fe^{3+} and ferrihydrite:

$$Fe^{2+} + \tfrac{1}{4}O_2 + 2\tfrac{1}{2}H_2O \rightarrow Fe(OH)_{3(fer)} + 2H^+$$

Thiobacillus ferrooxidans

The low solubility of ferric hydroxide or ferrihydrite [$Fe(OH)_{3(fer)}$] allows it to precipitate from solution at circumneutral pH. As acidity is generated, the pH of the aqueous setting drops and Fe^{3+} is released to the solution. Sulfide oxidation is then greatly accelerated by a second, abiotic oxidation pathway, where ferric and ferrous iron cycle under aerobic conditions:

$$FeS_2 + 14Fe^{3+} + 8H_2O \rightarrow 15Fe^{2+} + 2SO_4^{2-} + 16H^+$$

$$Fe^{2+} + \tfrac{1}{4}O_2 + H^+ \rightarrow Fe^{3+} + \tfrac{1}{2}H_2O$$

The result is a jump to pH levels as low as 1–2. Negative pH values have even been observed in some extreme cases. Sulfate concentrations increase from the ppm to the parts per thousand (permil) and even percent range. The transition metals, including Fe, Mn, Ni, Cu, and Zn, also become soluble at low pH and rise to high concentrations.

The high acidity of AMD accelerates the weathering of aluminosilicate minerals. However, unlike natural weathering reactions driven by carbonic acid at pH levels of 5–7, weathering by AMD precludes significant formation of kaolinite and other clays due to the high solubility of Al^{3+} at low pH.

Secondary Mineral Precipitates

The high acidity and elevated sulfate and metal concentrations in AMD evolve by mixing with natural watershed drainage. The most important initial reaction is dilution, which has an immediate effect on pH. However, each unit increase of pH requires a tenfold dilution of the acid drainage unless the receiving waters are highly buffered with bicarbonate.

An increase in pH affects pH-sensitive solutes, such as Fe and Al, which move into the stability fields of their less soluble hydroxides. The response is a precipitation of ferrihydrite and gibbsite. Ferrihydrite, with high surface area and charge, sorbs other metals from solution.

In the unsaturated environment of tailings and pyritic waste rock where AMD is generated, seasonal dry periods allow desiccation of acidic pore waters and the accumulation of sulfate salts. These evaporite minerals also form along the borders of AMD channels and surface water accumulations. However, rather than representing a mineralogical control or sink for metals and sulfate in the environment, they act only as a storage term in the watershed. Their dissolution following renewed precipitation is accompanied by high solute and acidity loading to runoff.

Soluble ferric iron and aluminum sulfates, such as jarosite [$KFe_3(SO_4)_2(OH)_6$], rhomboclase [$HFe(SO_4)_2{\cdot}4H_2O$], and alunogen [$Al_2(SO_4)_3{\cdot}17H_2O$], contribute high concentrations of SO_4^{2-} as well as Fe^{3+} and Al^{3+}. Hydration and dissociation of these trivalent cations also lower the pH. Rhomboclase contributes H^+ directly to the solution.

Remediation

A legacy of mining over the past 200 years can be observed as stockpiles of reactive waste rock and tailings sitting on the surface. Long-term solutions to the generation of AMD at these sites are expensive and focus on the isolation of the source materials under saturated conditions. This cuts the supply of oxygen and greatly reduces the rate of sulfide oxidation. Allowing vegetation to reestablish over the site as a wetland or soil provides the added security of an organic carbon loading, which reverses the oxidation reactions and raises pH. Bacterial iron and sulfate reduction during oxidation of organic carbon are key reactions. They can result in the neutralization of pH and the precipitation of sulfide minerals:

$$CH_2O + 4Fe(OH)_3 + 7H^+ \rightarrow 4Fe^{2+} + HCO_3^- + 10H_2O$$
$$2CH_2O + SO_4^{2-} + Fe^{2+} \rightarrow FeS_{mackinawite} + 2HCO_3^- + 2H^+$$

The reduction of ferric iron occurs under higher Eh conditions than sulfate reduction (see Chapter 7) and results in the mobility of iron and other metals, which may have been adsorbed onto ferric iron hydroxide precipitates in tailings or acidic waste rock. However, subsequent sulfate reduction will greatly lower their solubility as sulfide minerals. In neutral-pH waters, a black amorphous iron sulfide or mackinawite forms, whereas at lower pH, fine-grained biogenic pyrite [FeS_2] may also precipitate. The overall reaction is acid consuming, and the pH is buffered by the bicarbonate produced.

Engineered wetlands are not possible where topography, precipitation and bedrock drainage are constraints. Seasonal desiccation and remobilization of metal-laden dust is also a concern.

Short-term solutions include treating the effluent with lime [CaO] or hydrated (slaked) lime [$Ca(OH)_2$] or limestone [$CaCO_3$]. These buffer the acidity at neutral pH and precipitate most of the metals as hydroxides. However, this presents the problem of isolating the metal-laden precipitates so that subsequent leaching or wind erosion does not remobilize the metals.

$$CaO + H_2O + 2H^+ \rightarrow Ca^{2+} + 2H_2O$$
$$CaCO_3 + H^+ \rightarrow Ca^{2+} + HCO_3^-$$

The following two examples represent drainage from mine environments with differing levels of acidity. The Iron Mountain site in northern California represents perhaps the most acidic drainage ever recorded, with subsequent mineralization by a suite of unusual iron sulfate salts. By contrast, acid drainage from Nordic Mine in Elliot Lake, Ontario, shows the important role of carbonates for the attenuation of acid drainage.

Hyperacid Drainage at Iron Mountain, California

Mining of the Iron Mountain massive sulfide deposit in northern California began in the 1860s, with underground workings producing copper plus silver, gold, iron, and zinc in addition to sulfur for sulfuric acid. Mining activities ended in 1962, and remediation efforts were implemented following its designation as a Superfund environmental cleanup site. Until that time, several thousand tons of pyrite per year were being oxidized and contributing to acid drainage from the site. Along with the discharge of this acidity to the Sacramento River went hundreds of tons per year of dissolved Cd, Cu, and Zn.

The drainage from the old mines at Iron Mountain provides a good example of weathering, metal loading, and pH under extremely acidic conditions (Nordstrom and Alper 1999). At this site, phreatic drainage through fractures in the host rock and ore body (Figure 8.25) creates a highly oxygenated environment for pyrite oxidation. Heat liberated from these exothermic reactions raises the local groundwater temperatures to around 40°C, enhancing reaction. The sulfide oxidation reactions follow the reactions shown above, including the abiotic reaction based on reduction of Fe^{3+} to Fe^{2+}, which is immediately recycled by atmospheric oxygen back to Fe^{3+}. The net product is sulfuric acid, with few mineralogical controls under these extremely acidic conditions to limit the pyrite oxidation reaction.

The generation of sulfuric acid and dissolved iron from pyrite oxidation in this environment is illustrated in Figure 8.26. Using pH electrodes calibrated with sulfuric acid buffers, Nordstrom et al. (2000) measure pH values as low as –3.6, with sulfate concentrations at that pH of close to 800 g/L (800,000 mg/L) and total dissolved iron concentrations of 140 g/L (140,000 mg/L). The high sulfate and metal

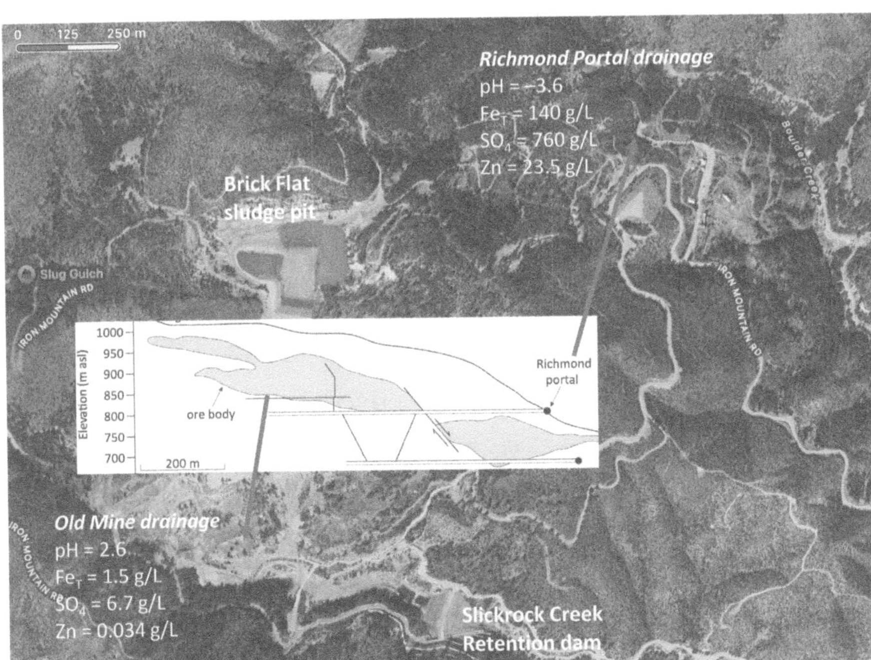

FIGURE 8.25 Iron Mountain acid mine drainage Superfund site location and cross section showing sites of drainage from Richmond mine portal. Host rock is fractured rhyolite encompassing the massive sulfide body. (Modified from Nordstrom and Alper 1999.)

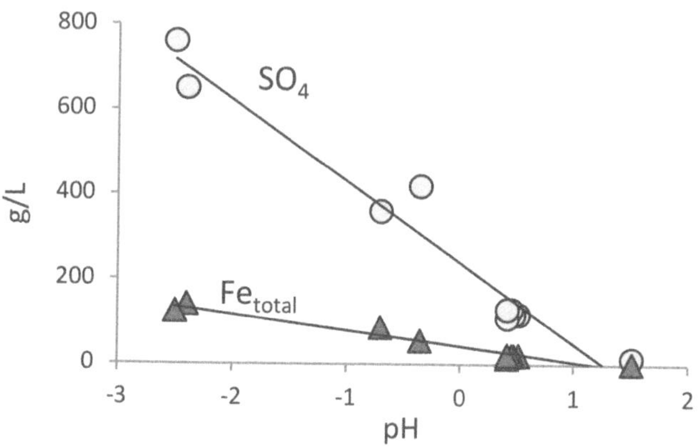

FIGURE 8.26 Plot of dissolved sulfate and iron in drainage from the Richmond Mine with measured pH. Negative pH values signify activities for H^+ that are greater than 1. (After Nordstrom et al. 2000.)

loadings give rise to some interesting and colorful mineralization. However, equally interesting are the negative pH values.

Because pH is a logarithmic expression, a pH value as low as –3.6 would suggest that, assuming $a_{H^+} \approx m_{H^+}$, some 390 kg of sulfuric acid was dissolved per kg of this AMD ($10^{-pH} = 3981\,\text{mol/kg}$), which is clearly wrong. To understand what these negative pH values represent, one needs to consider the activity–concentration relationship for hydrogen ions in these highly acidic conditions. To do this, these researchers calculate the hydrogen ion activities for sulfuric acid at different concentrations, using the Pitzer (1973) equations incorporated in PHRQPITZ (Plummer et al. 1988). These equations represent ion activities in strong electrolyte solutions, showing that the activity coefficient, γ_{H^+}, reverses exponentially from a fractional value less than 1 in lower-salinity solutions ($m_{H_2SO_4} < 1\,\text{mol/kg}$) to 10,000 in 10-molal solutions (Figure 8.27). For this reason, the measured H^+ activity is much greater than the associated concentration of H^+ ions based on the dissociation of the sulfuric acid at its given concentration ($m_{H_2SO_4}$):

$$H_2SO_4 \rightarrow H^+ + HSO_4^-$$

$$a_{H^+} = m_{H^+} \times \gamma_{H^+}$$

The sulfide mineralization at Iron Mountain is hosted by rhyolitic volcanic sequences dominated by albite plus phyllosilicates (chlinochlore and muscovite), quartz, and only minor calcite. The weathering of albite was shown in Chapter 5 to consume acidity and produce low-temperature clay minerals. However, under conditions of hyperacidity, these clays are no longer stable due to the high solubility of aluminum (Figure 2.3). Iron and other metals all become highly soluble at low pH, where hydroxide species ($Al(OH)_3$, $Fe(OH)_3$, $Cu(OH)_2$, etc.) are unstable. Only the silicate minerals, such as amorphous silica, opal, and chalcedony, are stable at low pH (Figure 5.7).

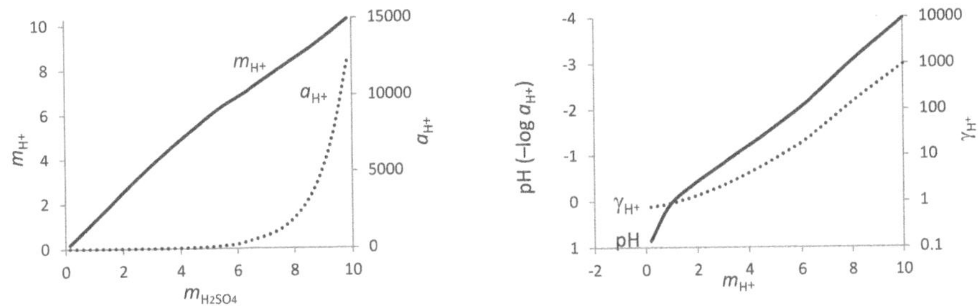

FIGURE 8.27 Activity–concentration behavior for H^+ in sulfuric acid solutions. Left: H^+ concentration (m_{H^+}) increases uniformly with sulfuric acid concentration ($m_{H_2SO_4}$) from the dissociation reaction ($H_2SO_4 \rightarrow H^+ + HSO_4^-$), but H^+ activity (a_{H^+}) increases logarithmically. This is due to the logarithmic rise in the H^+ activity coefficient (γ_{H^+}) in high-ionic-strength solutions, such as sulfuric acid. In low ionic strength solutions, γ is less than 1, but in strong electrolyte solutions (brines, acids) it increases exponentially (see Figure 2.4). Right: The strong increase in pH ($-\log a_{H^+}$) correlates with γ_{H^+} (both on log scales).

$$NaAlSi_3O_{8\,\text{albite}} + 4H_2O + 4H^+ \rightarrow Na^+ + Al^{3+} + 3H_4SiO_4$$

$$Al_2Si_2O_5(OH)_{4\,\text{kaolinite}} + 6H^+ \rightarrow 2Al^{3+} + 2H_4SiO_4 + H_2O$$

$$(Mg_5Al)AlSi_3O_{10}(OH)_{8\,\text{clinochlore}} + 16H^+ \rightarrow 5Mg^{2+} + 2Al^{3+} + 3H_4SiO_4 + 6H_2O$$

$$KAl_2(AlSi_3O_{10})(OH)_{2\,\text{muscovite}} + 10H^+ \rightarrow K^+ + 3Al^{3+} + 3H_4SiO_4$$

This series of acid-consuming mineral dissolution reactions can then be compared with the geochemistry of the acid discharge from Iron Mountain. Figure 8.28 shows the high concentrations produced by the sulfide (mainly pyrite) oxidation reactions, which release base metals and sulfuric acid (pH = −2.5) into solution. The base metal concentrations reflect their abundance in the sulfide ore, which is mainly pyrite [FeS_2]. The acid produced by sulfide oxidation is then consumed by mineral weathering, producing Al^{3+}, Mg^{2+}, Na^+ and K^+ found in the rhyolite. The relative

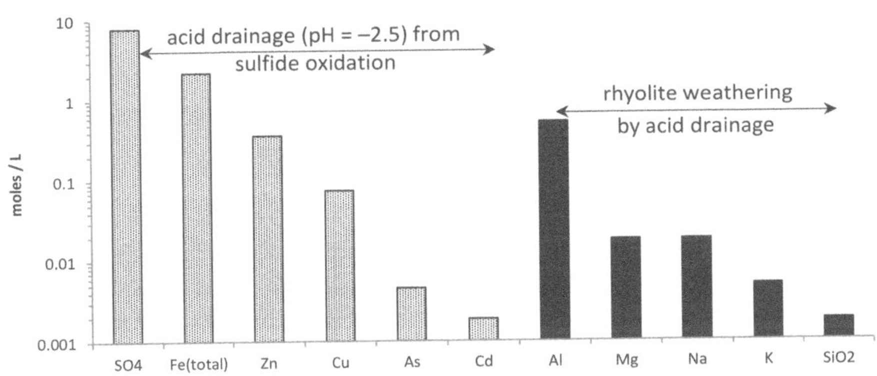

FIGURE 8.28 Geochemistry of the acid drainage waters with pH −2.5 from Iron Mountain. (Data from Nordstrom et al. 2000.)

abundances of these cations in the acid drainage reflect their relative abundances in the minerals above. Only silicic acid is anomalously low, due to precipitation of amorphous silica and its polymorphs at low pH.

The extreme geochemistry of the Iron Mountain discharge generates a range of rare efflorescent (evaporative) salts formed by evaporative concentration of the acid drainage at seepage points in the mine. Nordstrom and Alpers (1999) analyzed such precipitates collected at drips and seepage faces, showing them to be sulfate minerals with variable metal and water contents. Most were iron sulfates, including the following:

- Ferrous iron sulfates, such as melanterite [$FeSO_4{\cdot}7H_2O$] and rozenite [$FeSO_4{\cdot}4H_2O$]
- Ferrous-ferric sulfates, including copiapite [$Fe^{II}Fe_4^{III}(SO_4)_6(OH)_2 \cdot 20H_2O$] and voltaite [$K_2Fe_5^{II}Fe_4^{III}(SO_4)_{12} \cdot 18H_2O$]
- Ferric iron sulfates kornelite [$Fe_2(SO_4)_3{\cdot}7H_2O$] and rhomboclase [$HFe(SO_4)_2{\cdot}4H_2O$]

The ferrous iron sites in these minerals are readily substituted with divalent Cu^{2+} and Zn^{2+}, and the ferric iron sites with Al^{3+}. Dissolution of these minerals generates significant acidity, not only due to the release of H^+ in the structure of minerals such as rhomboclase, but also by the hydrolysis of Fe^{3+} and Al^{3+} in solution. The result from the dissolution of these efflorescent salts is the heavy loading of acidity and metals during early stream discharge following rainfall.

Nordic Mine Uranium-Tailings Impoundment, Elliot Lake, Ontario

This study by Morin et al. (1988) provides a great example of the acid generation and subsequent buffering reactions as the acid drainage moves through the subsurface. In this case, there is some acid-buffering capacity in the sandy aquifer, which contains a few percent calcite and dolomite. The acid plume movement and reaction in the aquifer are shown in cross section in Figure 8.29, together with the multilevel piezometer-monitoring network.

The core of the plume in the tailings was determined to be generated according to the classic reactions on residual pyrite:

$$FeS_2 + 3\tfrac{1}{2}O_2 + H_2O \rightarrow Fe^{2+} + 2SO_4^{2-} + 2H^+ \qquad \textit{Thiobacillus thiooxidans}$$
$$Fe^{2+} + \tfrac{1}{4}O_2 + 2\tfrac{1}{2}H_2O \rightarrow Fe(OH)_3 + 2H^+ \qquad \textit{Thiobacillus ferrooxidans}$$
$$Fe(OH)_{3(fer)} + 3H^+ \rightarrow Fe^{3+} + 3H_2O \qquad \textit{abiotic}$$

The tailings themselves are carbonate-free, and pH values below 3 are generated. At this pH, metals, including Fe, Cu, Pb, Zn, and U, are very soluble and are mobilized from the tailings. The low pH also reduced the CEC of the substrate, precluding sorption as a mechanism to slow contaminant migration. This allows high mobility for radium, which as a divalent base cation (Ra^{2+}, like Sr^{2+}) would normally have a high retardation factor. This allows the high activity of 5 Bq/L that was measured. Along with these metals is a high concentration of SO_4^{2-} from sulfide

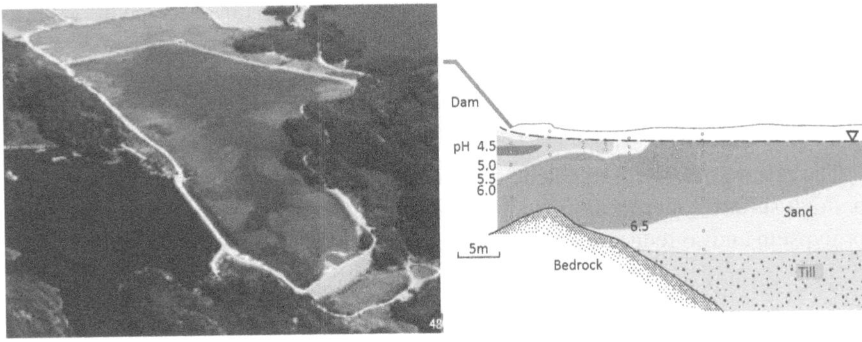

FIGURE 8.29 pH of acid drainage migrating below a tailings dam at the Nordic Mine, Elliot Lake, Ontario. pH values increase by reaction during flow through the carbonate-bearing sand. Multilevel piezometer tips shown. (After Morin et al. 1988.)

oxidation. Note that the high concentration of SO_4^{2-} is also favored at low pH because the second dissociation constant for sulfuric acid is 10^{-2}. At pH 2.7, over 20% of the dissolved sulfate is as bisulfate, HSO_4^-, and as pH decreases, the solubility of gypsum increases.

As this plume migrates in the subsurface, its velocity is retarded by reactions with native minerals in the aquifer. The principal attenuation process is the dissolution of carbonate minerals in the aquifer. At lower pH, this generates carbonic acid:

$$2H^+ + CaCO_{3\,(cal)} \rightarrow Ca^{2+} + H_2CO_3 \quad \text{below pH 6.4}$$

As the pH rises by consuming H^+ in the plume, the carbonic acid will dissociate, and so attenuation reactions at higher pH produce bicarbonate:

$$H^+ + CaCO_{3\,(cal)} \rightarrow Ca^{2+} + HCO_3^- \quad \text{above pH 6.4}$$

The production of bicarbonate and increase in pH begin to increase the activity of dissolved carbonate:

$$HCO_3^- \rightarrow CO_3^{2-} + H^+$$

which acts to attenuate the concentration of ferrous iron through precipitation of siderite:

$$Fe^{2+} + CO_3^{2-} \rightarrow FeCO_{3\,(sid)}$$

Siderite is less soluble than calcite, so ferrous iron is also attenuated by the incongruent dissolution of calcite:

$$Fe^{2+} + CaCO_{3\,(cal)} \rightarrow FeCO_{3\,(sid)} + Ca^{2+}$$

As these reactions raise the pH, ferric iron species in the plume become insoluble as ferrihydrite and precipitate as an additional reactive substrate for sorption of other metals:

$$Fe(OH)_2^+ + H_2O \rightarrow Fe(OH)_{3\,(fer)} + H^+$$

The release of Ca^{2+} from calcite dissolution and siderite formation, together with the high sulfate concentration from the tailings generate supersaturated conditions for gypsum and so reduce the very high dissolved SO_4^{2-}:

$$Ca^{2+} + SO_4^{2-} + 2H_2O \rightarrow CaSO_4 \cdot 2H_2O_{(gyp)}$$

Finally, the increased pH also greatly reduces the solubility of aluminum. Toward circumneutral pH values, Al^{3+} is stable as hydroxide and precipitates as amorphous aluminum hydroxide, gibbsite.

$$Al^{3+} + 3H_2O \rightarrow Al(OH)_{3\,(gib)} + 3H^+$$

These reactions will proceed until the buffering capacity of the aquifer is depleted. With measurements of the calcite content in the aquifer, mass balance calculations of the long-term acid generation capacity of the tailings versus the acid and metal buffering capacity of the aquifer can be made. If there is insufficient buffering in the subsurface for the long-term loading from the tailings, other remedial measures must be considered. In the case of the Nordic mine, reducing conditions in the tailings were established to limit the production of acid drainage. This was accomplished with a low-permeability cap and soil over the top, and monitoring since that time shows decreasing levels of acidity, iron, and other solutes.

Surface discharges from the tailings impoundment are treated differently at this mine, by diversion and treatment. In particular, the acidity and potential for ^{226}Ra contamination of the local surface waters is prevented by treatment with soluble barium chloride and lime [CaO]:

$$CaO + H_2O \rightarrow Ca^{2+} + 2OH^-$$

$$OH^- + H^+ \rightarrow H_2O$$

$$^{226}Ra^{2+} + Ba^{2+} + SO_4^{2-} \rightarrow Ba_{(1-x)}\,{}^{226}Ra_xSO_4$$

Subsequent flow through a settling pond prior to discharge into natural drainage waters allows the insoluble barium sulfate to precipitate and settle out, sequestering with it the trace concentrations of highly radioactive ^{226}Ra. The neutral pH also then sequesters ferric hydroxides (ferrihydrite) and other metals along with it.

Selenium in Mine Drainage

Like sulfur, which sits one row above, selenium is found in oxidized settings as Se^{VI} selenate, SeO_4^{2-} and Se^{IV} selenite, SeO_3^{2-} (not to be confused with the clear variety of gypsum [$CaSO_4{\cdot}H_2O_{selenite}$], both names derived from the Greek for *moon*). In low redox settings it is reduced to insoluble zero-valence selenium, $Se^0_{(s)}$ or further to Se^{-II} selenide, Se^{2-}. It is often chelated as a variety of soluble organoselenides. Its elemental state as crystalline selenium is not observed in the environment, but this insoluble form can be produced in mine water treatment systems and for industrial use.

As a trace non-metal, its participation in environmental cycling and uptake in vegetation is typically through substitution for sulfur. It is widely distributed in the environment through volcanic emissions and through weathering of sulfides, including pyrite and selenide [FeSe]. It is recovered commercially from sulfide ore during electrolytic refining of copper and is critical in metallurgy, in glass making, and more recently in photovoltaics. It is an essential trace nutrient for plants and humans at daily levels below 1 mg/kg, but toxic at levels only ten times greater than this (Naga Jyothi et al. 2020). The World Health Organization (WHO) drinking water limit of 10 mg/L. Selenium toxicity in surface waters is sensitive to its oxidation state, as selenite and organoselenides are more bioavailable.

Stable as sulfide minerals and crystalline $Se^0_{(s)}$, the greatest release of Se into groundwaters and surface waters is from oxidation of selenide-bearing minerals during coal mining. Oxidation and leaching from sulfides in exposed waste rock releases very soluble selenate (SeO_4^{2-}) and selenite (SeO_3^{2-}) anions (Figure 8.30). The extensive coal mining operations in the Elk River Valley of British Columbia, Canada, contribute to rising levels of Se in surface waters during low-flow seasons, where levels regularly exceed the surface water quality limit of 50 mg/L. Remediation strategies have been implemented to reduce leaching from waste rock piles (Teck 2019). Bioreactors have been established that remove selenate in mine waste rock drainage by reduction to either insoluble zero-valence selenium $S^0_{(s)}$ or ferrous selenide (Figure 8.30).

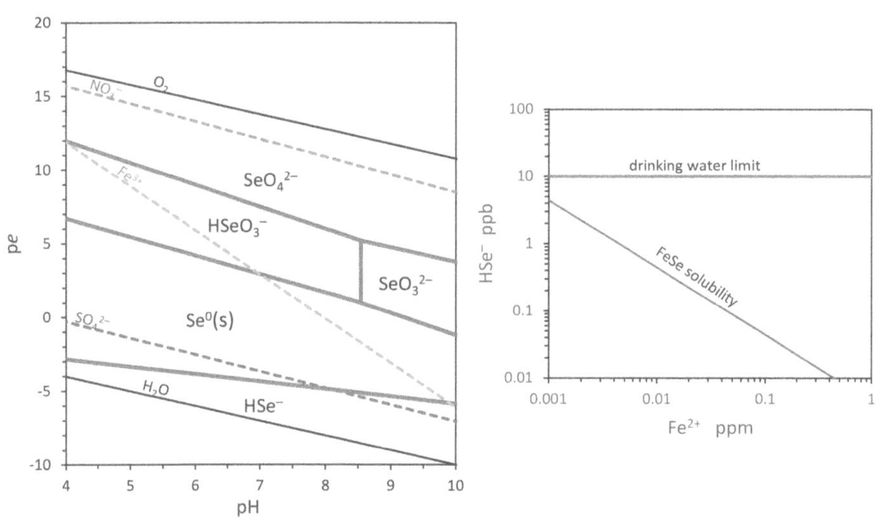

FIGURE 8.30 Left: Selenium stability diagram for 25°C and total dissolved selenium of 10^{-6} mol/kg (0.8 ppm) (from data in Séby et al. 2001). The zero-valence redox state is intermediary between selenite and selenide. Elemental selenium [$Se^0_{(s)}$] is essentially insoluble. Right: Solubility of Ferrous selenide [FeSe] with Fe^{2+}.

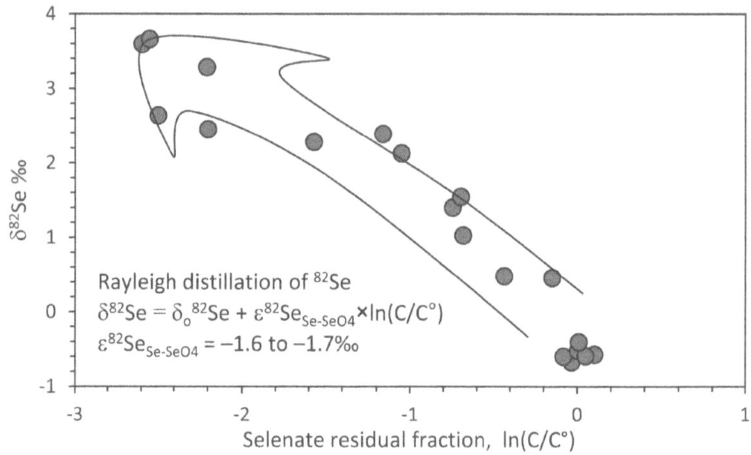

FIGURE 8.31 Rayleigh fractionation during sub-aerobic reduction of selenate to $Se^0_{(s)}$ in column bioreactor column experiments (Hendry et al. 2024).

$$SeO_4^{2-} + 2e^- + 2H^+ \rightarrow SeO_3^{2-} + H_2O$$
$$SeO_3^{2-} + 4e^- + 6H^+ \rightarrow Se^0 + 3H_2O$$
$$Fe^{2+} + HSe^- \rightarrow H^+ + FeSe \quad K = 10^8$$

Increases in more bioavailable $HSeO_3^-$ in the treatment effluent were further treated by sorption with iron chloride (Golder 2019). The reduction in selenate in treated mine effluent was demonstrated with stable isotopes of selenate during field and lab bioreactors as a verification of anoxic redox reaction (Hendry et al. 2024). The data (Figure 8.31) allowed calculation of the enrichment factor for the full reaction from selenate to zero-valence selenium, from the slope of the regression of the inverse relationship for $\delta^{82}Se_{SeO4}$ and $\ln(f)$ where $\ln(f)$ is expressed as ln(C/C°).

BASE METALS IN GROUNDWATER

The first series of transition metals include the important base metals, manganese, iron, cobalt, nickel, copper, and zinc, which are found as contaminants in mining camps and industrial settings. Iron, the fourth most abundant element in the earth's crust and the most abundant of the transition metals, is discussed in the context of redox evolution, along with Mn in Chapter 6. Like iron, most of the transition metals are multivalent cations that form hydroxide species over the range of pH of natural waters and have low solubilities at neutral pH. In circumneutral waters under oxidizing conditions, these metals are readily adsorbed onto Fe and Mn oxyhydroxides, complexed with organics, or precipitated as hydroxides or carbonates at elevated pH. All form sulfide or oxide minerals, and with the exception of Fe and Cu, are not found in their native state. Despite these similarities, variations in their oxidation states and their crustal abundances allow differences in their geochemical behavior in the environment.

Chromium (Cr)

Chromium is a trace element in the natural environment, occurring mainly as trivalent Cr^{III} and readily substituting for Fe^{III} in oxides and hydroxides. Toxic hexavalent Cr^{VI} is found only under highly oxidizing conditions, usually from anthropogenic activities. It is commonly mined as chromite ($FeCr_2O_4$), a refractory mineral associated mainly with ultramafic rocks and used to line blast furnaces. It is used for its corrosion resistance in chromium steel and metal plating, for its chromatographic characteristics in paints and dyes, in explosives, pesticides, and drilling mud, as well as in pressure-treated wood as chromated copper arsenate. Although trivalent Cr^{III} is an essential micronutrient, hexavalent Cr^{VI} is a neurotoxin and carcinogen at levels as low as 50 ppb. It gained public attention following the lethal contamination of Baechtel Creek in California by the dumping of plating liquids up to 1995.

Concentrations in natural waters are typically in the range of 1 to 20 ppb and average only generally below the WHO guideline of 50 ppb in drinking water. Seawater contains less than 1 ppb Cr. Cr^{III} has very low solubility in natural waters, where it exists in sixfold coordination with H_2O as $Cr(H_2O)_6^{3+}$) below pH 3.8 and as the hydroxide $Cr(OH)^{2+}$ up to neutral pH (Figure 8.32). Under alkaline conditions, $Cr(OH)_3$dominates. Only above pH 11.5 does Cr^{III} dominate as an anion, $Cr(OH)_4^-$. Under neutral pH conditions, solid chromium hydroxide [$Cr(OH)_{3(s)}$] maintains $Cr(OH)_3^\circ$ concentrations at less than about 5 ppb. In lower pH waters, the cation forms are readily sorbed onto clays or precipitate with iron oxyhydroxides.

Cr^{III} can be oxidized to Cr^{VI} under natural conditions by O_2 or other electron acceptors, including NO_3^- and Mn^{4+}. Unlike Cr^{III}, hexavalent Cr^{VI} forms chromic acid

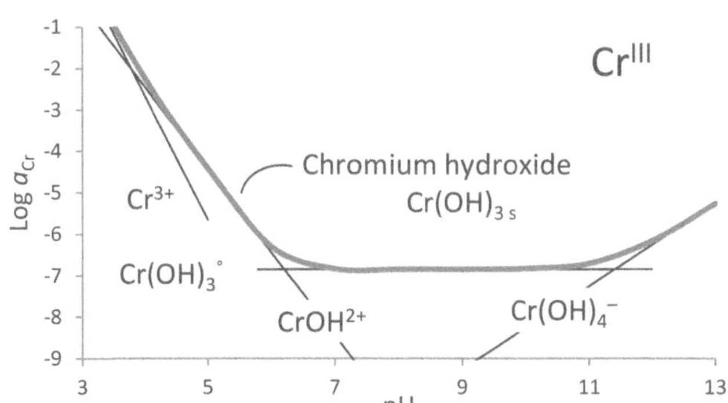

FIGURE 8.32 Cr^{III} speciation and solubility. (Based on Rai et al. 1987; Palmer and Wittbrodt 1991; Wu et al. 2008.)

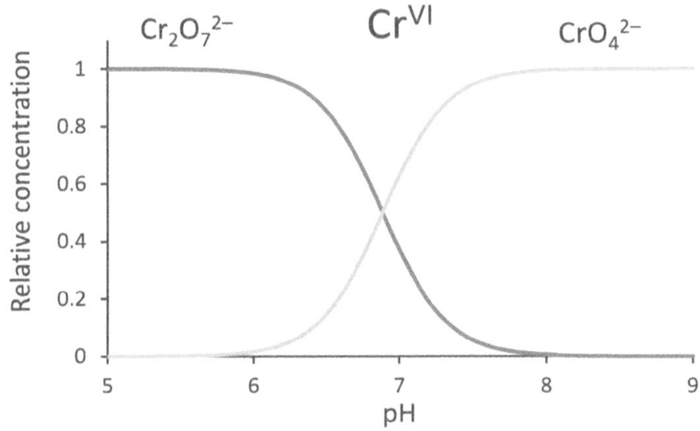

FIGURE 8.33 Cr^{VI} speciation with pH. Toxic chromate ions are highly soluble.

(H_2CrO_4). Dichromate ($Cr_2O_7^{2-}$) dominates in natural waters up to pH 6.5 where it dissociates to chromate (CrO_4^{2-}) (Figure 8.33). These anions are highly soluble but are readily reduced back to Cr^{III} by organic carbon and other elements of lower redox potential, including ferrous iron (Fe^{II}) and sulfide. Seldom do chromate concentrations exceed a few ppb in natural waters. Hem (1985) cites pH 9 groundwaters with up to 200 ppb Cr as CrO_4^{2-}. The hyperalkaline groundwaters in northern Jordan (see Example 6.11) that discharge at pH 12.5 are pale green with 7.02 ppm Cr (Khoury et al. 1992).

Trivalent Cr^{III} dominates in groundwater because it is slow to oxidize and does so only at a high electromotive potential (Figure 8.32). From this diagram, only manganese, nitrate, or oxygen can be used to oxidize trivalent chromium to soluble hexavalent Cr^{VI}.

The release of hexavalent chromate from industrial activities has led to reactive barrier technologies to mitigate its mobility (Blowes and Ptacek 1992). These make use of the low solubility of trivalent chromium hydroxide, providing it can be reduced to this form. From Figure 8.34, any electron donor below the Cr^{VI}/Cr^{III} line can potentially reduce chromate to precipitate trivalent chromium hydroxide ($Cr[OH]_{3(s)}$). Examples are reduction with ferrous iron and with carbon, although the most effective reduction rates were observed when using zero-valence iron as the electron donor with precipitation of ferric chromium oxyhydroxide [$CrFe(OOH)_2$]:

$$CrO_4^{2-} + 3Fe^{2+} + 5H_2O \rightarrow Cr(OH)_3 + 3Fe(OH)_2^+ + H^+$$

$$2Cr_2O_7^{2-} + 3CH_2O + 4H^+ + H_2O \rightarrow 4Cr(OH)_3 + 3CO_2$$

$$CrO_4^{2-} + Fe^0_{(s)} + 2H^+ \rightarrow CrFe(OOH)_{2(s)}$$

These reactions were traced using chromium isotopes in batch experiments to monitor the reduction of chromate to chromium hydroxide using fixed carbon (Jamieson-Hanes et al. 2012) and using ferrous iron (Kitchen et al. 2012) as the

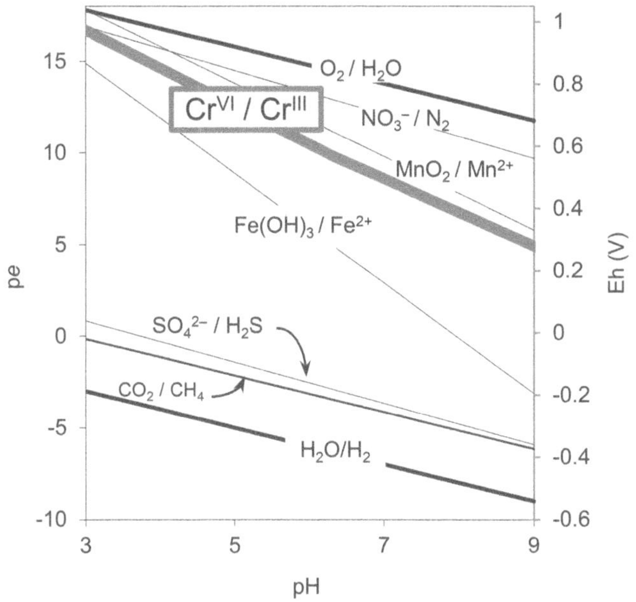

FIGURE 8.34 Electromotive potential for the Cr^{VI}/Cr^{III} redox reaction.

electron donor. As discussed for other systems, isotopes resolve reactive loss of a contaminant from attenuation by dispersion and mixing, which are not fractionating. Chromium has four stable isotopes, ^{50}Cr (4.35%), ^{52}Cr (83.8%), ^{53}Cr (9.5%), and ^{54}Cr (2.37%). The ^{53}Cr to ^{52}Cr ratio is typically measured using a multicollector inductively coupled plasma mass spectrometer (ICP-MS) and is referenced to a National Institute of Standards and Technology chromium nitrate solid (NIST SRM 979) and expressed in the standard delta-permil fashion as $\delta^{53}Cr‰$ SRM 979. The experimental data for chromate reduction by ferrous iron and organic carbon are shown in Figure 8.35. Both impart an enrichment on the residual chromate, suggesting that $\delta^{53}Cr$ can be a useful tool to track chromate reduction and attenuation in field sites of hexavalent chromium contamination.

Cobalt (Co)

Cobalt is a trace metal with a crustal abundance of only 25 ppm. Like iron, it can be magnetized, but its major use is to improve the strength and corrosion resistance of steel when alloyed with iron and nickel. A Co–Cr alloy is used for hip and knee joint replacements. Co-oxides provide blue tints in glass and ceramics. As a structural component of vitamin B-12, cobalt is an essential micronutrient for plants and animals and can be added as sulfate to phosphate fertilizers. It has one stable isotope, ^{59}Co, and one long-lived radionuclide, ^{60}Co, produced by neutron activation of ^{59}Co in nuclear reactors. It decays to ^{60}Ni by beta emission ($T_{½}$ = 5.27 years), with strong

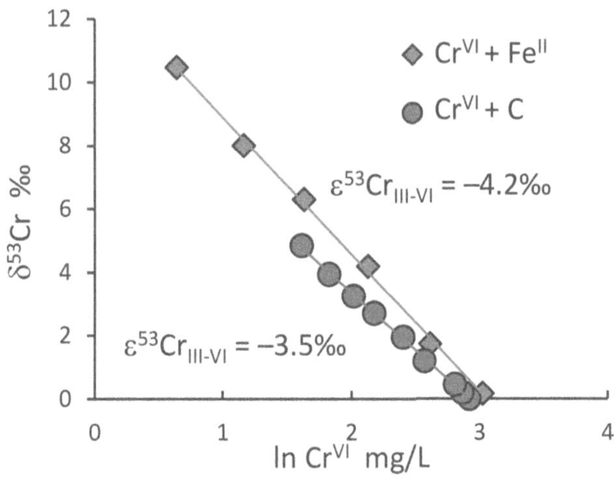

FIGURE 8.35 $\delta^{53}Cr$ fractionation and Rayleigh distillation during reduction of Cr^{VI} chromate to Cr^{III} using ferrous iron at pH 5 (gray diamonds; Kitchen et al. 2012) and with organic carbon (black discs; Jamieson-Hanes et al. 2012). Both reduction reactions show Rayleigh-type enrichments in ^{53}Cr, with enrichment factors of $\varepsilon^{53}Cr_{VI\text{-}III}$ = −4.2‰ for Cr^{VI} reduction with ferrous iron and $\varepsilon^{53}Cr_{VI\text{-}III}$ = −3.5‰ for reduction on carbon.

gamma radiation that is used in nuclear medicine. Stable cobalt is generally coproduced from copper–nickel deposits, where it occurs in a variety of sulfides and arsenides, with releases to the environment through smelting operations and coal-fired generating plants. Cobalt is soluble under lower pH conditions and generally follows Fe and Mn oxyhydroxides.

Freshwaters seldom exceed 1 ppb except where anthropogenic sources exist. Both divalent Co^{II} and trivalent Co^{III} exist in natural waters, although $Co^{III}$is a strong oxidant and readily converted to soluble divalent cobalt. Uncomplexed Co^{2+} and inorganic complexes such as $CoCO_3^{\circ}$ dominate in freshwaters. However, transport is dominantly in colloidal form complexed with organics or adsorbed to Mn and Fe oxyhydroxides. Like most transition metals, higher concentrations are favored at lower pH. Under anoxic conditions, cobalt sulfides are highly insoluble.

Cobalt is not biomagnified in ecosystems, although it is accumulated by both plants and organisms. It is not identified as a contaminant by either the WHO or U.S. EPA and has no drinking water limit. Environmental regulations for surface water quality limit Co to 0.1 ppm.

Nickel (Ni)

Like Co, nickel has similar geochemical characteristics to iron, with which it is alloyed in the core of the terrestrial planets and in iron meteorites. Its concentration is lower in the mantle and only 75 ppm in the crust, associated preferentially with mafic and ultramafic rocks, where it substitutes for Fe^{2+} and Mg^{2+}. Nickel is found in economic concentrations as pentlandite [$(Fe,Ni)_9S_8$] associated with copper sulfide, but also as iron–nickel oxyhydroxides and nickel arsenides, such as niccolite (NiAs). It is used primarily for its anticorrosion and high strength characteristics in the production of stainless steel and in nickel plating.

Like Co, divalent Ni^{II} dominates in minerals and waters. It is essentially insoluble as a sulfide. Under aerobic conditions, nickel tends to be relatively mobile. It becomes increasingly soluble as divalent Ni^{2+} below pH 6. At higher pH, nickel is readily transported as inorganic and organic chelates, or is adsorbed onto Fe and Mn oxyhydroxides. Above pH 9, solubility is limited by its hydroxide [$Ni(OH)_2$] ($K = 10^{-15.26}$) to less than 1 ppm. Although nickel has five stable isotopes, little development as tracers exists, given the relatively low reactivity and redox of nickel in the environment.

Natural surface waters typically carry low to sub-ppb concentrations of Ni, although values over 100 ppb–10 ppm can be found in high source-term regions and mining camps. As nickel partitions onto solid phases, freshwater sediments have concentrations averaging 25 ppm in background regions. Seawater has less than 1 ppb Ni, where scavenging by particulates and organics control solubility. Similarly, groundwaters seldom have concentrations over 10 ppb. Drinking water typically has less than 10 ppb Ni. The WHO provides a drinking water limit of 20 ppb, although no adverse health effects in humans from ingested nickel have been demonstrated.

Copper (Cu)

Copper has a crustal abundance of only 55 ppm and is one of the least abundant transition metals. Nonetheless, it is significantly concentrated during magmatic and hydrothermal activity as insoluble sulfides. Chalcopyrite ($CuFeS_2$) is the most common ore mineral, although Cu forms other sulfide minerals as well as a host of oxides, such as cuprite (Cu_2O) and carbonates, including azurite ($2CuCO_3 \cdot Cu[OH]_2$) and malachite ($CuCO_3 \cdot Cu[OH]_2$). It is also one of the few metals to be found and mined in its elemental state, as native Cu^0. It is an essential trace element for metabolism in plants and animals, but at higher concentrations, becomes toxic to microorganisms and so is used in pesticides, in antifouling paints for marine boat hulls, and in pressure-treated wood as chromated copper. Copper has two stable isotopes, ^{63}Cu (69.17%) and ^{65}Cu (30.83%), expressed as $\delta\,^{65}Cu$ referenced to the NIST 976 Cu standard, which have been shown to fractionate between Cu^I and Cu^{II} redox states during copper mineralization (Braxton and Mathur 2011). Partitioning during aqueous complexation and transport suggests species-dominated fractionation (Maréchal and Albarède 2002), although only minor in natural aqueous systems (Bigalke et al. 2010).

Weathering of copper mineralization is the dominant natural source of Cu in aquatic systems above background levels of several ppb. Runoff from mining camps and tailings, particularly if acidic, and fallout from smelting are sources of much higher concentrations in the environment. In minerals and waters, cupric Cu^{II} predominates. Monovalent or cuprous Cu^I is unstable under most pH conditions and rarely found. Like other transition metals, copper has a strong affinity for adsorption

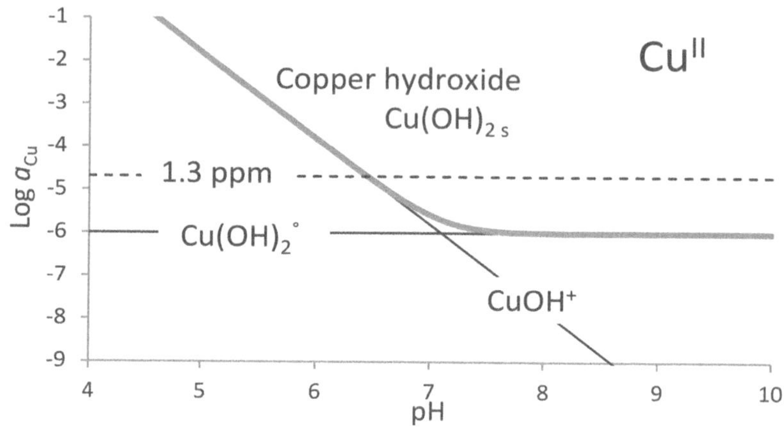

FIGURE 8.36 Solubility of copper hydroxide ($Cu[OH]_2$), with pH. (After Cravotta 2008.)

onto organic and inorganic substrates, including ferric oxyhydroxides and clays. Accordingly, its solubility as Cu^{2+} is favored only at low pH, allowing concentrations in the range of 100–1000 ppm in AMD.

Copper hydroxide [$Cu(OH)_2$] limits Cu^{2+} solubility to less than 1 ppm at circumneutral pH and in the low ppb range at higher pH (Figure 8.36). DIC further reduces copper solubility by precipitation of malachite [$Cu_2(OH)_2CO_3$]. Copper can be concentrated in reducing environments by reduction to monovalent Cu^I and precipitation as chalcocite [Cu_2S].

Natural surface waters have less than 0.1 ppm dissolved Cu. Similarly, groundwater concentrations are limited by sorption to less than 1 ppm. In natural waters, sorption seems to limit dissolved copper to concentrations below those predicted by pH-mineral stability calculations. The WHO recommends an upper limit of 2 ppm for copper in drinking water, based on potential gastrointestinal irritation at higher levels and aesthetic qualities of taste and staining. The U.S. EPA sets a limit for drinking water at 1.3 ppm. The Canadian water quality guideline for aquatic life is 4 ppb and for irrigation waters is 200 ppb.

Zinc (Zn)

Zinc has a crustal abundance of 70 ppm but is concentrated by hydrothermal solutions. Sphalerite [ZnS] is the principal ore-forming mineral, most commonly found in association with galena [PbS] filling dissolution voids in carbonate rocks. Zn replaces Fe and other divalent metals in a host of minerals, including magnetite and ferromagnesian silicates. The greatest industrial use of zinc is in galvanization of steel to reduce corrosion by Zn-electroplating or hot dipping into molten zinc. It is also used in alloys, particularly with copper to produce brass. Metallic zinc is rolled as malleable sheeting for batteries, roofing, and other applications requiring low-corrosion seals.

In the weathering environment, Zn retains its divalent state and is soluble under acidic conditions where Zn^{2+} is stable. Zn^{2+} is the dominant dissolved species below pH 7, whereas under alkaline conditions, $ZnHCO_3^+$ and $Zn(OH)^+$ become the important dissolved phases. At higher pH, zinc hydroxide [$Zn(OH)_2$] limits solubility (2 ppm at pH 8). Dissolved Zn^{2+} is also very sensitive to DIC, as smithsonite [$ZnCO_3$] ($K = 10^{-10}$) is a significant control when DIC exceeds about 10 ppm. At very high pH, anionic Zn hydroxides such as $Zn(OH)_4^{2-}$ enhance solubility, although such environments are restricted to cement pore waters, coal fly-ash deposits, and slags where Na and Ca hydroxides may leach.

Like other transition metals, Zn is preferentially sorbed onto humic substances and iron hydroxides in water. Under circumneutral to high pH conditions, this is the dominant control on Zn solubility and is responsible for the low concentration in most natural waters. Concentrations in natural groundwater seldom exceed 100 ppb, and less than 50 ppb in lakes, where sorption onto organics and uptake by microorganisms occur. Higher concentrations are commonly associated with wastewater discharges, particularly from sewage plants. AMD can be very significant point sources of Zn^{2+} in aquatic systems (Figure 8.37).

Accumulation by sorption onto particulates and solid organic carbon is reflected by the concentration of Zn in sediments, which averages about two times its crustal abundance. Zn has four stable isotopes, ^{64}Zn (48.6%), ^{66}Zn (27.9%), ^{67}Zn (4.1%), and ^{68}Zn (18.8%), and measurements are made of the ^{66}Zn to ^{64}Zn ratio, expressed as $\delta^{66}Zn$. The sorption and cycling of Zn in these aqueous systems suggest that isotope fractionation could be significant (Maréchal and Albarède 2002).

Zinc is an essential nutrient, required in biological systems for the synthesis of nucleic acids and enzymes. Drinking water standards have a relatively high limit of 3–5 ppm based on aesthetic considerations. Humans require about 10–20 mg of Zn daily. In aquatic systems, standards are more severe due to high bioaccumulation

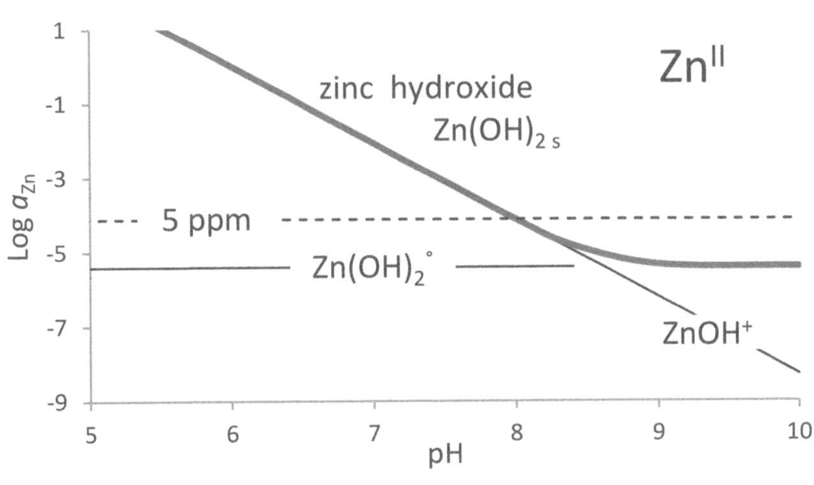

FIGURE 8.37 Zinc hydroxide speciation and solubility with pH.

factors in biota. Maximum allowable concentrations of dissolved Zn are set in the low ppb range (EPA indicates 17 ppb), although this limit rises progressively to over 300 ppb in high DIC waters due to complexation with carbonate.

ARSENIC

Arsenic has gained considerable interest over the past decades, as concentrations increase in a number of groundwater systems and its impact on human health becomes more prevalent, and the impacts of chronic ingestion of arsenic at low concentrations become more evident (Kapaj et al. 2006). For this reason, the WHO has recommended a safe drinking water limit of 10 μg/L. However, the redox chemistry of arsenic and its behavior in natural groundwater flow systems often make this limit difficult to achieve. It also makes it difficult to manage in legacy mine-tailings sites.

Arsenic Sources and Solubility

Arsenic sits in the fourth row of the periodic table, at atomic number 33 as a metalloid between germanium and selenium, and so has a variety of valences, including arsenide (As^{-III}), elemental (As^0), arsenite (As^{III}), and arsenate (As^V). As elemental arsenic, it has metallic properties and can be alloyed to strengthen metals, for example, lead in car batteries. In its reduced valence state, it can form metallic arsenides, such as GaAs semiconductors in the electronics industry. Arsenic trioxide (As_2O_3) is highly soluble (>10 g/L) and very toxic.

Its greatest use is perhaps the antibacterial properties of arsenate compounds with applications from fungicides and pesticides like sodium methyl arsenate to pressure-treated wood preservatives such as chromated copper arsenate, which is now restricted due to exposure to As and replaced in many applications by cupric ammonium and cupric chromium compounds.

Natural As minerals include the arsenite sulfide orpiment [As_2S_3] and realgar [As_4S_4] with As covalent bonding, but formally as As^{II}. Both are associated with hydrothermal deposits as low-temperature mineralization. Another substantial natural occurrence is as an arsenide replacing S in iron sulfides—arsenopyrite [FeAsS]. Weathering of arsenopyrite and other As minerals oxidizes and releases arsenic to the environment. This has led to some regions, such as Bangladesh, with significant concentrations in the drinking water. Arsenopyrite can host economic grades of gold as microinclusions. However, it cannot be routinely recovered by crushing and floatation with CN^- complexation and must be roasted, releasing As oxide gases that are recovered as arsenic trioxide [As_2O_3] powder. Past mining practices have led to significant As contamination of soils and tailings in some gold-mining camps.

In the aqueous environment, arsenic is generally found as either As^V, arsenate, or As^{III}, arsenite, with the speciation of either redox state a function of pH. Arsenate, like P^V sitting just above in column 15 of the periodic table, forms a weak acid as

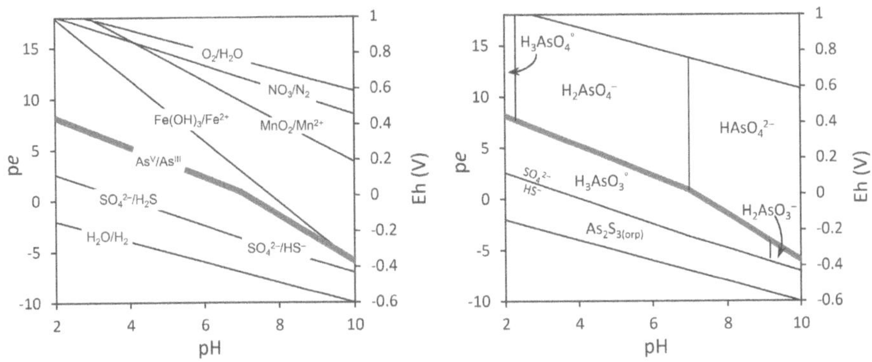

FIGURE 8.38 pH–redox stability fields for arsenic hydroxides and sulfides. Left: The As^V/As^{III} redox line with respect to other major redox pairs. Right: Stability fields for arsenate and arsenite hydroxides and arsenic sulfide. The redox line for conversion of $H_3AsO_3^\circ$ to orpiment, $As_2S_{3(orp)}$, is calculated for a total sulfur ($SO_4^{2-} + H_2S + HS^-$) concentration of 3.2 ppm or 10^{-4} mol/kg, and coincides with that for sulfate reduction to hydrogen sulfide, SO_4^{2-}/HS^-.

a hydroxide. Arsenate hydroxide species and their dissociation constants are as follows (from Nordstrom and Archer 2003):

$$H_3AsO_4^\circ \rightarrow H_2AsO_4^- + H^+ \quad K_{H_3AsO_4} = 10^{-2.3}$$

$$H_2AsO_4^- \rightarrow HAsO_4^{2-} + H^+ \quad K_{H_2AsO_4^-} = 10^{-6.99}$$

$$HAsO_4^{2-} \rightarrow AsO_4^{3-} + H^+ \quad K_{HAsO_4^{2-}} = 10^{-11.8}$$

while As^{III} has only two stable hydroxides within the pH range of natural waters:

$$H_3AsO_3^\circ \rightarrow H_2AsO_3^- + H^+ \quad K_{H_2AsO_3} = 10^{-9.17}$$

The speciation of the arsenate and arsenite hydroxides is shown in the left-hand chart of Figure 8.38.

The As^V hydroxide species $H_2AsO_4^-$ dominates below pH 7, and $HAsO_4^{2-}$ dominates above pH 7, while reduced As^{III} as the neutral species $H_3AsO_3^\circ$ dominates up to pH 9.17. The reduction of arsenate, As^V, to arsenite, As^{III}, occurs at a relatively low *pe*, well below NO_3^- and even after ferric iron (Figure 8.38; right chart).

Under oxidizing conditions, where both ferrihydrite [$Fe(OH)_3$] and As^V exist, arsenate is strongly sorbed onto this ubiquitous solid iron hydroxide phase, particularly at low pH, and dissolved arsenic concentrations are well below the drinking water standard of 10 ppb. As the speciation shifts to the $HAsO_4^{2-}$ field with increasing pH, As sorption decreases (Figure 8.39). As^{III} is also sorbed onto ferrihydrite, less so at low pH but significantly under alkaline conditions. Under strongly reducing conditions, where sulfate reduction produces hydrogen sulfide (H_2S and HS^-), As^{III} is highly insoluble as the arsenite sulfide, orpiment (As_2S_3).

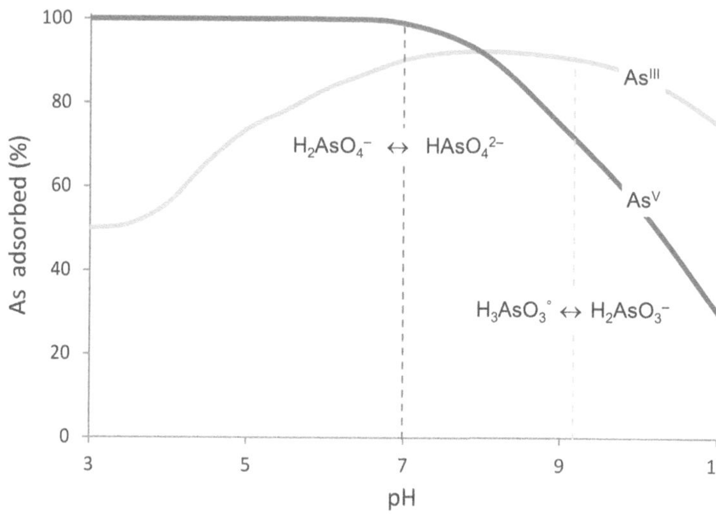

FIGURE 8.39 Sorption of arsenate (As^V) and arsenite (As^{III}) on ferric hydroxide, $Fe(OH)_3$, over a range of pH. (Modified from Masue et al. 2007.) The pH boundaries for the dominant arsenate and arsenite hydroxide species are shown with vertical lines. A reduction in sorption for both As^{III} and As^V occurs as these species become more anionic.

Arsenate reduction occurs above sulfate reduction and below iron reduction, which provides a window in the redox ladder where arsenic has very high solubility. Above the ferric/ferrous iron line, As solubility is limited by sorption onto ferrihydrite, and below the sulfate reduction line, by orpiment solubility. Between, only claudetite can exert any solubility control.

Although arsenic pentoxide [As_2O_5] can be synthesized under highly oxidizing conditions, arsenic trioxide [As_2O_3] is the only stable arsenic oxide, with two polymorphs—cubic arsenolite and the more stable monoclinic claudetite, which dissolve with very high solubilities:

$$As_2O_{3\,(claud)} + 3H_2O \rightarrow 2H_3AsO_{3(aq)} \quad K_{claud} = a^2_{H_3AsO_3} = 10^{-1.34}$$

This provides an arsenic activity of $10^{-0.67}$. Not accounting for ionic strength, this gives a dissolved concentration of 0.2 mol/kg which is 15,000 ppm or some 15 grams per L (Figure 8.40).

Bangladesh

Perhaps one of the most serious examples of natural arsenic contamination of groundwater is that of the Ganges delta in Bangladesh, which supplies millions of people with drinking water. Wells have been installed to replace surface water supplies

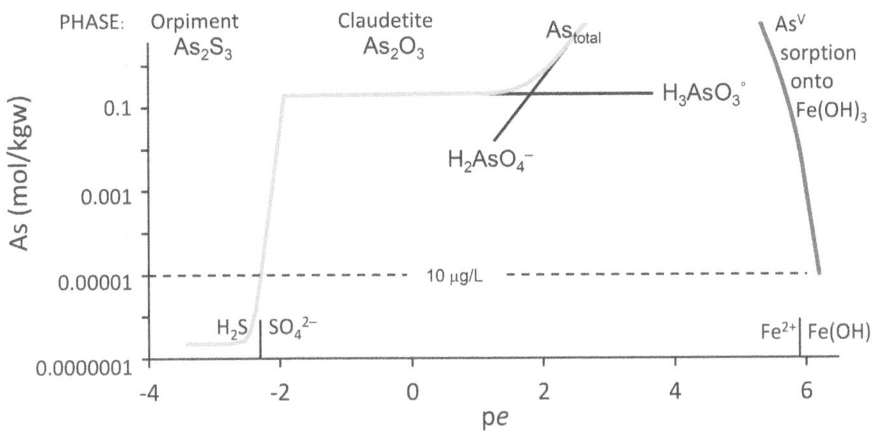

FIGURE 8.40 Mineral controls on arsenic solubility. Left: Redox control on arsenic solubility, constructed for pH 6, assuming 10^{-4} molar total sulfur as SO_4^{2-} and H_2S. At pe above about 1.5, mineralogical control of As can be achieved by AsV ($H_2AsO_4^-$) sorption onto ferric oxyhydroxides.

that are typically contaminated with nitrate and bacteria. However, the groundwaters have As concentrations up to and exceeding 300 μg/L (von Brömssen et al. 2007; Figure 8.41). The source and mobility in this case represent a good example of the solubility of arsenic under different redox conditions. Nickson et al. (1998) show that these deltas build from sediments, with over 10 ppm As sorbed on the

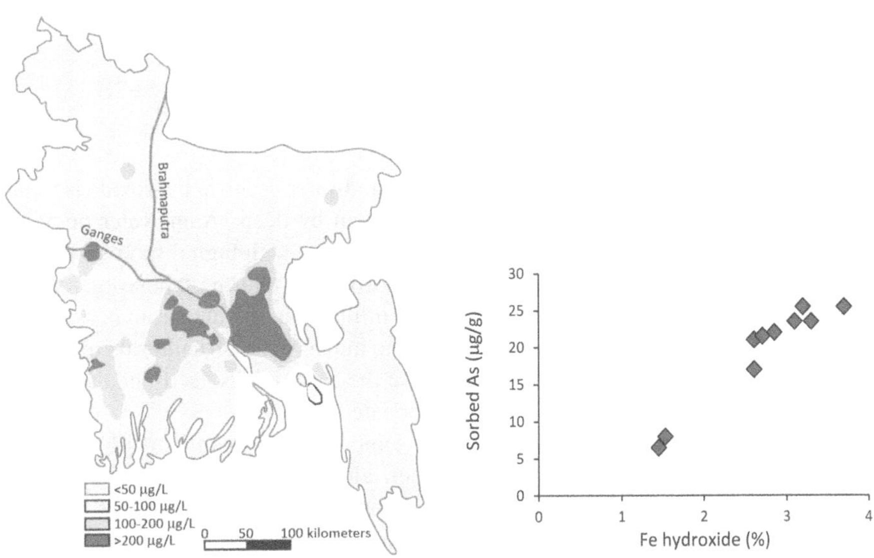

FIGURE 8.41 As in groundwater in the Ganges River delta of Bangladesh (left; von Brömssen et al. 2007). As sorbed on ferrihydrite in the sediments (right; Nickson et al. 1998) is released to groundwaters under reducing conditions.

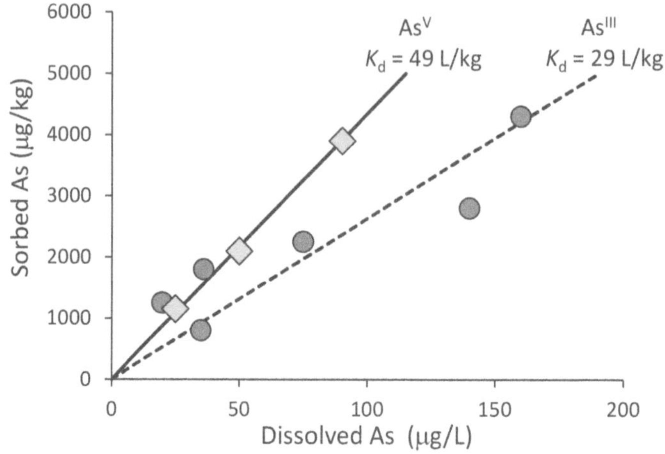

FIGURE 8.42 Sorption isotherms for As^{V} (as $H_2AsO_4^-$) and As^{III} (as H_3AsO_3) on deep (58 m) sands with ferric iron hydroxide, below the zone of As-contaminated groundwaters (Radloff et al. 2011), showing greater sorption for the more oxidized As^{V} than the more soluble As^{III}.

solid iron hydroxides derived from oxidation of base metal sulfides in the Himalayas (Figure 8.42). In the subsurface, organic material generates reducing conditions using the ferric iron hydroxide as an electron acceptor:

$$CH_2O + As^{V} - 2Fe(OH)_3 \rightarrow 2Fe^{2+} + H_3AsO_3^{\circ} + HCO_3^- + H_2O + 2H^+$$

The system then releases sorbed As into groundwater as ferric hydroxide is transformed to ferrous iron. Perturbation of the system by deep groundwater pumping then draws down these shallow, reducing groundwater with high dissolved arsenite into the exploited aquifer. Neumann et al. (2010) show that the labile organic carbon involved in the release of As comes with infiltration of water from constructed ponds, which are the major source of recharge to much of the agricultural area. Thus, the geochemical signatures of this process are the elevated As itself but also highly dissolved (ferrous) iron and increased bicarbonate.

The widespread contamination has been somewhat mitigated by drilling deeper wells that are exclusively for drinking water, although the downward infiltration of the high-As groundwaters from the overlying reducing zones remains a threat. Radloff et al. (2011) have measured the sorption of As^{III} and As^{V} onto the brown, ferric iron hydroxide-bearing sands, which today host groundwaters with less than 2 µg/L As. They show high sorption for arsenite, As^{III}, but even greater sorption for As^{V} (Figure 8.42), which will retard the migration of As from above due to groundwater pumping from the deeper sands.

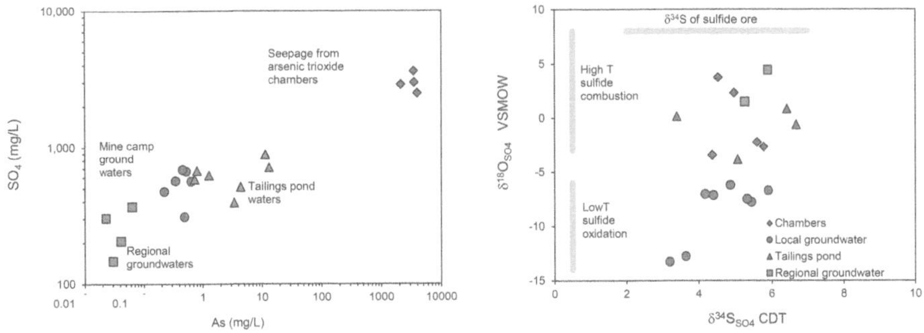

FIGURE 8.43 Arsenic and sulfate in Giant mine, Yellowknife, mine waters (Clark and Raven 2003). The association between SO_4^{2-} and As (left) allows sulfate isotopes to be used As a proxy for the source of As at this mine (right), showing contributions from both high-temperature roasting of the ore (chambers, tailings, and regional groundwaters) and low-temperature oxidation of arsenopyrite in the mining camp (local groundwaters).

Arsenic in Mine Waters

Arsenic in mine drainage is common in gold camps working ore that is hosted with arsenopyrite. Determining its origin is important for remediation. At the Giant Mine gold mining camp in Yellowknife in the Canadian Arctic, arsenic emissions from roasting of the arsenopyrite ore in the past have impacted the regional landscape. More modern procedures condensed arsenic gases and stockpiled the arsenic trioxide [As_2O_3] powder in underground chambers. Leaching of this has raised As concentrations in some mine waters to over 4 g/L. Mine tailings backfilled in the underground workings also host residual arsenopyrite in tailings from which low-temperature oxidation releases both SO_4^{2-} and As. Here, sulfate isotopes distinguish between these two sources of SO_4^{2-} and As in the mine waters. The high roasting temperature and exclusive involvement of atmospheric O_2 ($\delta^{18}O_{O_2}$ = 23.5‰) in combustion are evident in the sulfate from the arsenic trioxide chambers, as well as in the regional sulfate from ore-roasting emissions in the past and in the tailings mine water that comes from mine-water discharge. By contrast, the local groundwaters, which have As from 100 ppb to 1 ppm, derived their arsenic and sulfate from the aqueous (low temperature) oxidation of sulfides in the underground (Figure 8.43).

PERMEABLE REACTIVE BARRIERS

The construction of passive reaction cells in aquifers to intercept localized, point-source groundwater contaminant plumes is often favored over long-term active pump-and-treat approaches. They are designed to be more permeable than the aquifer to ensure the plume moves through the cell rather than around it and host media formulated to react with and attenuate the contaminant of interest. The first PRB was installed in a sand aquifer to intercept a plume of chlorinated solvents (TCE and tetrachloroethene) using granulated zero-valent metallic iron mixed with sand in roughly a 20% mixture (O'Hannesin and Gillham 1998) (Figure 8.44).

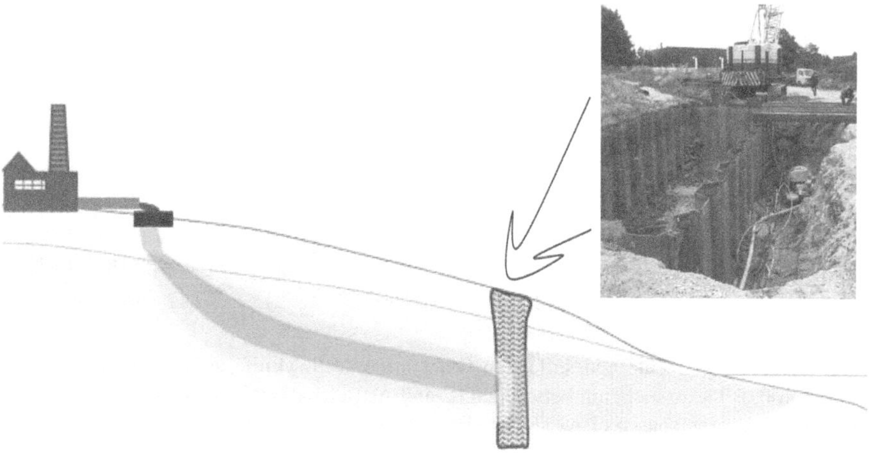

FIGURE 8.44 Permeable reactive barriers are constructed in unconsolidated sediments to intercept contaminant plumes in groundwater and with a suitable mix of reactants transform the contaminants to benign forms.

NUCLEAR WASTE

Uranium, the heaviest of all naturally occurring elements, was first identified in pitchblende [UO_2] in 1789 by Martin Klaproth. Just over a century later the development of photography led to Antoine Becquerel's discovery of radioactivity from observations that uranium salts would expose photographic plates. By the turn of the century, Marie Curie had isolated polonium and radium from uranium ore, demonstrating the occurrence of radioactive daughter elements neighboring U on the periodic table. Lord Rutherford showed that this decay occurred with a constant half-life, providing geologists with the basis for geochronology. However, fission was not discovered until 1938, when Hahn and Meitner exposed uranium to neutrons and found that they produced barium, an element with atomic number 56 and far from uranium on the periodic table. This rapidly led to the dramatic effort to exploit the associated energy release from the conversion of a small amount of mass to energy, as predicted by Einstein in 1905.

Initial efforts to develop fission as a weapon led to later efforts to harness it as an energy source. Both have left a considerable environmental legacy, from the mining, refining, and enrichment of uranium to the radioactive fission products of nuclear waste. Understanding the mobility of these radionuclides in the environment and finding sites for their long-term isolation present some of the most compelling hydrogeological challenges we face.

URANIUM GEOCHEMISTRY

Uranium has a very effective geochemical mechanism for accumulation, based on its high solubility as oxidized hexavalent U^{VI} but almost insoluble nature as tetravalent

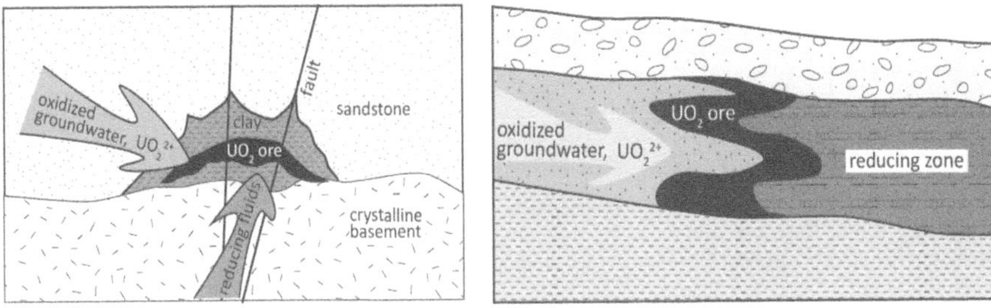

FIGURE 8.45 Typical redox *traps* for the formation of uranium deposits. Left: Cigar Lake (after Cramer 1994), Canada, mixing between reducing hydrothermal fluids and U-bearing oxidizing groundwaters. Right: South Australia, flow of oxidizing, U-bearing groundwaters into an aquifer zone with reduced compounds such as sulfides and organics.

U^{IV}. Most uranium ores form at a redox interface in groundwater or hydrothermal systems (Figure 8.45).

Oxidizing waters leach uranium from primary materials, such as granite or granitic sediments. Oxidation to hexavalent U^{VI} makes uranium soluble as the uranyl cation UO_2^{2+} or as a carbonate-complexed species, such as $UO_2CO_3^{\circ}$ and $UO_2(CO_3)_2^{2-}$. When uraniferous groundwaters encounter reducing zones in the aquifer, the uranium is reduced back to the tetravalent state and precipitated from solution as insoluble uraninite [UO_2] and pitchblende, which can have forms ranging from UO_2 to U_3O_7. Such *roll-front* deposits, so-called for their slow forward migration, occur where reduced species, such as organic carbon, graphite, and sulfides, are found within the aquifer, or at a zone of mixing with reduced hydrothermal fluids. As such settings are typically found with in clastic sediments, porosity is high, and accumulations of UO_2 ore can reach high percentages. Major uranium ore-producing regions are often found on the periphery of sedimentary basins in shield terrain.

Uranium is the heaviest of the naturally occurring elements and has no stable isotopes. It is dominantly ^{238}U with a half-life of 4.5 billion years. The fissionable isotope ^{235}U ($T_{½}$ = 704 My) makes up only 0.72% of natural uranium—insufficient for sustained fission in light water and graphite reactors. Decay of ^{238}U by release of α particles (2p2n which is a ^{4}He nucleus) generates a series of daughters, including ^{230}U, ^{230}Th, ^{226}Ra, and ^{222}Rn, eventually ending with stable lead, ^{206}Pb (Figure 7.27). It is a lithophilic element that is enriched in granitic rocks. Its average crustal abundance of about 2 ppm varies up to 10 ppm in U-rich granites and to less than 50 ppb in mafic rocks.

Uranium is found as tetravalent U^{IV} and hexavalent U^{VI} redox states. In the reduced state, uranium solubility is limited by uraninite [UO_2] and its microcrystalline form, pitchblende, which are the principal U-ore-forming minerals (Figure 8.46). Only below pH 3.5 will UO_2 support dissolved U concentrations above 100 ppb (as $U[OH]^{3+}$). At neutral pH, uraninite limits U solubility to less than 1 ppb.

Hexavalent uranium is soluble as the uranyl ion, UO_2^{2+}, up to pH 5.8, at which point the uranium hydroxide $UO_2(OH)^+$ becomes the dominant species (Figure 8.46). The low solubility of $UO_2(OH)_2$ maintains dissolved uranium concentrations at less than

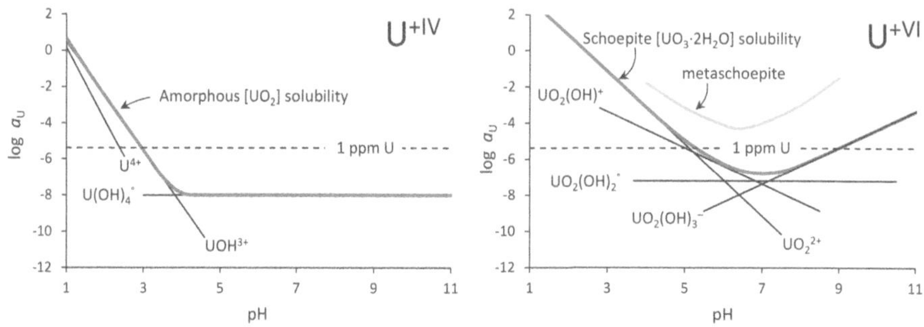

FIGURE 8.46 Solubility of reduced (U^{IV}) uranium oxide (amorphous) (UO_2) (left) and oxidized (U^{VI}) uranium oxyhydroxides, schoepite, [$UO_3{\cdot}2H_2O$] (right). Also shown is the solubility of metaschoepite (Gorman-Lewis et al. 2008). Crystalline uranium oxide, uraninite, has a solubility some eight orders of magnitude lower than amorphous UO_2. Lines drawn using $\Delta G°$ values compiled by Langmuir (1997) from various sources.

10 ppb at pH 6 and less than 1 ppb at pH 7 (Figure 8.47). The uranyl ion, however, readily complexes with the carbonate anion, greatly enhancing uranium solubility as $UO_2(CO_3)_2^{2-}$ and making uranium carbonate, Rutherfordine [UO_2CO_3] one of the most soluble uranium minerals in low-ionic-strength water. By contrast, vanadium, V, complexes the uranyl ion, limiting uranium solubility to less than 1 ppm with the uranium vanadate mineral carnotite [KUO_2VO_4].

Surface waters and groundwaters have, on average, less than 1 ppb, although concentrations of several ppb can be observed. Low concentrations of soluble U^{VI} species in natural waters are due to its low crustal abundance and sorption of the uranyl ion on clays. It is also limited by reaction with iron oxyhydroxide at higher pH. The

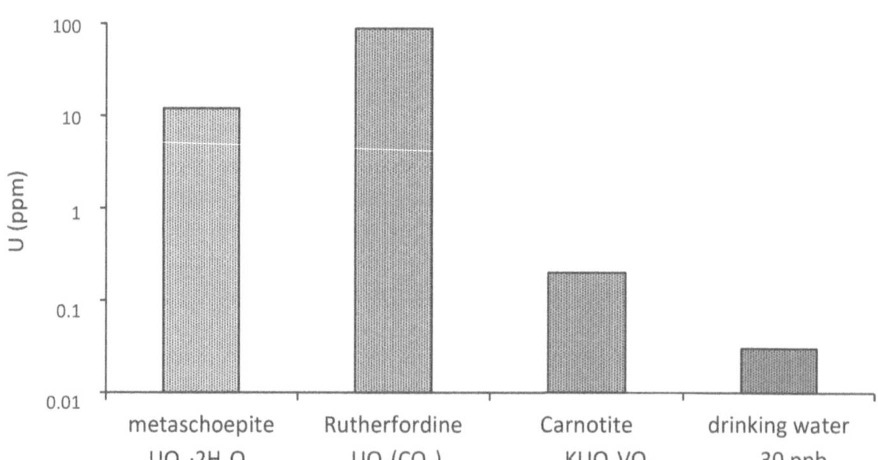

FIGURE 8.47 Solubility of U^{VI} minerals in fresh water showing the importance of carbonate complexed uranium, $UO_2(CO_3)_2^{2-}$, to increase the solubility of the uranyl ion, UO_2^{2+}, and vanadium, V^V as vanadate, $H_2VO_4^-$, to decrease UO_2^{2+} solubility in fresh water.

U.S. EPA places a maximum concentration limit of 30 ppb for total U in drinking water. The WHO uses a radiologically derived limit of 130 ppb.

The long half-life of ^{238}U gives it only a minimal radiological hazard. However, it is a toxic heavy metal that accumulates in the kidneys, creating long-term health problems. Important sources of uranium in the environment are from the uranium enrichment process used to increase the $^{235}U/^{238}U$ ratio for light-water nuclear reactors and for nuclear weapons (uranium depleted in ^{235}U or DU being the waste product). The use of depleted uranium for armor and armor-piercing shells has also increased environmental exposure.

Nuclear Fuel and Radioactive Waste

Most reactors *burn* purified uranium dioxide (UO_2) that has been fired to produce a fine-grain porcelain structure. Light-water reactors require enriched fuel, where the ratio of fissionable ^{235}U to common ^{238}U is increased from the natural level of 0.72% to over 3% in the fuel. This is done by the gaseous diffusion of uranium hexafluoride (UF_6), which generates large quantities of ^{235}U-depleted uranium hexafluoride or DUF_6. A solid at room temperature, UF_6 readily hydrates to produce soluble uranium oxyfluoride (UO_2F_2) and highly corrosive hydrofluoric acid (HF), both of which are toxic. Spent uranium fuel can also be reprocessed to re-enrich the residual ^{235}U content, recover plutonium for fuel, and reduce the mass of radioactive waste. Reprocessing, however, is accompanied by the release of short-lived noble gas radioisotopes, as well as volatile radionuclides, including ^{129}I and tritium.

The natural radioactivity in uranium fuel generates a neutron flux that propagates fission of ^{235}U, producing new high-energy neutrons and two lighter-weight nuclides. In a nuclear weapon, a supercritical mass of pure ^{235}U, the fission neutrons released propagate an uncontrolled chain reaction. In a nuclear reactor, the much lower ^{235}U-to-^{238}U ratio sustains a lower neutron flux that is controlled by rods of neutron-absorbing elements, usually boron. A moderator is also used to slow the high-energy neutrons (by multiple neutron-hydrogen impacts) to thermal range, where they are more likely to be captured by other ^{235}U nuclides to sustain fission. Reactors burning enriched uranium fuel use light water as a moderator, although H_2O will capture neutrons and so is an inefficient moderator. The deuterium in heavy water (D_2O) has a small neutron capture cross section and so is more efficient at slowing neutrons, allowing nonenriched fuel to be used. This is the basis for the Canadian Deuterium Uranium or CanDU reactor design launched in 1944 by C.J. Mackenzie as director of the Canadian National Research Council.

Fission generates one to two fast neutrons and two lighter nuclides or fission products, with masses of about 90 and 140. The kinetic energy of these products accounts for over 80% of the total energy release and is harnessed as heat by the cooling water to drive steam turbines. Subsequent β and γ decay of the daughter nuclides accounts for the remainder of the energy released. Examples of possible decay schemes are as follows:

$$^{235}_{92}U_{143} + n \rightarrow {}^{90}_{38}Sr_{52} + {}^{143}_{60}Nd_{83} + 2n$$
$$^{235}_{92}U_{143} + n \rightarrow {}^{96}_{40}Zr_{56} + {}^{137}_{55}Cs_{82} + 2n$$
$$^{235}_{92}U_{143} + n \rightarrow {}^{104}_{44}Ru_{60} + {}^{129}_{53}I_{76} + 2n$$

Excess neutrons in the fission products convert to protons by β decay. This produces elements of higher atomic number and lower neutron number, but with essentially the same mass. There is no loss of nucleons (one neutron with mass 1.6749×10^{-24} g decays to one proton with mass 1.6726×10^{-24}, with the difference of 2.3×10^{-27} or 0.14% attributed to the β particle and energy), and so the decay chains are essentially isobaric, stepping left to right from element to element along the rows of the periodic table until a stable or long-lived radionuclide is reached. Some fission products decay rapidly to stable nuclides and are not of concern:

$$^{140}\mathrm{Xe}_{(13.6\,s)} \rightarrow {}^{140}\mathrm{Cs}_{(63.7\,s)} \rightarrow {}^{140}\mathrm{Ba}_{(12.8\,d)} \rightarrow {}^{140}\mathrm{La}_{(1.68\,d)} \rightarrow {}^{140}\mathrm{Ce}_{(\mathrm{stable})}$$

Others decay by β emission to long-lived radionuclides that pose a long-term radio-hazard:

$$^{90}\mathrm{Se}_{(140\,ms)} \xrightarrow{\beta} {}^{90}\mathrm{Br}_{(1.92\,s)} \xrightarrow{\beta} {}^{90}\mathrm{Kr}_{(32\,s)} \xrightarrow{\beta} {}^{90}\mathrm{Rb}_{(158\,s)} \xrightarrow{\beta} {}^{90}\mathrm{Sr}_{(28.78y)}$$

$$^{129}\mathrm{In}_{(0.61\,s)} \xrightarrow{\beta} {}^{129}\mathrm{Sn}_{(2.23\,m)} \xrightarrow{\beta} {}^{129}\mathrm{Sb}_{(4.4\,h)} \xrightarrow{\beta} {}^{129}\mathrm{Te}_{(69.6\,m)} \xrightarrow{\beta} {}^{129}\mathrm{I}_{(15.7My)}$$

In addition to the suite of fission products in spent uranium fuel is a second group of radionuclides produced by neutron activation of other elements in the reactor core. Tritium and radiocarbon are both produced by neutron impacts on nitrogen, present as ubiquitous N_2 gas in the fuel rods, the coolant water, and the moderator water.

$$^{14}_{7}\mathrm{N}_7 + \mathrm{n} \rightarrow {}^{3}_{1}\mathrm{H}_2 + {}^{12}_{6}\mathrm{C}_6$$

$$^{14}_{7}\mathrm{N}_7 + \mathrm{n} \rightarrow {}^{14}_{6}\mathrm{C}_8 + \mathrm{p}$$

Trace amounts of chloride in the fuel and water are also neutron-activated:

$$^{35}_{17}\mathrm{Cl}_{18} + \mathrm{n} \rightarrow {}^{36}_{17}\mathrm{Cl}_{19} + \gamma$$

Finally, the tremendous neutron flux in the fuel also activates ^{238}U to form the transuranic or superheavy elements. Activation of abundant ^{238}U produces ^{239}Np, an isotope of the first transuranic element neptunium. This decays with a half-life of 2.4 days to more stable plutonium, ^{239}Pu:

$$^{238}_{92}\mathrm{U} + \mathrm{n} \rightarrow {}^{239}_{92}\mathrm{U}_{147(24\,m)} \xrightarrow{\beta} {}^{239}_{92}\mathrm{Np}_{146(2.4\,d)} \xrightarrow{\beta} {}^{239}_{94}\mathrm{Pu}_{146(24,100y)}$$

The long half-life of ^{239}Pu allows it to accumulate in the fuel rods, where further activation produces other long-lived radionuclides of the actinides, including

$$^{240}\mathrm{Pu}_{(6,570y)},\ {}^{242}\mathrm{Pu}_{(376,000y)},\ {}^{241}\mathrm{Am}_{(432y)},\ {}^{243}\mathrm{Am}_{(7,370y)},\ \text{and}\ {}^{244}\mathrm{Cm}_{(18.1y)}.$$

The radioactivities contributed by the main radionuclides in spent fuel are listed in Table 8.3, together with their half-life, their final stable decay product, and the geochemical form. Also calculated here are their contribution to the mass (based on $Aa = \lambda\ Nn$) of spent fuel. It is noteworthy that the most radioactive constituents, ^{90}Sr and ^{137}Cs, represent only 100 mg and 240 mg, respectively, in 1 kg of spent fuel (i.e., 100 and 240 ppm).

Spent fuel consists of fission-microfractured, porcelain-like UO_2 pellets in corrosion-resistant zircaloy tubes (98% zirconium with ~1.5% tin and minor iron). Radionuclides mobile as anions or gas, including T, ^{14}C, ^{36}Cl, and ^{129}I, represent the early release group, followed by the soluble alkali, Cs^+, and the alkaline earth, Sr^{2+}. The long-term release group is essentially the actinides present as insoluble tetravalent oxides. Only extensive oxidation of the uranium fuel pellets will release them to groundwater.

With the high radioactivity of these fission and activation products, spent fuel is considered *high-level waste* (HLW). Once removed from the reactor, it is stored in pools to dissipate the heat from β decay of the early fission products. However, long-term isolation is required for the decay of nuclides in Table 8.4. In addition to HLW is the class of *low- and intermediate-level waste* (LILW), consisting of ^{14}C extracted onto resin columns from reactor moderator water plus irradiated

TABLE 8.4
The Long-lived Radionuclides of Concern in Spent Uranium Fuel Waste

Nuclide	Half-life (years)	Radioactivity (million Bq per kg spent UO_2 fuel)	Mass (g per kg spent UO_2 fuel)	Stable decay product	Geochemic species
Activation products					
^{3}H (or T)	12.32	0.16	4.5E-10	^{3}He	H_2O, H_2
^{14}C	5730	42.6	0.00026	^{14}N	CO_2, CH_4
^{36}Cl	301,000	0.46	0.00038	^{36}Ar	Cl^-
Fission products					
^{90}Sr	28.78	514,000	0.10	^{90}Zr	Sr^{2+}
^{99}Tc	211,100	128.4	0.20	^{99}Ru	TcO_2
^{93}Zr	1,530,000	15.6	0.17	^{93}Nb	ZrO_2
^{129}I	15,700,000	5.18	0.79	^{129}Xe	I^-
^{135}Cs	2,300,000	1.01	0.024	^{135}Ba	Cs^+
^{137}Cs	30.07	766,000	0.24	^{137}Ba	Cs^+
Actinides					
^{238}U	4500 My	12.2	987	^{206}Pb	UO_2
^{239}Pu	24,100	6360	2.8	^{207}Pb	PuO_2
^{240}Pu	6570	8700	1.0	^{208}Pb	PuO_2
^{242}Pu	376,000	8.81	0.061	^{206}Pb	PuO_2
^{241}Am	432	11,100	0.087	^{209}B	AmO_2
^{243}Am	7370	19.4	0.0026	^{207}Bi	AmO_2
^{244}Cm	18.10	448	0.00015	^{208}Pb	CmO_2

equipment, decommissioning waste and materials associated with the handling of nuclear compounds, laboratory clothing and liquids, and wastes from hospitals and industries using nuclear radiation in specific applications. Although low in overall radioactivity, LILW generally has high volume and, with the long half-life of some radionuclides like ^{14}C, requires a long-term disposal solution.

Radioactive Waste Disposal

Burial of nuclear waste in deep geological repositories (DGRs) is considered by most countries to be the most favorable option to provide secure isolation for the period of time required decay to low levels of radioactivity. Figure 8.48 shows the rate of decay for the radionuclides in spent fuel listed in Table 8.4. The time periods for decay range from decades for tritium to millions of years for ^{129}I. Both HLW and LILW require isolation on a time scale of 10,000 years or longer to allow decay of highly radioactive and soluble ^{90}Sr and ^{137}Cs and of very mobile ^{14}C. Beyond this time frame, only ^{129}I, mobile as an anion, is considered of concern.

DGRs include packaging of the waste in copper canisters emplaced in holes drilled in adit floors and packed with dry bentonite clay buffer. Adits and shafts are then backfilled with clay and sand, and/or cement. This engineered barrier, or *near field*, provides security over a long period of time, but conservative models accept

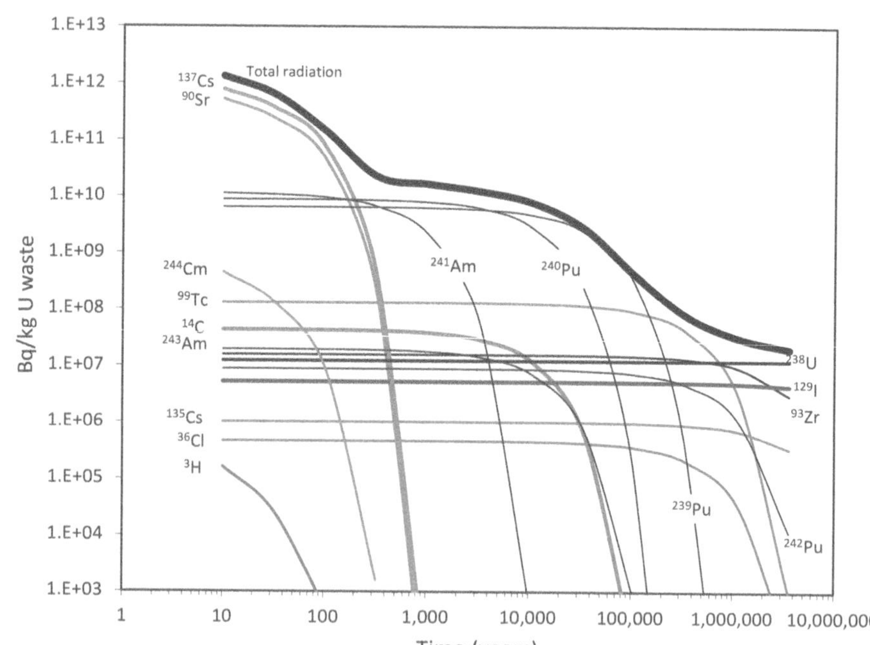

FIGURE 8.48 Contribution of major radionuclides to overall radiation of spent uranium fuel and their decay over time. Both ^{14}C and ^{129}I are of concern as C and I are mobile as anions (HCO_3^-, I^-, and IO_3^-) and gases (CO_2, CH_4, and CH_3I) and are biophylic.

that it will at one point be breached. Containment of the radionuclides in Table 9.5 after this time relies on the integrity of the *far field*—the hydrogeological and geochemical environment beyond the engineered barrier of the DGR.

Migration through the far field to the surface could occur by advection with groundwater along fractures and permeable horizons or by diffusion through porewater. In either case, mobility will be affected by geochemical reaction with the surfaces of minerals and organics along the pathway. Redox and pH, of course, play significant roles. The following discusses the important geochemical considerations for migration of the radionuclides of concern.

^{3}H or T—Hydrogen is highly mobile as H_2O. However, the risk posed by tritium is not from its presence in HLW but in releases from power plants during operation. Releases occur as tritiated water (HTO), and tritiated hydrogen gas (HT). Most tritium produced by neutron activation of H and D in moderator water is stripped and stored as solid titanium hydride (TiT_2) to decay. Some is sealed as T_2 gas in glass tubes with a phosphor coating that emits light by radioluminescence for emergency lighting, gun sights, and other applications that demand nonelectrical lighting. The high volatility of HTO and its accumulation in vegetation as organically bound tritium creates halos around nuclear generating facilities that are up to about 30 Bq/L compared to regional backgrounds of about 3 Bq/L. Nonetheless, levels are well below those posing a radio-hazard.

^{14}C—Carbon in natural groundwaters is largely present as DIC but is particularly mobile in its reduced form as CH_4. Organic resins used to strip ^{14}C from moderator water represent most of the radiocarbon inventory of LILW. Conservative models consider that the organic wastes with ^{14}C will decompose to $^{14}CO_2$ and, with H_2 from corrosion of steel containers in the vaults, would produce highly mobile $^{14}CH_4$ by CO_2 reduction. Methane in the far field may then be oxidized by methanotrophic bacteria in the near surface, representing a potential loss of ^{14}C to the biosphere. Radiocarbon transported as DIC would experience some attenuation by precipitation and exchange with carbonate minerals along fractures. Radiocarbon as DIC would also be retained by reaction with any concrete used as backfill material and bulkheads in the decommissioned repository, as well as cement grout to fill fractures and fault zones. Reardon and Dewaele (1990) and Dayal and Reardon (1994) show that DIC species react strongly with the calcium–silica–hydrate minerals in cement, producing more stable calcite.

^{36}Cl—Chloride is ubiquitous in natural waters as the conservative anion, Cl^-, and particularly in groundwaters, where it can originate from saline sources such as crustal brines, seawater and accumulated salts. Almost always present as an anion, $^{36}Cl^-$ is very conservative and does not adsorb onto minerals or organics.

^{90}Sr—Strontium is always present as a divalent alkaline earth cation, Sr^{2+}. It is not redox sensitive and replaces Ca^{2+} in feldspars and carbonate minerals. Sr^{2+} is retarded by adsorption on clay minerals. The bentonite buffer around the canisters, while creating a very low permeability barrier after hydration, would also retain strontium by cation exchange.

^{99}Tc—Technetium is a most interesting element in that it has no stable isotope and is not found in nature. Its longest-lived isotope, ^{98}Tc, has a half-life of only 4.2 million years, and so any technetium generated during nucleosynthesis at the creation of our solar system some 4.5 billion years ago has long since decayed away. This makes understanding its geochemical behavior somewhat problematic, and we must rely on close analogues, such as its heavier Group VIIA neighbor, rhenium, Re. Under reducing conditions in the spent fuel repository, tetravalent TcO_2 should be stable, but, like ReO_2, is weakly soluble, with concentrations greater than 40–100 ppb. If oxidized to Tc^{VII}, it is highly mobile as the TcO_4^- anion (Kim and Boulègue 2003).

^{135}Cs and ^{137}Cs—Cesium is an alkali metal, found exclusively in the hydrosphere as Cs^+. Like K^+, it is soluble in water, but in aquifers it is highly retarded by cation exchange on clays. Some 50 to 100 × 10^{15} Bq of ^{137}Cs (about 25 kg) out of a total of 1 to 2 × 10^{18} Bq of radioactivity (4% to 8%) was released from Chernobyl Reactor 4 following the steam explosions and ignition of the graphite core in April 1986 (NEA 2002). ^{137}Cs remains today the radionuclide of greatest concern, although monitoring shows that local deposition was largely retained on clays within the upper soil horizon.

^{129}I—Iodine is a halogen and is stable in the environment as the reduced form, iodide, I^-, and also as the oxidized species, iodate, IO_3^- (Figure 8.49). As an anion, it has higher potential for mobility from nuclear waste than the cations, Cs+, Sr2+, and the actinides. However, it can be readily complexed by organics and so has a strong potential for adsorption in soils (Alvarado et al. 2002) and on ferric iron oxides (Kennedy et al. 2011). It has high volatility as methyl iodide, ICH_3. The risk from ^{129}I is its uptake by the thyroid gland, coupled with its high mobility and long half-life. The short-lived isotope (^{131}I, T½ = 8 days) is a hazard from immediate releases from accidents, but is not considered in the waste inventory.

^{239}Pu, ^{240}Pu, ^{242}Pu, ^{241}Am, ^{243}Am, and ^{244}Cm—The transuranic radionuclides produced in the fuel by neutron capture with ^{238}U are all stable as tetravalent oxides, PuO_2, AmO_2, and CmO_2. Like UO_2, they will be essentially immobile under reducing conditions. Under oxidizing conditions, they can gain mobility with higher valence states as anions, such as PuO_4^{2-}. However, unlike hexavalent uranium, these actinides are readily reduced back to their insoluble tetravalent state.

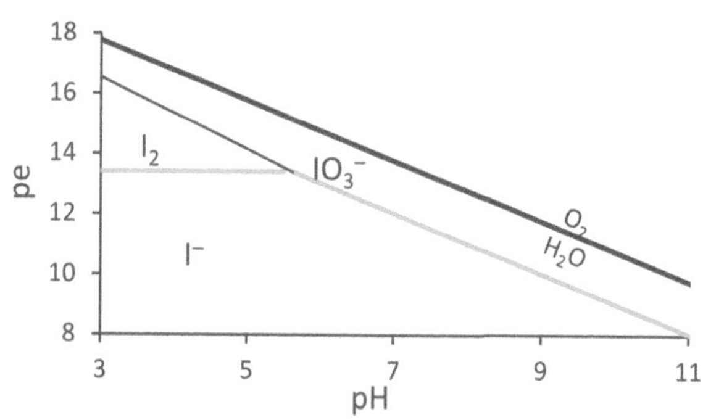

FIGURE 8.49 Aqueous speciation of iodine with redox and pH, showing the dominant species and the upper stability field of water. Iodide can also form methyl iodide, ICH_3 at low *pe* and with available organic matter.

Tritium in Groundwater, *Pembroke, Ontario*

Although tritium is a naturally occurring radionuclide in fresh waters produced by cosmic radiation, great quantities are produced by neutron activation of hydrogen in reactors. The Canadian designed CanDU heavy water (D_2O) reactor produces a high level of tritium through activation of deuterium. Experimental fusion reactors use tritium as fuel ($D + T \rightarrow {}^4He + n$) and will gain predominance in power generation in decades to come, contributing to tritium losses to the environment. The short half-life and low energy of tritium decay make it useful in new products, including non-electrical ones, for emergency lighting or illumination where spark hazards are of concern (Figure 8.51).

Encapsulation of HT, as shown in the photo in Figure 8.50, is accompanied by emissions to the atmosphere. Hydrogenase enzymes in soils convert the HT gas to HTO, which recharges groundwater with precipitation. This is closely monitored at a site in Pembroke Ontario (SRB Technologies 2016). Improvements to the encapsulation procedures reduced emissions after 2000 by over an order of magnitude. The record of emissions was recovered in porewaters from a 5-m core of silty clay sediments, where the 2000 peak has migrated to the base of the profile overlying the limestone (Shkarupin et al. 2025).

The migration of this tritium plume has been monitored for compliance, with data shown in Figure 8.51 for over a decade, and thus close to a full half-life for tritium. The monitoring data for the top of bedrock wells (left chart) record the declining limb of the 2000 emissions peak at the well 10 m from the stack. By about 2009–2010, the attenuated peak is passing the 50-m well and has reached the 115-m monitoring well as a much attenuated peak after 2014. The significant attenuation

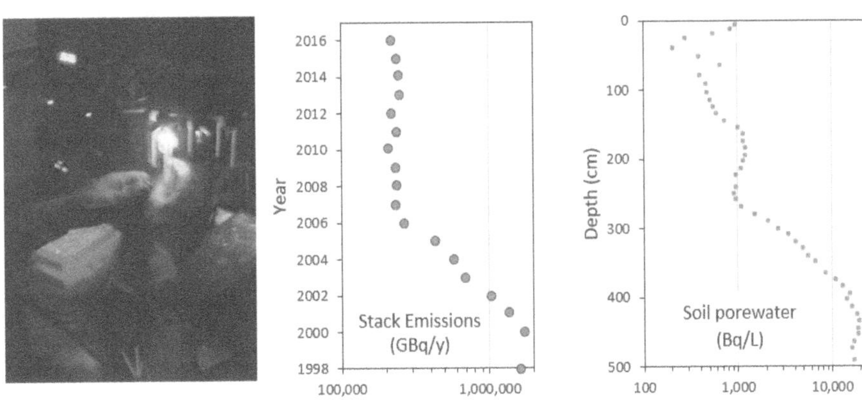

FIGURE 8.50 Tritium as hydrogen gas (HT) is encapsulated by the rig operator into glass tubes that glow (green) from the beta-irradiation of the ZnS lining (left photo). Center: Rig stack emissions of HT are oxidized to tritiated water (HTO) and enter groundwater with precipitation. Improvements in the encapsulation rig in 2000 reduced emissions by over an order of magnitude. Right: Soil porewater tritium profile produced in 2017 from a 5-m core to bedrock of silty clay, showing the 2000 peak reached the base of the sediments and entered the bedrock aquifer (Shkarupin et al. 2025).

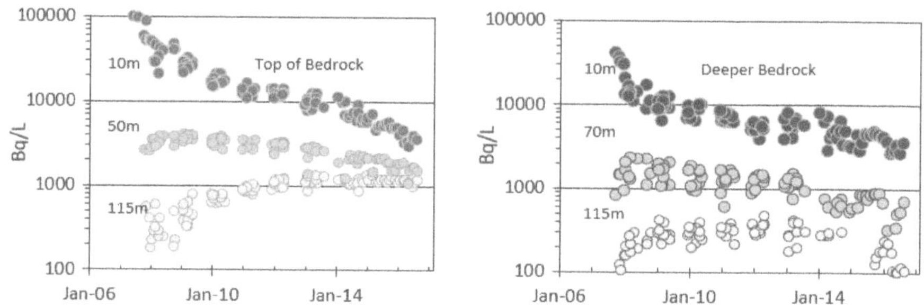

FIGURE 8.51 Plume of tritium in groundwater in a limestone aquifer below 5 m of low permeability silty clay, monitored from January 2006 to October 2016. (Data from SRB Technologies 2016; Schmidt 2017).

is due to mixing, matrix diffusion, and decay (50% over this period), as sorption of tritium does not occur. Monitoring in the deeper bedrock wells over this same period shows similar trends, although with earlier arrival times and much greater attenuation. These trends suggest that the 2000 plume will be well below the 7000 Bq/L drinking water limit before reaching the Muskrat River drainage channel.

Strontium-90 in Groundwater, *Chalk River, Ontario*

In November 1954, the accidental loss of process solutions from an experimental spent fuel reprocessing plant at the Chalk River Nuclear Laboratories released some 26 TBq (26×10^{12} Bq) of ^{90}Sr and 37 TBq of ^{137}Cs into a shallow sand aquifer, along with much lower activities of other fission products, ^{154}Eu ($T_{½}$ = 8.6 years), ^{106}Ru ($T_{½}$ = 374 days), and ^{144}Ce ($T_{½}$ = 285 days). Some 35 years later, it was found that the undecayed ^{137}Cs and ^{154}Eu remained sorbed within the unsaturated zone and that the short-lived ^{106}Ru and ^{144}Ce radionuclides had decayed to background. However, the ^{90}Sr had generated a plume extending up to 350 m from the source (Killey and Munch 1987), with potential discharge into a wetland before the ^{90}Sr activity had decayed to below the standard of 5 Bq/L.

The sand aquifer has an average linear groundwater velocity of about 70 m/y, giving a circulation time for groundwater along the 350 m flow path of about 5 years. The ^{90}Sr plume, however, was found to be moving at a much slower rate due to sorption onto ferric iron oxyhydroxide coatings on the sand (Killey and Munch 1987). Distribution coefficients for ^{90}Sr in the core of the plume were calculated from concentrations in the porewater compared with the sorbed fraction and found to be high, at 12.9 mL/g. However, the K_d values measured at the leading edge of the plume as low as 0.8 mL/gd, due to the existence of preferential pathways with less ferric iron substrate for sorption. Using the calculation for a retardation factor discussed above, the values for porosity, $\theta = 0.33$, and bulk density, $\rho = 2$ g/cm^3, velocities for the core of the plume and for the leading edge can be determined. Both are important, as the latter determines when the receptor wetland will be impacted and the former determines whether decay to background will occur before the bulk of the ^{90}Sr reaches the wetland.

Core

$$R = 1 + \frac{\rho}{\theta} \times K_d$$
$$= 1 + \frac{2}{0.33} \times 12.9$$
$$= 79$$

Leading edge

$$R = 1 + \frac{\rho}{\theta} \times K_d$$
$$= 1 + \frac{2}{0.33} \times 0.8$$
$$= 5.8$$

Using the relationship for $v_{Sr} = \frac{v_{gw}}{R}$ and groundwater velocity of 70 m/y allows calculation of the plume:

$$v_{Sr} = \frac{v_{gw}}{R}$$
$$= 70/79$$
$$= 0.88\,\text{m/y}$$

Time to wetland: 395 years

$$v_{Sr} = \frac{v_{gw}}{R}$$
$$= 70/5.8$$
$$= 12.1\,\text{m/y}$$

29 years

The velocity of the leading edge of the plume predicts arrival at the wetland within about one half-life, while the core of the plume would not arrive before 14 half-lives elapsed, bringing ^{90}Sr activities to background levels. Accordingly, an in situ wall-and-curtain PRB was designed to remove ^{90}Sr as the leading edge reached the wetland, but not for the full load of ^{90}Sr in the aquifer (Lee et al. 1998). A sheet piling wall was driven to the base of the aquifer, and a permeable curtain in front was filled with a zeolite, clinoptilolite ($K_3Al_3[Al,Si]_2Si_{13}O_{36}\cdot 12H_2O$), with a high surface area and high CEC. The groundwaters have a Ca^{2+} concentration of only 10 mg/L, whereas higher ionic-strength water would decrease ^{90}Sr sorption due to competition from other cations (Hull and Schafer 2008).

The K_d for $^{90}Sr^{2+}$ sorption on clinoptilolite is reported to be as high as 2600 mL/g, providing sorption for several decades (Lee et al. 1998). Lee and Hartwig (2005) report the following performance of the wall and curtain:

Contaminated groundwater treated: 15 million L/y (15,000 m^3/y)
Uncontaminated groundwater diverted: 10 million L/y (10,000 m^3/y)
^{90}Sr activity of contaminated groundwater: 85 Bq/L
^{90}Sr activity of treated groundwater: <0.6 Bq/L
^{90}Sr removed from groundwater: 850 million Bq/y or 99.5%

Core Meltdown, *Fukashima, Japan*

The magnitude 9.0 Tōhoku earthquake of March 11, 2011, generated a tsunami that killed over 15,000 people and displaced 4.4 million households. The tsunami also breached the seawall at the Fukushima Dai'ichi nuclear power plant, leading to meltdowns and releases of radionuclides to the atmosphere. Extensive monitoring of short- and long-lived radionuclides has been undertaken in air, soils, surface waters, and groundwaters (MEXT 2012; Kinoshita et al. 2011). The greatest activities measured were contributed by short-lived ^{131}I ($T_{½}$= 8 days), with levels exceeding 10^6 Bq/m^2 along a northwest trajectory generated by the prevailing winds at that time from the power plant to the 30-km perimeter and beyond into the extended evacuation zone (Kinoshita et al. 2011). However, more emphasis was placed on ^{137}Cs and ^{90}Sr as long-lived radioisotopes that contribute the greatest amount to the long-term effective radiation dose. Figure 8.52 shows that ^{137}Cs activities exceeded 1 million Bq/m^2 along the northwest trajectory and over 500,000 Bq/m^2 within the extended evacuation zone, whereas ^{90}Sr activities were largely below 10,000 Bq/m^2.

Contributions of these long-lived radionuclides to soil and groundwater remain as ongoing questions. Subsequent monitoring of well water in the region, however, found activities for ^{137}Cs below 0.72 Bq/kg in groundwater and one detectable level of ^{90}Sr at 0.0014 Bq/kg (Iijima 2013). Surface waters draining the high-activity corridor had ^{137}Cs activities on the order of 1–2 Bq/kg and ^{90}Sr less than 0.02 Bq/kg. The disparity between the high levels found in soils and those found in runoff and infiltrating waters is attributed to sorption. The distribution coefficients measured for highly organic paddy and upland soils in Japan are on the order of 100–1000 mL/g for ^{90}Sr and an order of magnitude greater for ^{137}Cs (Kamei-Ishikawa et al. 2008). The associated retardation factors, R, and velocities (v_{Sr} and v_{Cs}) then make ^{137}Cs essentially immobile and ^{90}Sr highly retarded in the soils and aquifers of the Fukushima region.

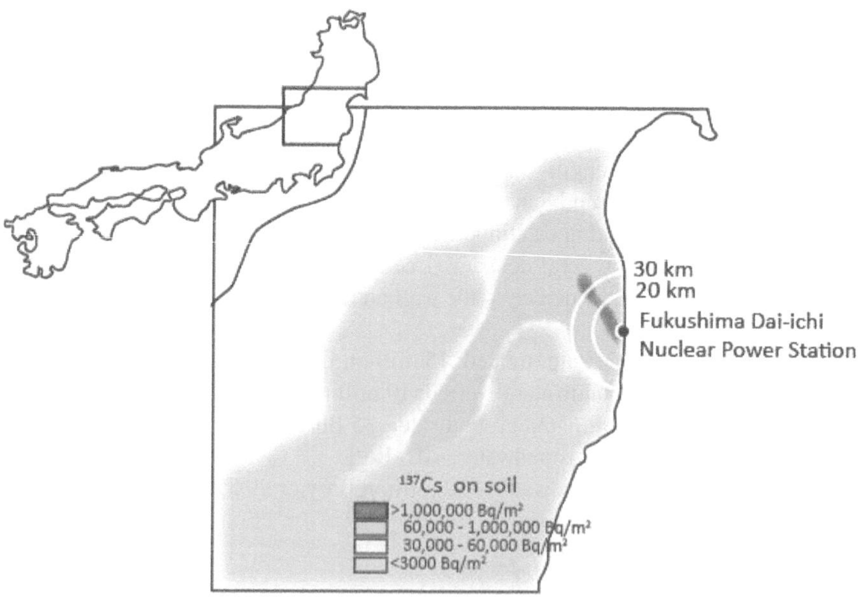

FIGURE 8.52 ^{137}Cs fallout on soils of the Fukushima region (after MEXT 2012) in northeastern Japan (inset).

Aquitards and Nuclear Waste Isolation

Most countries with a nuclear power generation capability are now looking for appropriate aquitards in which to bury their LILW and the spent fuel that constitutes HLW in a DGR. Safety assessment of a DGR considers that the engineered barrier will fail at some point and release radionuclides into the groundwater. The hydrogeological *far-field* beyond the repository then becomes the long-term barrier to the migration of radionuclides. Site investigations must demonstrate that groundwaters in the far-field are ancient and that solute transport is by diffusion only.

Geological settings that satisfy these two conditions fall into two categories, crystalline bedrock and argillaceous sedimentary formations. Crystalline terrains, generally granites, have very low hydraulic conductivities, providing the rock mass has no significant fracturing. However, finding a sufficiently large volume of unfractured crystalline rock is a challenge. In contrast, sedimentary formations in undeformed basins have predictable geology with few fractures. Clay minerals in shales, argillaceous carbonates, and unlithified clay units reduce permeability and preclude advective transport. Both the advective and diffusive regimes can be assessed with geochemistry and isotopes in the porewaters. Advective flow in permeable formations represents the boundaries to the site, and a diffusive regime is the geological environment being evaluated for its barrier properties.

Solute transport by diffusion in aquitards is most clearly demonstrated by the porewater profiles for naturally occurring conservative tracers, such as ^{18}O, D, and Cl^-, as well as highly diffusive species like helium (Mazurek et al. 2009). Sampling porewaters for analysis of these natural tracers can be challenging, with methods ranging from slowly responding piezometers to extraction from core samples by squeezing, direct equilibration, or vacuum distillation. The shape of the profile that these tracers establish is a function of (1) their initial concentrations at the aquitard boundaries and in the aquitard porewaters, (2) their rate of diffusion, and (3) time. Using measurements of effective diffusion coefficients in the aquitard (see Chapter 5 in the section on diffusive mixing in aquitards) together with an understanding of the initial boundary conditions, the time to establish the observed profile can be determined. This provides an estimate of the time afforded by the aquitard to retain radionuclides migrating from the DGR.

Hendry and Wassennaar (1999) use δD measured in porewaters collected from piezometer nests installed through an 80-m-thick glacial clay till aquitard overlying a 76-m-thick Cretaceous marine clay. This profile is shown in Figure 4.27 (Left profile) as an example of diffusive mixing of groundwater. The profile was used to calibrate a 1-D groundwater flow and transport model to constrain the time to establish the δD profile (Figure 8.53). The lower profile is considered to have developed over the last 30,000 years, following emplacement of the till during the last advance of the Laurentide Ice Sheet, while the upper profile developed over the Holocene after the glacier retreated some 10,000 years ago.

A diffusional profile developed over a longer period of time was studied in porewater of the Opalinus Clay at Benken, Switzerland (Gimmi et al. 2007). Here, the aquitard has two bounding aquifers, with well-established timing for meteoric groundwaters replacing initially marine water. Diffusional modeling indicates

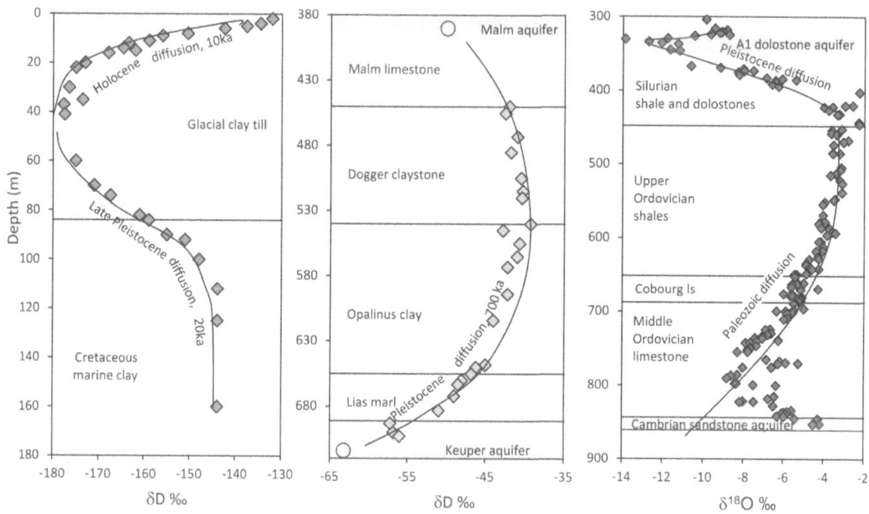

FIGURE 8.53 Isotope profiles and 1D diffusional modeling in aquitards. Left: δD profile in porewaters from the Pleistocene Battleford till and underlying Cretaceous Bearpaw marine clay, southern Saskatchewan (after Hendry and Wassenaar 1999), center: $\delta^{18}O$ profile through the Opalinus Clay in Switzerland (after Gimmi et al. 2007), and right: $\delta^{18}O$ profile through Ordovician to lower Silurian shales and carbonates of the Michigan Basin (Al et al. 2015).

that the initially marine porewaters evolved to the observed profile over some 700,000 years (Figure 8.53, center profile).

A final example comes from a site in the Ordovician shales and limestones on the eastern flank of the Michigan Basin in central North America considered for a DGR for LILW in southern Ontario (Figure 8.53, right profile). Here, the proposed host Cobourg Fm is overlain by a 250 m shale aquiclude with porewaters originating as evaporated Silurian seawater (Intera Engineering Ltd. 2011; Clark et al. 2013). The $\delta^{18}O$ profile (Figure 8.53) shows a diffusional profile through the Ordovician shales established over a 300 million year period (Al et al. 2015). Above, the profile in the Silurian section shows Pleistocene glacier effects.

An additional perspective on the dominance of diffusive transport in aquitards is given by the analysis of helium concentrations and isotopes in porewaters. Sampling this highly diffusive gas can be done from piezometers but more often requires encapsulation of core samples immediately following recovery followed by outgassing and analysis under high vacuum. The resulting profiles of the $^3He/^4He$ ratio give insights into the source of the geogenic component, and the helium concentration in the porewaters allows calculation of diffusion rates and time.

Helium profiles for the three sites discussed above in Figure 8.53 are shown in Figure 8.54. For the southern Saskatchewan clay till site, Hendry et al. (2005) use the $^3He/^4He$ ratio to show radiogenic production below the profile and the concentration profile to show upward diffusion toward the surface over the past 15,000–25,000 years. By contrast, helium in the Opalinus Clay in Switzerland (Mazurek et al. 2009) originates largely within the aquitard by in situ radiogenic production over a

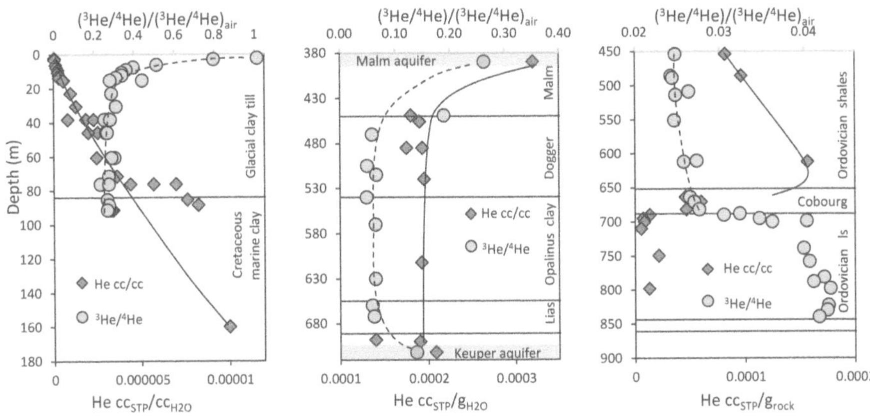

FIGURE 8.54 Helium concentration and isotope profiles in aquitards. Left: Clay till, southern Saskatchewan with diffusion trend toward atmospheric concentration and isotope ratio at surface (after Hendry et al. 2005); Opalinus clay, Switzerland, with linear distribution for concentration and isotopes from in situ production in Opalinus and Dogger clays (after Mazurek et al. 2009); and Ordovician shales and carbonates, Michigan basin, southern Ontario, with upward diffusion from in situ production in the shales but isolated below the Cobourg Formation from basin-derived helium (Intera Engineering Ltd. 2011; Clark et al. 2013).

17-million-year period with a minor component of deep (mantle) He with enriched $^3He/^4He$. The helium isotope profile from the Ordovician strata of the Michigan Basin shows a zone of in situ He production in the Ordovician shales, which is isolated at the base of the Cobourg Formation from a deeper zone of regionally sourced helium with a much more enriched $^3He/^4He$ signature. The concentrations of helium in this aquiclude proposed for a DGR suggest an age for the porewaters of some 260 million years.

References

Aeschbach-Hertig, W., 2014. Radiokrypton dating finally takes off. *PNAS*, 111: 6856–6857.

Aeschbach-Hertig, W., Peeters, F., Beyerle, U. and Kipfer, R., 1999. Interpretation of dissolved atmospheric noble gases in natural waters. *Water Resources Research*, 35: 2779–2792.

Aeschbach-Hertig, W., Peeters, F., Beyerle, U. and Kipfer, R., 2000. Palaeotemperature reconstruction from noble gases in ground water taking into account equilibration with entrapped air. *Nature*, 405: 1040–1044.

Aggarwal, P. K., Alduchov, O. A., Froehlich, K. O., Araguas-Araguas, L. J., Sturchio, N. C. and Kurita, N., 2012. Stable isotopes in global precipitation: A unified interpretation based on atmospheric moisture residence time. *Geophysical Research Letters*, 39: L11705.

Aggarwal, P. K., Longstaffe, F. J. and Schwartz, F. W., 2023. Ice core ^{17}O reveals past changes in surface air temperatures and stratosphere to troposphere mass exchange. *Atmosphere*, 14: 1268.

Aggarwal, P. K., Romatschke, U., Araguas-Araguas, L., Belachew, D., Longstaffe, F. J., Berg, P., Schumacher, C. and Funk, A., 2016. Proportions of convective and stratiform precipitation revealed in water isotope ratios. *Nature Geoscience*, 9: 624–630.

Al, T. A., Clark, I. D., Kennell, L., Jensen, M. and Raven, K. G., 2015. Geochemical evolution and residence time of porewater in low-permeability rocks of the Michigan Basin, southwest Ontario. *Chemical Geology*, 404: 1–17.

Alvaredo Quiroz, N. G., Kotzer, T. G., Milton, G., Clark, I. D. and Bottomley, D., 2002. Partitioning of ^{127}I and ^{129}I in an unconfined glaciofluvial aquifer on the Canadian Shield. *Radiochimica Acta*, 90: 469–478.

Anderson, K. K. and Hooper, A. B., 1983. O_2 and H_2O are each the source of one O in produced from NH3 by nitrosomas: ^{15}N–NMR evidence. *FEBS Letters*, 64: 236–240.

Andrews, J. A. and Fontes, J.-C., 1992. Importance of the in situ production of ^{36}Cl, ^{36}Ar and ^{14}C in hydrology and hydrogeochemistry. In *Isotope Techniques in Water Resources Development 1991*, IAEA Symposium 319, March 1991, Vienna: 245–269.

Andrews, J. N., Davis, S. N., Fabryka-Martin, J., Fontes, J.-C., Lehmann, B. E., Loosli, H. H., Michelot, J.-L., Moser, H., Smith, B. and Wolf, M., 1989. The in situ production of radioisotopes in rock matrices with particular reference to the Stripa granite. *Geochimica et Cosmochimica Acta*, 53: 1803–1815.

Araoka, D., Kawahata, H., Takagi, T., Watanabe, Y., Nishimura, K. and Nishio, Y., 2013. Lithium and strontium isotopic systematics in playas in Nevada, USA: Constraints on the origin of lithium. *Mineralium Deposita*, 49: 371–379.

Aravena, R. and Wassenaar, L. I., 1993. Dissolved organic carbon and methane in a regional confined aquifer: Evidence for associated subsurface sources. *Applied Geochemistry*, 8: 483–493.

Aravena, R., Wassenaar, L. I. and Barker, J. F., 1995. Distribution and isotopic characterization of methane in a confined aquifer in southern Ontario. Canada. *Journal of Hydrology*, 173: 51–70.

Aravena, R., Wassenaar, L. I. and Plummer, L. N., 1995. Estimating ^{14}C groundwater ages in a methanogenic aquifer. *Water Resources Research*, 31: 2307–2317.

Artinger, R., Buckau, G., Kim, J. I., Geyer, S. and Wolf, M., 1996. Influence of sedimentary organic matter on dissolved fulvic acids in groundwater: Significance for groundwater dating with ^{14}C in dissolved organic matter. In *Isotopes in Water Resources Management*, vol. 1, IAEA Symposium 336, March 1995, Vienna, pp. 57–72.

Bäckström, M., Karlsson, S., Bäckman, L., Folkeson, L. and Lind, B., 2004. Mobilisation of heavy metals by deicing salts in a roadside environment. *Water Research*, 38: 720–732.

Bajjali, W., Clark, I. D. and Fritz, P., 1997. The artesian thermal groundwaters of northern Jordan: Insights to their recharge history and age. *Journal of Hydrology*, 192: 355–382.

Bao, H. and Gu, B., 2004. Natural perchlorate has a unique oxygen isotope signature. *Environmental Science and Technology*, 38: 5073–5077.

Barker, J. F. and Pollock, S. J., 1984. The geochemistry and origin of natural gases in southern Ontario. *Bulletin of Canadian Petroleum Geology*, 32: 313–326.

Barnes, I. and O'Neil, J. R., 1978. Present day serpentinization in New Caledonia, Oman and Yugoslavia. *Geochimica Cosmochimica Acta*, 42: 144–145.

Bayari, S., 2002. TRACER: An EXCEL workbook to calculate mean residence time in ground water by use of tracers CFC-11, CFC-12 and tritium. *Computers & Geosciences*, 28: 621–630.

Benetti, P., Calaprice, F., Calligarich, E., Cambiaghi, M., Carbonara, F., Cavanna, F. and Cocco, A. G. et al., 2007. Measurement of the specific activity of ^{39}Ar in natural argon. *Nuclear Instruments and Methods in Physics Research A*, 574: 83–88.

Bentley, H. W., Phillips, F. M. and Davis, S. N., 1986. Chlorine-36 in the terrestrial environment. In *Handbook of Environmental Isotope Geochemistry, Volume 2, The Terrestrial Environment, B*, edited by P. Fritz and J.-Ch. Fontes, Elsevier, Amsterdam, the Netherlands, pp. 427–480.

Bernat, M., Church, T. and Allegre, C., 1972. Barium and strontium concentrations in Pacific and Mediterranean sea water profiles by direct isotope dilution mass spectrometry. *Earth and Planetary Science Letters*, 16: 75–80.

Bester, M. L., Frind, E. O., Molson, J. W. and Rudolph, D. L., 2006. Numerical investigation of road salt impact on an urban wellfield. *Ground Water*, 44: 165–175.

Bieringer, J., Schlosser, C., Sartorius, H. and Schmid, S., 2009. Trace analysis of aerosol bound particulates and noble gases at the BfS in Germany. *Applied Radiation and Isotopes*, 67: 672–677.

Bigalke, M., Weyer, S. and Wilcke, W., 2010. Copper isotope fractionation during complexation with insolubilized humic acid. *Environmental Science & Technology*, 44: 5496–5502.

Birkham, T. K., Hendry, M. J., Wassenaar, L. I. and Mendoza, C. A., 2007. A transient model of vadose zone reaction rates using oxygen isotopes and carbon dioxide. *Vadose Zone Journal*, 6: 67–76.

Blowes, D. W. and Ptacek, C. J., 1992. Geochemical remediation of groundwater by permeable reactive walls: Removal of chromate by reaction with iron-bearing solids. *Proceedings of the Subsurface Restoration Conference, 3rd International Conference on Ground Water Quality Research* Dallas, TX, June 21–24, 1992, pp. 214–216.

Blume, O., Guitard, E., Crann, C., Orekhov, M., Amos, R., Clark, I. D., Lapen, D., Blowes, D., Ptacek, C., Craiovan, E. and Sunohara, M., 2022. Relationships between carbon age and CO_2 efflux in agricultural and drainage ditch soils using the thermonuclear bomb pulse. *Vadose Zone Journal*. doi: 10.1002/vzj2.20208.

Böhlke, J. K., Ericksen, G. E. and Revesz, K. M., 1997. Stable isotope evidence for an atmospheric origin of desert nitrate deposits in northern Chile and Southern California, USA. *Chemical Geology*, 136: 135–152.

Böhlke, J. K., Smith, R. L. and Miller, D. N., 2006. Ammonium transport and reaction in contaminated groundwater: Application of isotope tracers and isotope fractionation studies. *Water Resources Research*, 42. doi: 10.1029/2005WR004349.

Bollhöfer, A., Schlosser, C., Schmid, S., Konrad, M., Purtschert, R. and Krais, R., 2019. Half a century of Krypton-85 activity concentration measured in air over Central Europe: Trends and relevance for dating young groundwater. *Journal of Environmental Radioactivity*, 205–206: 7–16.

Bortolami, G. C., Ricci, B., Susella, G. F. and Zuppi, G. M., 1979. Isotope hydrology of the Val Corsaglia, Maritime Alps, Piedmont, Italy. In *Isotope Hydrology 1978*, vol. 1, IAEA Symposium 228, Nurenberg, Germany, June 1978, pp. 327–350.

Böttcher, J., Strebel, O., Voerkelius, S. and Schmidt, H.-L., 1990. Using isotope fractionation of nitrate-nitrogen and nitrate-oxygen for evaluation of microbial denitrification in a sandy aquifer. *Journal of Hydrology*, 114: 413–424.

Bottomley, D. J., Chan, L. H., Katz, A., Starinsky, A. and Clark, I. D., 2003. Lithium isotope geochemistry and origin of Canadian Shield Brines. *Ground Water*, 41: 847–856.

Bottomley, D. J., Clark, I. D., Batty, N. and Kotzer, T., 2005. Geochemical and isotopic evidence for a genetic link between Canadian Shield brines, dolomitization in the Western Canada Sedimentary Basin, and Devonian calcium-chloridic seawater. *Canadian Journal of Earth Sciences*, 42: 2059–2071.

Bottomley, D. J., Gregoire, D. C. and Raven, K. G., 1994. Saline groundwater and brines in the Canadian Shield: Geochemical and isotopic evidence for a residual evaporite brine component. *Geochimica et Cosmochimica Acta*, 54: 1483–1498.

Bottomley, D. J., Katz, A., Chan, L. H., Starinsky, A., Douglas, M., Clark, I. D. and Raven, K., 1999. The origin and evolution of Canadian Shield brines: Evaporation or freezing of seawater? New lithium isotope and geochemical evidence from the Slave craton. *Chemical Geology*, 155: 295–320.

Bottomley, D. J., Renaud, R., Kotzer, T. and Clark, I. D., 2002. Iodine-129 constraints on residence times of deep marine brines in the Canadian Shield. *Geology*, 30: 587–590.

Bottomley, D. J., Ross, J. D. and Clarke, W. B., 1984. Helium and neon isotope geochemistry of some groundwaters from the Canadian Precambrian Shield. *Geochimica et Cosmochimica Acta*, 48: 1973–1985.

Boyle, D. R. and Chagnon, M., 1995. An incidence of skeletal fluorosis associated with groundwaters of the maritime carboniferous basin, Gaspé region, Quebec, Canada. *Environmental Geochemistry and Health*, 17: 5–12.

Braxton, D. and Mathur, R., 2011. Exploration applications of copper isotopes in the supergene environment: A case study of the Bayugo porphyry copper-gold deposit, southern Philippines. *Economic Geology*, 106: 1447–1463.

Brown, D. J., Helmke, P. A. and Clayton, M. K., 2003. Robust geochemical indices for redox and weathering on a granitic laterite landscape, in Central Uganda. *Geochimica et Cosmochimica Acta*, 67: 2711–2723.

Brown, R. M., 1961. Hydrology of tritium in the Ottawa Valley. *Geochimica et Cosmochimica Acta*, 21: 199–216.

Buffle, J., 1977. Les substances humiques et leurs interactions avec les ions minéraux. In Conference Proceedings, La Commission d'Hydrologie Appliquée de l'A. G. H. T. M., L'Université de Paris-Sud, Orsay, France, pp. 3–10.

Burland, S. M. and Edwards, E. A., 1999. Anaerobic benzene biodegradation linked to nitrate reduction. *Applied and Environmental Microbiology*, 65: 529–533.

Burns, R. C. and Hardy, R. W. F., 1975. *Nitrogen Fixation in Bacteria and Higher Plants*, Springer, New York.

Canadell, J. G., Monteiro, P. M. S., Costa, M. H., Cotrim da Cunha, L., Cox, P. M., Eliseev, A. V., Henson, S., Ishii, M., Jaccard, S., Koven, C., Lohila, A., Patra, P. K., Piao, S., Rogelj, J., Syampungani, S., Zaehle, S. and Zickfeld, K., 2021. Global carbon and other biogeochemical cycles and feedbacks. In *Climate Change 2021: The Physical Science Basis. Contribution of Working Group I to the Sixth Assessment Report of the Intergovernmental Panel on Climate Change*, edited by V. Masson-Delmotte, P. Zhai, A. Pirani, S.L. Connors, C. Péan, S. Berger, N. Caud, Y. Chen, L. Goldfarb, M.I. Gomis, M. Huang, K. Leitzell, E. Lonnoy, J.B.R. Matthews, T.K. Maycock, T. Waterfield, O. Yelekçi, R. Yu, and B. Zhou, Cambridge University Press, Cambridge, United Kingdom and New York, NY, USA, pp. 673–816, doi:10.1017/9781009157896.007.

Chambers, L. A., Gooddy, D. C. and Binley, A. M., 2019. Use and application of CFC-1, CFC-12, CFC-113 and SF_6 as environmental tracers of groundwater residence time: A review. *Geoscience Frontiers*, 10: 1643–1652.

Chan, L. H., 1987. Lithium isotope analysis by thermal ionization mass spectrometry of lithium tetraborate. *Analytical Chemistry*, 59: 2662–2665.

Chan, L.-H. and Edmond, J. M., 1988. Variation of lithium isotope composition in the marine environment: A preliminary report. *Geochimica et Cosmochimica Acta*, 52: 1711–1717.

Chanton, J. P., Bauer, J. E., Glaser, P. A., Siegel, D. I., Kelley, C. A., Tyler, S. C., Romanowicz, E. H. and Lazrus, A., 1995. Radiocarbon evidence for the substrates supporting methane formation within northern Minnesota peatlands. *Geochimica et Cosmochimica Acta*, 59: 3663–3668.

Chapelle, F. H. and Lovely, D. R., 1992. Competitive exclusion of sulfate reduction by Fe(III)-reducing bacteria: A mechanism for producing discrete zones of high-iron ground water. *Ground Water*, 30: 29–36.

Chebotarev, I. I., 1955. Metamorphism of natural waters in the crust of weathering. *Geochimica et Cosmochimica Acta*, 8, Part 1 22–48; Part II 137–170; Part III 198–212.

Chen, C., Riley, R. J., Prentice, I. C. and Keenan, T. F., 2022. CO_2 fertilization of terrestrial photosynthesis inferred from site to global scales, *PNAS* 119 (10) e2115627119

Christensen, T. H., Kjeldsen, P., Bjerg, P. L., Jensen, D. L., Christensen, J. B., Baun, A., Albrechtsen, H.-J. and Heron, G., 2001. Biogeochemistry of landfill leachate plumes. *Applied Geochemistry*, 16: 659–718.

Church, C. D., Isabelle, L. M., Pankow, J. F., Rose, D. L. and Tratnyek, P. G., 1997. Method for determination of methyl-tert-butyl ether (MTBE) and its degradation products in water. *Environmental Science & Technology*, 31: 3723–3726.

Ciais, P., Sabine, C., Bala, G., Bopp, L., Brovkin, V., Canadell, J., Chhabra, A., DeFries, R., Galloway, J., Heimann, M., Jones, C., Le Quéré, C., Myneni, R. B., Piao, S. and Thornton, P., 2013: Carbon and other biogeochemical cycles. In *Climate Change 2013: The Physical Science Basis. Contribution of Working Group I to the Fifth Assessment Report of the Intergovernmental Panel on Climate Change*, edited by T. F. Stocker, D. Qin, G.-K. Plattner, M. Tignor, S. K. Allen, J. Boschung, A. Nauels, Y. Xia, V. Bex and P. M. Midgley, Cambridge University Press, Cambridge, United Kingdom and New York, NY, USA, pp. 465–570.

Clark, I. D., Al, T., Jensen, M., Kennell, L., Mazurek, M., Mohapatra, R. and Raven, K. G., 2013. Paleozoic-aged brine and authigenic helium preserved in an Ordovician shale aquiclude. *Geology*, 41: 951–954.

Clark, I. D., Bajjali, W. T. and Phipps, G. C., 1995. Constraining ^{14}C ages in sulphate reducing groundwaters: Two case studies from arid regions. In: *Isotopes in Water Resources Management*, IAEA Symposium 336, March 1995, Vienna, 43–56.

Clark, I. D., Douglas, M., Raven, K. and Bottomley, D. J., 2000. Recharge and preservation of glacial meltwater in the Canadian Shield. *Ground Water*, 38: 735–742.

Clark, I. D. and Fontes, J.-C., 1990. Palaeoclimatic reconstruction in northern Oman based on carbonates from hyperalkaline groundwaters. *Quaternary Research*, 33: 320–336.

Clark, I. D., Fontes, J.-C. and Fritz, P., 1992. Stable isotope disequilibria in travertine from high pH waters: Laboratory investigations and field observations from Oman. *Geochimica Cosmochimica Acta*, 56: 2041–2050.

Clark, I. D. and Fritz, P., 1997. *Environmental Isotopes in Hydrogeology*, Lewis Publishers, Boca Raton, FL, 328 pp.

Clark, I. D., Fritz, P., Quinn, O. P., Rippon, P. W., Nash, H., Sayyid Barghash, B. G. and Said, A., 1987. Modern and fossil groundwater in an arid environment: A look at the hydrogeology of Oman. In *Isotope Techniques in Water Resources Development*, International Atomic Energy Agency, Vienna, pp. 167–187.

Clark, I. D., Fritz, P., Seidlitz, H. K., Trimborn, P., Milodowski, A. E., Pearce, J. M. and Khoury, H. N., 1993. Recarbonation of metamorphosed marls in Jordan. *Applied Geochemistry*, 8: 473–481.

Clark, I. D., Fritz, P. and Souther, J. G., 1989. Geochemistry and isotope hydrogeology of the Mount Edziza–Mess Creek geothermal area. *Canadian Journal of Earth Sciences*, 26: 1160–1171.

Clark, I. D., Henderson, L., Chappellaz, J., Fisher, D. A., Koerner, R., Worthy, D. E. J., Kotzer, T., Norman, A.-L. and Barnola, J.-M., 2007. CO_2 isotopes as tracers of firn air diffusion and age in an Arctic ice cap with summer melting, Devon Island, Canada. *Journal of Geophysical Research*, 112: D01301. doi: 10.1029/2006jD007471.

Clark, I. D. and Lauriol, B., 1992. Kinetic enrichment of stable isotopes in cryogenic calcites. *Chemical Geology (Isotope Geoscience Section)*, 102: 217–228.

Clark, I. D., Lauriol, B., Marschner, M., Sabourin, N., Chauret, Y. and Desrochers, A., 2004. Endostromatolites from permafrost karst, Yukon, Canada: Paleoclimatic proxies for the Holocene hypsithermal. *Canadian Journal of Earth Sciences*, 41: 387–399.

Clark, I. D. and Raven, K. G., 2003. Arsenic sources in Giant Mine groundwaters, Yellowknife, NWT, Canada. *Isotopes in Environmental and Health Studies*, 40: 115–128.

Clark, I. D., Timlin, R., Bourbonnais, A., Jones, K., Lafleur, D. and Wickens, K., 2008. Origin and fate of industrial ammonia in municipal groundwaters: Tracing anaerobic oxidation (anammox) and apportionment with $^{15}N_{NH_4^+}$. *Ground Water Monitoring and Remediation*, 28: 73–82.

Claypool, G. E., Holser, W. T., Kaplan, I. R., Sakai, H. and Zak, I., 1980. The age curves of sulfur and oxygen isotopes in marine sulphate and their mutual interpretation. *Chemical Geology*, 28: 199–260.

Clayton, R. N., Friedman, I., Graff, D. L., Mayeda, T. K., Meents, W. F. and Shimp, N. F., 1966. The origin of saline formation waters, 1. Isotopic composition. *Journal of Geophysical Research*, 71: 3869–3882.

Clayton, R. N., Grossman, L. and Mayeda, T. K., 1973. A component of primitive nuclear composition in carbonaceous meteorites. *Science*, 182: 485–488.

Coleman, D. D., Liu, C.-L. and Riley, K. M., 1988. Microbial methane in the shallow Paleozoic sediments and glacial deposits of Illinois, U.S.A. *Chemical Geology*, 71: 23–40.

Coplen, T. B., 2011. Report of stable isotopic composition, reference material LSVEC (Carbon and lithium isotopes in lithium carbonate). United States Geological Survey, Reston Stable Isotope Laboratory, Reston Virginia, 20192.

Coplen, T. B., Brand, W. A., Gehre, M., Gröning, M., Meijer, H. A. J., Toman, B. and Verkouteren, R. M., 2006. New guidelines for ^{13}C measurements. *Analytical Chemistry*, 78: 2439–2441.

Coplen, T. B., Hopple, J. A., Böhlke, J. K., Peiser, H. S., Rieder, S. E., Krouse, H. R., Rosman, K. J. R., Ding, T., Vocke, R. D., Jr., Revesz, K. M., Lamberty, A., Taylor, P. and Bievre, P. D., 2002. Compilation of minimum and maximum isotope ratios of selected elements in naturally occurring terrestrial materials and reagents. *Water-Resources Investigations* Report 2001–4222.

Coplen, T. B., Kendall, C. and Hopple, J., 1983. Comparison of stable isotope reference samples. *Nature*, 302: 236–238. doi: 10.1038/302236a0.

Cornett, R. J., Cramer, J., Andrews, H. R., Chant, L. A., Davies, W., Greiner, B. F. and Imahori, Y. et al., 1996. In situ production of ^{36}Cl in uranium ore: A hydrogeological assessment tool. *Water Resources Research*, 32: 1511–1518.

Cox, K. G. and Hawkesworth, C. J., 1985. Geochemical stratigraphy of the Deccan traps at Mahabaleshwar, Western Ghats, India, with implications for open system magmatic processes. *Journal of Petrology*, 26: 355–377.

Craig, H., 1961. Isotopic variations in meteoric waters. *Science*, 133: 1702–1703.

Craig, H., 1963. The isotopic geochemistry of water and carbon in geothermal areas. In *Nuclear Geology on Geothermal Areas*, edited by E. Tongiorgi, Spoleto, 1963, Consiglio Nazionale delle Ricerche, Laboratorio di Geologia Nucleare, Pisa, pp. 17–53.

Cramer, J. J. and Whiteshell Laboratories, 1994. Final Report of the AECL/SKB Cigar Lake Analog Study, Whiteshell Laboratories, Pinawa, Canada.

Cravotta, Ch. A., III, 2008. Dissolved metals and associated constituents in abandoned coalmine discharges, Pennsylvania, USA. Part 2: Geochemical controls on constituent concentrations. *Applied Geochemistry*, 23: 203–226.

Cruz-san, J., Araguas, L., Rozanski, K., Cardenal, J., Hidalgo, M. C., Garcia-Lopez, S., Martinez-Garrido, J. C., Moral, F. and Olias, M., 1992. Sources of precipitation over southeastern Spain and groundwater recharge. An isotopic study. *Tellus*, 44: 226–236.

Curtis, J., 2002. Fractured shale-gas systems. *American Association of Petroleum Geologists Bulletin*, 86: 1921–1938.

Dansgaard, W., 1964. Stable isotopes in precipitation. *Tellus*, 16: 436–468.

Daoust, P., Clark, I. D. and Desrochers, A., 2023. Origin of gases and waters from a hyper-saline, carbonate spring on Anticosti Island, Québec, Canada. *Canadian Journal of Earth Sciences*, 60: 1530–1543.

Darling, W. G. and Gooddy, D. C., 2007. Assessing the applicability of global CFC and SF_6 input functions to groundwater dating in the UK. *Science of the Total Environment*, 387: 353–362.

Darling, W. G., Gooddy, D. C., MacDonald, A. M. and Morris, B. L., 2012. The practicalities of using CFCs and SF_6 for groundwater dating and tracing. *Applied Geochemistry*, 27: 1688–1697.

Das, A., Krishnaswami, S., Sarin, M. M. and Pande, K., 2005. Chemical weathering in the Krishna Basin and Western Ghats of the Deccan Traps, India: Rates of basalt weathering and their controls. *Geochimica et Cosmochimica Acta*, 69: 2067–2084.

Dayal, R. and Reardon, E. J., 1994. Carbon-14 behaviour in a cement-dominated environment: Implications for spend CANDU resin waste disposal. *Waste Management*, 14: 457–466.

Desaulniers, D. E., Kaufmann, R. S., Cherry, J. A. and Bentley, H. W., 1986. ^{35}Cl-^{37}Cl variations in a diffusion-controlled groundwater system. *Geochimica et Cosmochimica Acta*, 50: 1757–1746.

Desaulty, M.-A., Climent Monfort, D., Lefebvre, G., Cristiano-Tassi, A., Peralta, D., Perret, S., Urban, A. and Guerrot, C., 2022. Tracing the origin of lithium in Li-ion batteries using lithium isotopes. *Nature Communications*, 13: 4172.

Dimitrakopoulos, R. and Muehlenbachs, K., 1987. Biodegradation of petroleum as a source of ^{13}C-enriched carbon dioxide in the formation of carbonate cement. *Chemical Geology (Isotope Geoscience Section)*, 65: 283–29.

Ding, T., Valkiers, S., Kipphardt, H., De Bievre, P., Taylor, P. D. P., Gonfiantini, R. and Krouse, H., 2001. Calibrated sulfur isotope abundance ratios of three IAEA sulfur isotope reference materials and VCDT with a reassessment of the atomic weight of sulfur. *Geochimica et Cosmochimica Acta*, 65: 2433–2437.

Domenico, P. A. and Schwartz, F. W., 1998. *Physical and Chemical Hydrogeology*, Wiley, New York, NY, p. 506.

Doney, S. C., Glover, D. M. and Jenkins, W. J., 1992. A model function of the global bomb tritium distribution in precipitation, 1960–1986. *Journal of Geophysical Research-Oceans*, 97(C4): 5481–5492.

Douglas, M., Clark, I. D., Raven, K. and Bottomley, D., 2000. Groundwater mixing dynamics at a Canadian Shield mine. *Journal of Hydrology*, 235: 88–103.

Drever, J. I., 1997. *The Geochemistry of Natural Waters, Surface and Groundwater Environments*, 3rd ed., Prentice Hall, Upper Saddle River, NJ, 436 pp.

Drever, J. I., 1997. The carbonate system and pH control. In *The Geochemistry of Natural Waters*, Prentice Hall, Upper Saddle River, NJ, p. 436.

Dunn, P. J. H., Malinovsky, D., Ogrinc, N., Potočnik, D., Flierl, L., Rienitz, O., Dipayan, P., and Meijer, H. A. J., 2024. Re-determination of R($^{13}C/^{12}C$) for Vienna Peedee belemnite (VPDB). *Rapid Communications in Mass Spectrometry*, 38; e9773.

Dwight, B., LeeAnn, M., Hillary, J., Scott, H. and Keith, L., 2013. A Preliminary Deposit Model for Lithium Brines, U. S. Geological Survey Open-File Report 2013–1006, p. 6.

Eastoe, C. J., Peryt, T. M., Petrychenko, O. Y. and Geisler-Cussey, S., 2007. Stable chlorine isotopes in Phanerozoic evaporites. *Applied Geochemistry*, 22: 575–588.

Edmunds, W. M. and Bogush, A. A., 2012. Geochemistry of natural waters: The legacy of V. I. Vernadsky and his students. *Applied Geochemistry*, 27: 1871–1886.

Edwards, T. W. D., Aravena, R. O., Fritz, P. and Morgan, A. V., 1985. Interpreting paleoclimate from ^{18}O and ^{2}H in plant cellulose: Comparison with evidence from fossil insects and relict permafrost in southwestern Ontario. *Canadian Journal of Earth Sciences*, 22: 1720–1726.

Eggenkamp, H. G. M. and Coleman, M. L., 2000. Rediscovery of classical methods and their application to the measurement of stable bromine isotopes in natural samples. *Chemical Geology*, 167: 393–402.

Eiler, J. M., 2007. Clumped-isotope" geochemistry – The study of naturally-occurring, multiply-substituted isotopologues. *Earth and Planetary Science Letters*, 262: 309–327.

Einsiedl, F. and Mayer, B., 2005. Sources and processes affecting sulfate in a karstic groundwater system of the Franconian Alb, Southern Germany. *Environmental Science and Technology*, 39: 7118–7125.

Eleazer, W. E., Odle, W. S., III, Wang, Y.-S. and Barlaz, M. A., 1997. Biodegradability of municipal solid waste components in laboratory-scale landfills. *Environmental Science and Technology*, 31: 911–917.

Emblanch, C., Zuppi, G. M., Mudry, J., Blavoux, B. and Batiot, C., 2003. Carbon-13 of TDIC to quantify the role of the unsaturated zone: The example of the vaucluse karst systems (Southeastern France). *Journal of Hydrology*, 279: 262–274.

EPA, 2013. Municipal Solid Waste Generation, Recycling, and Disposal in the United States: Facts and Figures for 2011. EPA530-F-13-001 Solid Waste and Emergency Response (5306P) Washington, DC 20460. www.epa.gov/wastes.

EPICA Community Members, 2004. Eight glacial cycles from an Antarctic ice core. *Nature*, 429(6992): 623–628.

Epstein, J. A., Feist, E. M., Zmora, J. and Marcus, Y., 1981. Extractions of lithium from the Dead Sea. *Hydrometallurgy*, 6: 269–275.

Erda, L., Wei, X., Hui, J., Yinlong, X., Yue, L., Liping, B. and Liyong, X., 2005. Climate change impacts on crop yield and quality with CO_2 fertilization in China. *Philosophical Transactions of the Royal Society Biological Sciences*, 360: 2149–2154.

Eriksson, E., 1952. Composition of atmospheric precipitation. II Sulfur, chloride, iodine compounds. *Tellus*, 4: 280–303.

FAO, 2008. Current World Fertilizer Trends and Outlook to 2011/12, Food and Agriculture Organization of the United Nations, Viale delle Terme di Caracalla, 00153 Rome. www.fao.org.

Faure, G., 1997. *Principles and Applications of Geochemistry*, 2nd ed., Prentice Hall, Upper Saddle River, NJ, 600 pp.

Fennell, D. E. and Gossett, J. M., 1997. Comparison of butyric acid, ethanol, lactic acid, and propionic acid as hydrogen donors for the reductive dechlorination of tetrachloroethene. *Environmental Science & Technology*, 31: 918–926.

Ferguson, P. and Veizer, J., 2007. Coupling of water and carbon fluxes via the terrestrial biosphere and its significance to the Earth's climate system. *Journal of Geophysical Research*, 112: D24S06. doi: 10.1029/2007JD008431.

Fetter, Ch. W., 2001. *Applied Hydrogeology*, Prentice Hall, Upper Saddle River, NJ, 598 pp.

Feurdean, V., Feurdean, L. and Lari, I., 1997. Deuterium distribution in crude oil. *Isotopes in Environmental and Health Studies*, 33: 389–398.

Field, C. B., Behrenfeld, M. J., Randerson, J. T. and Falkowski, P., 1998. Primary production of the biosphere: Integrating terrestrial and oceanic components. *Science*, 281: 237–240. doi: 10.1126/science.281.5374.237.

Finneran, K. T. and Lovley, D. R., 2001. Anaerobic degradation of methyl tert-butyl ether (MTBE) and tert-butyl alcohol (TBA). *Environmental Science & Technology*, 35: 1785–1790.

Fontes, J.-Ch. and Garnier, J.-M., 1979. Determination of the initial ^{14}C activity of total dissolved carbon: A review of existing models and a new approach. *Water Resources Research*, 15: 399–413.
Fortin, D., Roy, M., Rioux, J.-P. and Thibault, P.-J., 2000. Occurrence of sulfate-reducing bacteria under a wide range of physico-chemical conditions in sulfidic mine tailings. *FEMS Microbiological Ecology*, 33: 197–208.
Frape, S. K. and Fritz, P., 1982. The chemistry and isotopic composition of saline groundwaters from the Sudbury basin, Ontario. *Canadian Journal of Earth Sciences*, 19: 645–661.
Freeze, R. A. and Cherry, J. A., 1979. *Groundwater*, Prentice Hall, Upper Saddle River, NJ, 604 pp.
Fritz, P., Basharmal, G. M., Drimmie, R. J., Ibsen, J. and Qureshi, R. M., 1989. Oxygen isotope exchange between sulphate and water during bacterial reduction of sulphate. *Chemical Geology (Isotope Geoscience Section)*, 79: 99–105.
Fritz, P., Clark, I. D., Fontes, J.-Ch., Whiticar, M. J. and Faber, E. 1992. Deuterium and ^{13}C evidence for low temperature production of hydrogen and methane in a highly alkaline groundwater environment in Oman. In *Proceedings of the 7th International Symposium on Water–Rock Interaction*, Park City, Utah, 13–18 July, 1992, eds. Y. Kharaka and A.S. Maest, p. 793–796.
Fritz, P., Drimmie, R. J., Frape, S. K. and O'Shea, O., 1987a. The isotopic composition of precipitation and groundwater in Canada. In *Isotope Techniques in Water Resources Development*, IAEA Symposium 299, March 1987, Vienna, pp. 539–550.
Fritz, P. and Fontes, J.-Ch., 1980. *Handbook of Environmental Isotope Geochemistry*, Volume 1, The Terrestrial Environment, A, Elsevier, Amsterdam, The Netherlands, 558 pp.
Fritz, P. and Frape, S. K., 1982. Saline groundwaters in the Canadian Shield – A first overview. *Chemical Geology*, 36: 179–190.
Fröhlich, K., Franke, Th., Gellermann, R., Hebert, D. and Jordan, H., 1987. Silicon-32 in different aquifer types and implications for groundwater dating. International Symposium on the Use of Isotope Techniques in Water Resources Development, Isotope Techniques in Water Resources Development. IAEA Vienna 1987, 149–163.
Frost, C. D. and Toner, R. N., 2004. Strontium isotopic identification of water-rock interaction and groundwater mixing. *Ground Water*, 42: 418–432.
Gaillardet, J., Dupre, B., Louvat, P. and Allegre, C. J., 1999. Global silicate weathering and CO_2 consumption rates deduced from the chemistry of large rivers. *Chemical Geology*, 159: 3–30.
Gardner, P. and Solomon, D. K., 2009. An advanced passive diffusion sampler for the determination of dissolved gas concentrations. *Water Resources Research*, 45: W06423. doi: 10.1029/2008WR007399.
Garrels, R. M. and Christ, C. L., 1965. Carbonate equilibria. In *Solutions, Minerals and Equilibria*, Freeman, Cooper & Company, San Francisco, CA, p. 450.
Gat, J. R. and Carmi, I., 1970. Evolution of the isotopic composition of atmospheric waters in the Mediterranean Sea area. *Journal of Geophysical Research*, 75: 3039–3048.
Gault, A. G., Ibrahim, A., Langley, S., Renaud, R., Takahashi, Y., Boothman, C., Lloyd, J. R., Clark, I. D., Ferris, F. G. and Fortin, D., 2011. Microbial and geochemical features suggest iron redox cycling within bacteriogenic iron oxide-rich sediments. *Chemical Geology*, 281: 41–51.
General Electric Co. and KAPL Inc., 2002. *Chart of the Isotopes*, 16th ed., General Electric Co., Knolls Atomic Power Laboratory, Inc., Schenectady, NY.
Ghosh, P., Adkins, J., Affek, H., Balta, B., Guo, W., Schauble, E. A., Schrag, D. and Eiler, J. M., 2006. ^{13}C–^{18}O bonds in carbonate minerals: A new kind of paleothermometer. *Geochimica et Cosmochimica Acta*, 70: 1439–1456.
Gillham, R. W. and O'Hannesin, S. F., 1992. Metal-catalyzed abiotic degradation of halogenated organic compounds. In *Proceedings of the IAH Conference: Modern Trends in Hydrogeology*, Ontario, Canada, 10–13 May, 1992, p. 94.

Gillham, R. W. and O'Hannesin, S. F., 1994. Enhanced degradation of halogenated aliphatics by zero-valent iron. *Ground Water*, 32: 958.

Gillham, R. W., O'Hannesin, S. F. and Orth, W. S., 1993. Metal enhanced abiotic degradation of halogenated aliphatics: Laboratory tests and field trials. In *Proceedings of the Sixth Annual Environmental Management and Technology Conference*, edited by W. R. Bunner, Advanstar Expositions, Glen Ellyn, IL, p. 440.

Gillon, M., Barbecot, F., Gilbert, E., Corcho Alvarado, J. A., Marlin, C. and Massault, M., 2009. Open to closed system transition traced through TDIC isotopic signature at the aquifer recharge stage, implications for groundwater ^{14}C dating. *Geochimica Cosmochimica Acta*, 73: 6488–6501.

Gimmi, T., Waber, H. N., Gautschi, A. and Rübel, A., 2007. Stable water isotopes in pore water of Jurassic argillaceous rocks as tracers for solute transport over large spatial and temporal scales. *Water Resources Research*, 43: W04410. doi: 10.1029/2005WR004774.

Golder, 2019. Bioaccumulation analysis in support of the 2019 implementation plan adjustment. Unpublished technical report to Teck Coal. February 2019. https://www.teck.com/media/Annex-G-Bioaccumulation-Analysis.pdf

Goldschmidt, V. M., 1954. *Geochemistry*, edited posthumously by A. Muir, Oxford University Press, London.

Gonfiantini, R., 1965. Effetti isotopici nell'evaporazione di acque salate. *Atti Soc. Toscana Sci. Nat. Pisa, Ser. A*, 72: 550–569.

Gonfiantini, R. 1976. Consultant's meeting on stable isotope standards and intercalibration in hydrology and in geochemistry. Final Report, I.A.E.A., Vienna, 8–10 September 1976.

Gonfiantini, R., 1986. Environmental isotopes in lake studies. In *Handbook of Environmental Isotope Geochemistry, Vol. 2, The Terrestrial Environment, B*, edited by P. Fritz and J.-Ch. Fontes, Elsevier, Amsterdam, The Netherlands, pp. 113–168.

Gonfiantini, R. and Zuppi, G. M., 2003. Carbon exchange rate of DIC in karst groundwater. *Chemical Geology*, 197: 319–336.

Gori, Y., Wehrens, R., Greule, M., Keppler, F., Ziller, L., La Porta, N. and Camin, F., 2013. Carbon, hydrogen and oxygen stable isotope ratios of whole wood, cellulose and lignin methoxyl groups of *Picea abies* as climate proxies. *Rapid Communications in Mass Spectrometry*, 27: 265–275.

Gorman-Lewis, D., Fein, J. B., Burns, P. C., Szymanowski, J. E. S. and Converse, J., 2008. Solubility measurements of the uranyl oxide hydrate phases metaschoepite, compreignacite, Na–compreignacite, becquerelite, and clarkeite. *Journal of Chemical Thermodynamics*, 40: 980–990.

Gosse, J. C., Lal, D., Schaeffer, J., Ivy-Ochs, S. and Granger, D., 2006. Terrestrial cosmogenic nuclide exposure methods. In *The Encyclopedia of Quaternary Science*, edited by S. Elias, Blackwell Publishers, Oxford, UK, pp. 34–60.

Grasby, S. E. and Chen, Z., 2005. Subglacial recharge into the Western Canada Sedimentary Basin: Impact of Pleistocene glaciation on Basin hydrodynamics. *Geological Society of America Bulletin*, 117: 500–514.

Greene, S., Battye, N., Clark, I. D., Kotzer, T. and Bottomley, D., 2008. Canadian Shield brine from the Con Mine, Yellowknife, NT, Canada: Noble gas evidence for an evaporated Palaeozoic seawater origin mixed with glacial meltwater and Holocene recharge. *Geochimica et Cosmochimica Acta*, 72: 4008–4019.

Grim, R. E., 1968. *Clay Mineralogy*, McGraw-Hill, New York, 596 pp.

Gu, L., Post, W. M. and King, A. W., 2004. Fast labile carbon turnover obscures sensitivity of heterotrophic respiration from soil to temperature: A model analysis. *Global Biogeochemical Cycles*, 18: GB1022. doi: 10.1029/2003GB002119.

Hackley, K. C., Liu, C. L. and Coleman, D. D., 1996. Environmental isotope characteristics of landfill leachates and gases. *Ground Water*, 34: 827–836.

Hardie, L. A. and Eugster, H. P., 1980. Evaporation of seawater: Calculated mineral sequences. *Science*, 208: 498–500.

Harper, D. A., Longstaffe, F. J., Wadleigh, M. A. and McNutt, R. H., 1995. Secondary K-feldspar at the Precambrian-Paleozoic unconformity, southwestern Ontario. *Canadian Journal of Earth Science*, 32: 1432–1450.

Hem, J. D., 1985. *Study and Interpretation of the Chemical Characteristics of Natural Water.* U.S. Geological Survey Water-Supply Paper 2254. U.S. Department of the Interior, Washington, DC.

Henderson, A. D. and Demond, A. H., 2007. Long-term performance of zero-valence iron permeable reactive barriers: A critical review. *Environmental Engineering Science*, 24: 401–424.

Hendry, J. M., Johnson, Th. M., Schmeling, E., Shaw, S. and Kirk, L., 2024. Application of selenium isotopes to define selenium bioreduction in coal waste rock: Elk Valley, British Columbia. *Science of the Total Environment* 20:957:177838. doi: 10.1016/j.scitotenv.2024.177838.

Hendry, M. J., Kotzer, T. G. and Solomon, D. K., 2005. Sources of radiogenic helium in a clay till aquitard and its use to evaluate the timing of geologic events. *Geochimica et Cosmochimica Acta*, 69: 475–483.

Hendry, M. J. and Schwartz, F. W., 1990. The chemical evolution of groundwater in the Milk River aquifer, Canada. *Ground Water*, 28: 253–261.

Hendry, M. J. and Wassenaar, L. I., 1999. Implications of the distribution of δD in pore waters for groundwater flow and the timing of geologic events in a thick aquitard system. *Water Resources Research*, 35: 1751–1760.

Hill, R. D., Rinker, R. G. and Dale Wilson, H., 1980. Atmospheric nitrogen fixation by lightning. *Journal of the Atmospheric Sciences*, 37: 179–192.

Hogan, J. F. and Blum, J. D., 2003. Boron and lithium isotopes as groundwater tracers: A study at the Fresh Kills Landfill, Staten Island, New York, USA. *Applied Geochemistry*, 18: 615–627.

Holland, G., Sherwood Lollar, B., Li, L., Lacrampe-Couloume, G., Slater, G. F. and Ballentine, C. J., 2013. Deep fracture fluids isolated in the crust since the Precambrian era. *Nature*, 497: 357–362.

Howard, K. W. F., Boyce, J. I., Livingston, S. J. and Salvatori, S. L., 1993. Road salt impacts on ground water quality—The worst is still to come! *GSA Today*, 3: 319–321.

Howard, K. W. F. and Haynes, J., 1993. Urban geology 3. Groundwater contamination due to road de-icing chemicals: Salt balance implications. *Geoscience Canada*, 20: 1–8.

Hua, Q., Barbetti, M. and Rakowski, A. Z. (2013). Atmospheric radiocarbon for the period 1950–2010. *Radiocarbon*, *55*(4), 2059–2072.

Hughes, C. E. and Crawford, J., 2012. A new precipitation weighted method for determining the meteoric water line for hydrological applications demonstrated using Australian and global GNIP data. *Journal of Hydrology*, 464: 344–351.

Hull, L. and Schafer, A. L., 2008. Accelerated transport of ^{90}Sr following a release of high ionic strength solution in vadose zone sediments. *Journal of Contaminant Hydrology*, 97: 135–157.

Humlum, O., Stordahl, K. and Solheim, J.-E., 2013. The phase relation between atmospheric carbon dioxide and global temperature. *Global and Planetary Change*, 100: 51–69.

Hunkeler, D., Aravena, R. and Butler, B. J., 1999. Monitoring microbial dechlorination of tetrachloroethene (PCE) in groundwater using compound-specific stable carbon isotope radios: Microcosm and field studies. *Environmental Science and Technology*, 33: 2733–2738.

Hyvönen, R., Ågren, G. I., Linder, S., Persson, T., Cotrufo, M. F., Ekblad, A. and Freeman, M. et al., 2007. The likely impact of elevated [CO_2], nitrogen deposition, increased temperature and management on carbon sequestration in temperate and boreal forest ecosystems: A literature review. *New Phytologist*, 173: 463–480.

IAEA, 1981. *Stable Isotope Hydrology: Deuterium and Oxygen-18 in the Water Cycle*, Technical Report Series No. 210, International Atomic Energy Agency, Vienna.

IAEA, 2006. *Use of Chlorofluorocarbons in Hydrology: A Guidebook*, Atomic Energy Agency STI/PUB/1238. http://www-pub.iaea.org/MTCD/publications/PDF/Pub1238_web.pdf.

IAEA, 2020. Reference Sheet for LSVEC, Rev 02. IAEA, Vienna, 2020, 5 pp.

IAEA/WMO, 2001 and 2013. Global Network of Isotopes in Precipitation. *The GNIP Database*. http://www.iaea.org/water.

Iijima, K., 2013. Background and current status of understanding at both local (Fukushima NPP) and regional (Fukushima Predecture). International Atomic Energy Agency Workshop on Evaluating Groundwater Pathways and Residence Times as part of Site Investigations and Post-Closure Safety Assessments for Geological Repositories, Trieste, Italy, June 17–21, 2013.

Intera Engineering Ltd., 2011. *Descriptive Geosphere Site Model*, Intera Engineering Ltd. report for the Nuclear Waste Management Organization NWMO DGR-TR-2011-24 R000, Toronto, Canada.

IPCC, 2007. *Climate Change 2007: The Physical Science Basis. Contribution of Working Group I to the Fourth Assessment Report of the Intergovernmental Panel on Climate Change*, Cambridge University Press, Cambridge, United Kingdom and New York, NY.

Jackson, R. B., Vengosh, A., Darrah, T. H., Warner, N. R., Down, A., Poreda, R. J., Osborn, S. G., Zhao, K. and Karr, J. D., 2013. Increased stray gas abundance in a subset of drinking water wells near Marcellus shale gas extraction. *Proceedings National Academy of Science*, www.pnas.org/cgi/doi/10.1073/pnas.1221635110.

Jackson, R. E., Dusseault, M. B., Frape, S., Illman, W., Phan, Th. and Steelman, C., 2020. Investigating the origin and elevated H_2S in groundwater discharge from abandoned gas wells, Norfolk County, Ontario. *Proceedings, IAH GeoConvention, Virtual Event, September 21-23, 2020.*

Jackson, R. E. and Patterson, R. J., 1982. Interpretation of pH and Eh trends in a fluvial-sand aquifer system. *Water Resources Research*, 18: 1255–1268.

Jamieson-Hanes, J. H., Gibson, B. D., Lindsay, M. B. J., Kim, Y., Ptacek, C. J. and Blowes, D. W., 2012. Chromium isotope fractionation during reduction of Cr(VI) under saturated flow conditions. *Environmental Science & Technology*, 46: 6783–6789.

Jautzy, J. J., Petts, D., Clark, I. D., Al, T. A., Stern, R. and Jensen, M., 2020. Diagenetic evolution of a sedimentary system (Michigan Basin): insights from petrography and S-isotope micro-analysis of pyrite. *Chemical Geology* 541. doi: 10.1016/j.chemgeo.2020.119580.

Jetten, M. S. M., 1999. New pathways for ammonia conversion in soil and aquatic systems. *Plant and Soil*, 230: 9–19.

Jiang, W., Bailey, K., Lu, Z.-T., Mueller, P., O′Connor a, T. P., Cheng, C.-F. and Huc, S.-M. et al., 2012. An atom counter for measuring ^{81}Kr and ^{85}Kr in environmental samples. *Geochimica et Cosmochimica Acta*, 91: 1–6.

Jiang, W., Williams, W., Bailey, K., Davis, A. M., Hu, S.-M., Lu, Z.-T. and O'Connor, T. P. et al., 2011. ^{39}Ar detection at the 10^{-16} isotope abundance level with atom trap trace analysis. *Physical Review Letters*, 106: 103001.

Jiao, J. J., 2007. A 5,600-year-old wooden well in Zhejiang Province, China. *Hydrogeology Journal*, 15: 1021–1029.

Jones, C. D., Cox, P., and Huntingford, C., 2003. Uncertainty in climate–carbon-cycle projections associated with the sensitivity of soil respiration to temperature. *Tellus B*, 55, 642–648.

Jones, D. M., Head, I. M., Gray, N. D., Adams, J. J., Rowan, A. K., Aitken, C. M., Bennett, B. et al., 2008. Crude-oil biodegradation via methanogenesis in subsurface petroleum reservoirs. *Nature*, 451: 176–180.

Junge, C.E. and Werby, R.T., 1958. The concentration of chloride, sodium, potassium, calcium and sulfate in rainwater over the United States. *Journal of the Atmospheric Sciences*, 15: 417–425.

Jurgens, B. C., Böhlke, J. K., and Eberts, S. M., 2012, Tracer LPM (Version 1): An Excel® Workbook for Interpreting Groundwater Age Distributions from Environmental Tracer Data. U. S. Geological Survey Techniques and Methods Report 4-F3, p. 60.

Kagabu, M., Matsunaga, M., Ide, K., Momoshima, N. and Shimada, J., 2017. Groundwater age determination using ^{85}Kr and multiple age tracers (SF_6, CFCs and ^{3}H) to elucidate regional groundwater flow systems.

Kaiser, J., Hastings, M. G., Houlton, B. Z., Röckmann, T., and Sigman, D. M., 2007. Triple oxygen isotope analysis of nitrate using the denitrifier method and thermal decomposition of N_2O. *Analytical Chemistry*, 79: 599–607.

Kamei-Ishikawa, N., Uchida, S. and Tagami, K., 2008. Distribution coefficients for ^{85}Sr and ^{137}Cs in Japanese agricultural soils and their correlations with soil properties. *Journal of Radioanalytical and Nuclear Chemistry*, 277: 433–439.

Kapaj, S., Peterson, H., Liber, K. and Phattacharya, P., 2006. Human health effects from chronic arsenic poisoning: A review. *Journal of Science and Health Part A*, 41: 2399–2428.

Kaste, J. M., Lauer, N. E., Spaetzel, A. B. and Goydan, C., 2016. Cosmogenic ^{22}Na as a steady-state tracer of solute transport and water age in first-order catchments. *Earth and Planetary Science Letters*, 456: 78–86.

Kendall, C. and Aravena, R., 2000. Nitrate isotopes in groundwater systems. In *Environmental Tracers in Subsurface Hydrology*, edited by Peter Cook and Andrew L. Herczeg, Springer Science+Business Media, New York, NY, pp. 261–297.

Kendall, C. and Caldwell, E. A., 1998. *Isotope Tracers in Catchment Hydrology*, edited by C. Kendall and J. J. McDonnell, Elsevier, Amsterdam, the Netherlands, 839 pp.

Kennedy, C. B., Gault, A. G., Clark, I. D., Fortin, D. and Ferris, F. G., 2011. Retention of iodide by bacteriogenic iron oxides. *Geomicrobiology Journal*, 28: 387–395.

Kharaka, Y. K., Gunter, W. D., Aggarwal, P. K., Perkins, E. H. and DeBraal, J. D., 1988. SOLMINEQ.88: A computer program for geochemical modeling of water–rock interactions. *USGS Water-Resources Investigations Report*: 88–4227.

Khoury, H. N., 2019. Industrial rocks and minerals of Jordan: A review. *Saudi Society for Geosciences*. https://doi.org/10.1007/s12517-019-4750-0.

Khoury, H. N., Salameh, E., Clark, I. D., Fritz, P., Bajjali, W., Milodowski, A. E., Cave, M. R. and Alexander, W., 1992. A natural analogue of high pH cement pore waters from the Maqarin site of northern Jordan, 1: Introduction to the site. *Journal of Geochemical Exploration*, 46: 117–132.

Kieser, W. E., Zhao, X.-L., Clark, I. D., Kotzer, T. and Litherland, A. E., 2011. The New Accelerator Mass Spectrometry Facility at the University Of Ottawa. Applications of Accelerators in Research and Industry, AIP Conference Proceedings, vol. 1336, pp. 60–66. doi: 10.1063/1.3586057.

Killey, R. W. D. and Munch, J. H., 1987. Radiostrontium migration from a 1953–54 liquid release to a sand aquifer. *Water Pollution Research Journal of Canada*, 22: 107–128.

Kim, E. and Boulègue, J., 2003. Chemistry of rhenium as an analogue of technetium: Experimental studies of the dissolution of rhenium oxides in aqueous solutions. *Radiochimica Acta*, 91: 211–216.

Kim, J. and Rees, D. C., 1994. Nitrogenase and biological nitrogen fixation. *Biochemistry*, 33: 389–97. doi: 10.1021/bi00168a001.

Kim, J.-H., Ferguson, G., Person, M., Jiang, W., Lu, Z.-T., Ritterbusch, F., Yang, G.-M., Tyne, R., Bailey, L., Ballentine, C., Reiners, P. and McIntosh, J., 2022. Krypton-81 dating constrains timing of deep groundwater flow activation. *Geophysical Research Letters*, https://doi.org/10.1029/2021GL097618

Kim, S.-T., Coplen, T. B. and Horita, J., 2015. Normalization of stable isotope data for carbonate minerals: Implementation of IUPAC guidelines. *Geochimica et Cosmochimica Acta*, 158: 276–289. doi: 10.1016/j.gca.2015.02.011.

Kinoshita, N., Sueki, K., Sasa, K., Kitagawa, J., Ikarashi, S., Nishimura, T. and Wong, Y.-S. et al., 2012. Assessment of individual radionuclide distributions from the Fukushima nuclear accident covering central-east Japan. *Proceedings of the National Academy of Sciences* 108. www.pnas.org/cgi/doi/10.1073/pnas.1111724108.

Kinsley, C. and Crolla, A., and Joy, D., 2005. Impact of Water Softeners on Septic Tanks Field Evaluation Study. Final Report to Canadian Mortgage and Housing Corporation by Ontario Rural Wastewater Centre, University of Guelph, pp. 19 + appendices.

Kipfer, R., Aeschbach-Hertig, W., Peeters, F. and Stute, M., 2002. Noble gases in lakes and ground waters (in Noble gases in geochemistry and cosmochemistry). *Reviews in Mineralogy and Geochemistry*, 47: 615–700.

Kitchen, J. W., Johnson, T. M., Bullen, T. D., Zhu, J. and Raddatz, A. L., 2012. Chromium isotope fractionation factors for reduction of Cr (VI) by aqueous Fe (II) and organic molecules. *Geochimica et Cosmichimica Acta*, 89: 190–201.

Klass, D. L., 1984. Methane from anaerobic fermentation. *Science*, 223: 1021–1028.

Kloppmann, W., Chikurel, H., Picot, G., Guttman, J., Pettenati, M., Aharoni, A. and Guerrot, C. et al., 2009. B and Li isotopes as intrinsic tracers for injection tests in aquifer storage and recovery systems. *Applied Geochemistry*, 24: 1214–1223.

Kokelj, S. V., Lacelle, D., Lantz, T. C., Tunnicliffe, J., Malone, L., Clark, I. D. and Chin, K. S., 2013. Thawing of massive ground ice in mega slumps drives increases in stream sediment and solute flux across a range of watershed scales. *Journal of Geophysical Research: Earth Surface*, 118: 1–12.

Kolhatkar, R., Kuder, T., Philp, P., Allen, J. and Wilson, J. T., 2002. Use of compound-specific stable carbon isotope analyses to demonstrate anaerobic biodegradation of MTBE in groundwater at gasoline release site. *Environmental Science & Technology*, 26: 5139–5146.

Kousa, A., Komulainen, H., Hatakka, T., Backman, B. and Hartikainen, S., 2020. Variation in groundwater manganese in Finland. *Nature*, 43: 1193–1211.

Krapac, I. G., Dey, W. S., Roy, W. R., Smyth, C. A., Storment, E., Sargent, S. L. and Steele, J. D., 2002. Impacts of swine manure pits on groundwater quality. *Environmental Pollution*, 120: 475–492.

Krauskopf, K. and Bird, D. K., 1995. *Introduction to Geochemistry*, third edition, McGraw Hill, Inc., New York, NY, 647 pp.

Kraynov, S. R., Mer'kov, A. N., Petrova, N. G., Baturinskaya, I. V. and Zharikova, V. M., 1969. Highly alkaline (pH 12) fluorosilicate waters in the deeper zones of the lovozero massif. *Geochemistry. International*, 6: 635–640.

Lacelle, D., Bjornnson, J., Lauriol, B., Clark, I. D. and Troutet, Y., 2004. Segregated-intrusive ice of subglacial meltwater origin in retrogressive thaw flow headwalls, Richardson Mountains, NWT, Canada. *Quaternary Science Reviews*, 23: 681–696.

Lacelle, D. and Léveillé, R., 2010. Acid drainage generation and associated Ca–Fe–SO4 minerals in a periglacial environment, Eagle Plains, Northern Yukon, Canada: A potential analogue for low-temperature sulfate formation on Mars. *Planetary and Space Science*, 58: 509–521.

Landwehr, J. M. and Coplen, T. B., 2004. Line-conditioned excess: a new method for characterizing stable hydrogen and oxygen isotope ratios in hydrologic systems. In *Isotopes in Environmental Studies, Aquatic Forum 2004*. Monaco 25–29 October 2004. International Atomic Energy Agency, Vienna, Austria. IAEA-CN-118/56.

Langmuir, D., 1997. *Aqueous Environmental Geochemistry*, Prentice Hall, Upper Saddle River, NJ, p. 600.

Langmuir, D., 1997. Carbonate chemistry. In *Aqueous Environmental Geochemistry*, Prentice Hall, Upper Saddle River, NJ, pp. 193–230.

Lee, D. R. and Hartwig, D. S., 2005. Zeolite prevents discharge of strontium-90-contaminated groundwater. Canadian Nuclear Society, Waste Management Decommissioning and Environmental Restoration for Canada's Nuclear Activities: Current Practices and Future Needs. Ottawa, Ontario, Canada, May 8–11, 2005.

Lee, D. R., Smyth, D. J. A., Shikaze, S. G., Jowett, R., Hartwig, D. S. and Milloy, C., 1998. Wall-and-curtain for passive collection/treatment of contaminant plumes. In *Designing and Applying Treatment Technologies: Remediation of Chlorinated and Recalcitrant Compounds*, edited by G. B. Wickramanayake and R. E. Hinchee, Battelle Press, Columbus, OH, pp. 77–84.

Lee, K. Y., Syakir, M. I., Clark, I. D. and Veizer, J., 2013. Isotope constraints on the aquatic carbon budget: Langat watershed, Malaysia. *Aquatic Geochemistry*, 19: 443–475. doi: 10.1007/s10498-013-9198-3.

Lemarchand, E., Chabaux, F., Vigier, N., Millot, R. and Pierret, M. C., 2010. Lithium isotope systematics in a forested granitic catchment (Strengbach, Vosges Mountains, France). *Geochimica et Cosmochimica Acta*, 74: 4612–4628.

Leybourne, M. I. and Goodfellow, W. D., 2010. Geochemistry of surface waters associated with an undisturbed Zn-Pb massive sulfide deposit: Water-rock reaction, solute sources and the role of trace carbonate. *Chemical Geology*, 279: 40–54.

Li, T., 2013. Geochemistry and runoff processes in Wolf Creek, Whitehorse, Yukon. M.Sc. Thesis, Department of Earth Sciences, University of Ottawa, Ontario, Canada, p. 100.

Li, Y.-H., 1992. Seasalt and pollution inputs over the continental United States. *Water Air and Soil Pollution*, 64: 561–573.

Liaw, Y. P., 1990. Comparison of field, laboratory and theoretical estimates of global nitrogen fixation by lightning. *Journal of Geophysical Research*, 95: 489–494.

Liu, Q., Tossell, J. A. and Liu, Y., 2010. On the proper use of the Bigeleisen–Mayer equation and corrections to it in the calculation of isotopic fractionation equilibrium constants. *Geochimica et Cosmochimica Acta*, 74:6965–6983.

Loosli, H. H., Lehmann, B. E. and Smethie, W. M., Jr., 2000. Noble gas radioisotopes: ^{37}Ar, ^{85}Kr, ^{39}Ar, ^{81}Kr. In *Environmental Tracers in Subsurface Hydrology*, edited by P. Cook and A. L. Herczeg, Kluwer Academic, Dordrecht, the Netherlands, pp. 379–396.

Loosli, H. H. and Oeschger, H., 1969. ^{37}Ar and ^{81}Kr in the atmosphere. *Earth Planetary Science Letters*, 7: 67–71.

Loosli, H. H. and Oeschger, H., 1980. Use of ^{39}Ar and ^{14}C for groundwater dating. *Radiocarbon*, 22: 863–870.

Lowenstein, T. K., Hardie, L. A., Timofeeff, M. N. and Demicco, R. V., 2003. Secular variation in seawater chemistry and the origin of calcium chloride basinal brines. *Geology*, 31: 857–860.

Lowry, M. E., Reichenbaugh, R. C., Montgomery, J. S. L., Al-Mahrooqi, H. S. and Courtney, G. E., III., 1986. *Blue Pools of Oman*, Council for Conservation of Environment and Water Resources, Oman, CCEWR Report 86-8.

Lu, Z.-T., Schlosser, P., Smethie, W. M., Jr., Sturchio, N. C., Fischer, T. P., Kennedy, B. M. and Purtschert, R.et al., 2013. Tracer applications of noble gas radionuclides in the geosciences. *Earth Science Reviews* http://dx.doi.org/10.1016/j.earscirev.2013.09.002.

Lucas, L. L. and Unterweger, M. P., 2000. Comprehensible review and critical evaluation of the half-life of tritium. *Journal of Research of the National Institute of Standards and Technology*, 105: 541–549.

Mackay, D. M. and Cherry, J. A., 1989. Groundwater contamination: Pump-and-treat remediation. *Environmental Science and Technology*, 23: 630–636.

Majoube, M., 1971. Fractionnement en oxygène–18 et en deutérium entre l'eau et sa vapeur. *Journal of Chemical Physics*, 197: 1423–1436.

Maloszewski, P. and Zuber, A., 1982. Determining turnover time of groundwater systems with the aid of environmental tracers. I. Models and their applicability. *Journal of Hydrology*, 57: 207–231.

Maloszewski, P. and Zuber, A., 1996. Lumped parameter models for interpretation of environmental tracer data (Supplementary Annex). In *Manual on Mathematical Models in Isotope Hydrogeology, IAEA-TECHDOC-210*, IAEA, Vienna.

Mancini, S. A., Ulrich, A. C., Lacrampe-Couloume, G., Sleep, B., Edwards, E. A. and Sherwood-Lollar, B., 2003. Carbon and hydrogen isotopic fractionation during anaerobic biodegradation of benzene. *Applied Environmental Microbiology*, 69: 191–198.

Marchetti, A. A., Gu, F., Robl, R. and Straume, T., 1997. Determination of total iodine and sample preparation for AMS measurement of ^{129}I in environmental matrices. *Nuclear Instrumentation and Methods in Physics Research B*, 123: 352–355.

Maréchal, C. and Albarède, F., 2002. Ion-exchange fractionation of copper and zinc isotopes. *Geochimica Cosmochimica Acta*, 66: 1499–1509.

Martini, A. M., Walter, L. M., Ku, T. C. W., Budai, J. M., McIntosh, J. C. and Schoell, M., 2003. Microbial production and modification of gases in sedimentary basins: A geochemical case study from a Devonian shale gas play, Michigan basin. *AAPG Bulletin*, 87: 1355–1375.

Marty, B., Torgersen, T. and Meynier, V., 1993. Helium isotope fluxes and groundwater ages in the Dogger Aquifer, Paris Basin. *Water Resources Research*, 29: 1025–1035.

Masue, Y., Loeppert, R. H. and Kramer, T. A., 2007. Arsenate and arsenite adsorption and desorption behavior on coprecipitated aluminum: Iron hydroxides. *Environmental Science and Technology*, 41: 837–842.

Matsumoto, T., Han, L.-F., Jaklitsch, M. and Aggarwal, P. K., 2012. A portable membrane contractor sampler for analysis of noble gases in groundwater. *Ground Water*. doi: 10.1111/j.1745-6584.2012.00983.x.

Mazurek, M., Alt-Epping, P., Bath, A., Gimmi, T. and Waber, H. N., 2009. Natural Tracer Profiles across Argillaceous Formations: The CLAYTRAC Project. Organization for Economic Cooperation and Development, Nuclear Energy Agency, NEA No. 6253. http://www.oecd-nea.org/rwm/pubs/2009/6253-claytract-project-2009.pdf.

McCaffrey, M. A., Lazar, B. and Holland, H. D., 1987. The evaporation path of seawater and the coprecipitation of Br^- and K^+ with halite. *Journal of Sedimentary Petrology*, 57: 928–937.

McDonald, J. D., 2012. *Electric Power Substations Engineering The Electric Power Engineering Handbook*, 3rd ed., CRC Press, Boca Raton, FL.

McIntosh, J., Walter, L. and Martini, A., 2002. Pleistocene recharge to midcontinent basins: Effects on salinity structure and microbial gas generation. *Geochimica et Cosmochimica Acta*, 66: 1681–1700.

Meckenstock, R. U., Morasch, B., Griebler, C. and Richnow, H. H., 2004. Stable isotope fractionation analysis as a tool to monitor biodegradation in contaminated acquifers. *Journal of Contaminant Hydrology*, 75: 215–255.

MEXT, 2012. Analysis Results Concerning (i) Gamma-Emitting Nuclides and (ii) Sr-89 and Sr-90 (Second distribution survey) by MEXT. Ministry of Education, Culture, Sports, Science and Technology, Japan. http://radioactivity.nsr.go.jp/en/contents/6000/5636/24/338_Sr_0912018_e.pdf.

Michalski, G., Böhlke, J. K. and Thiemens, M., 2004. Long term atmospheric deposition as the source of nitrate and other salts in the Atacama Desert, Chile: New evidence from mass-independent oxygen isotopic compositions. *Geochimica et Cosmochimica Acta*, 68: 4023–4038.

Miller, M. F. and Pack, A., 2021. Why measure ^{17}O? Historical perspective, triple-isotope systematics and selected applications. *Reviews of Mineralogy and Geochemistry*, 86: 1–34.

Mohammadzadeh, H. and Clark, I. D., 2008. Degradation pathways of dissolved carbon in landfill leachate traced with compound-specific ^{13}C analysis of DOC. *Journal of Isotopes in Environmental and Health Studies*, 44: 1–28.

Mohammadzadeh, H., Clark, I. D., Marschner, M. and St-Jean, G., 2005. Compound specific isotopic analysis (CSIA) of landfill leachate DOC components. *Chemical Geology*, 218: 3–13.

Momoshima, N., Inoue, F., Ohta, T., Mahara, Y., Shimada, J., Ikawa, R. and Kagabu, M. et al., 2011. Application of ^{85}Kr dating to groundwater in volcanic aquifer of Kumamoto Area, Japan. *Radioanal Nuclear Chemistry*, 287: 761–767.

Mook, W., 1976. *The Dissolution-Exchange Model for Dating Groundwater with ^{14}C*, International Atomic Energy Agency (IAEA), IAEA, Vienna, pp. 213–225.

Mook, W. G., Bommerson, J. C. and Staverman, W. H., 1974. Carbon isotope fractionation between dissolved bicarbonate and gaseous carbon dioxide. *Earth and Planetary Science Letters*, 22: 169–176.

Moore, T., Xing, Y., Lazenby, B., Lynch, M., Schiff, S., Robertson, W. and Timlin, R. et al., 2011. Prevalence of anaerobic ammonium-oxidizing bacteria in contaminated groundwater. *Environmental Science & Technology*, 45: 7217–7225.

Morgenstern, U., 2000. Silicon-32. In *Environmental Tracers in Subsurface Hydrology*, edited by P. G. Cook and A. L. Herczeg, Kluwer Academic, Boston, pp. 499–502.

Morin, K. A., Cherry, J. A., Dave, N. A., Lim, T. P. and Vivyurka, A. J., 1988. Migration of acidic groundwater seepage from uranium-tailings impoundments, 1. Field study and conceptual hydrogeochemical model. *Journal of Contaminant Hydrology*, 2: 271–303.

Morin, K. A. and Hutt, N. M., 1997. *Environmental Geochemistry of Minesite Drainage: Practical Theory and Case Studies*, MDAG Publishing, Vancouver, Canada.

Moser, H. and Stichler, W., 1970. Deuterium measurements on snow samples from the Alps. *Isotope Hydrology 1970*, IAEA Symposium 129, March, 1970, Vienna, pp. 43–57.

Moses, C. O., 1999. Water. In *Encyclopedia of Geochemistry*, edited by C. Marshall and R. W. Fairbridge, Encyclopedia of Earth Sciences Series, Kluwer Academic, Boston, MA, pp. 661–666.

Mulder, A., Van de Graaf, A. A., Robertson, L. A. and Kuenen, J. G., 1995. Anaerobic ammonium oxidation discovered in a denitrifying fluidized bed reactor. *FEMS Microbiology and Ecology*, 16: 177–183.

Murphy, S., Ouellon, T., Ballard, J.-M., Lefebvre, R. and Clark, I. D., 2010. Tritium-helium groundwater age used to constrain a groundwater flow model of a valley-fill aquifer contaminated with trichloroethylene (Quebec, Canada). *Hydrogeology Journal*, 19: 195–207.

Musy, S., Hinsby, K., Troldborg, L., Delottier, H., Guillon, S., Brunner, P. and Purtschert, R., 2023. Evaluating the impact of muon-induced cosmogenic ^{39}Ar and ^{37}Ar underground production on groundwater dating with field observations and numerical modelling. *Science of the Total Environment*, 903: 166588.

Naga Jyothi, M. S. V., Ramaiah, B. J. and Maliyekkal, S. M. (2020). Occurrence, contamination, speciation and analysis of selenium in the environment. In *Measurement, Analysis and Remediation of Environmental Pollutants*. Energy, Environment, and Sustainability, edited by Gupta, T., Singh, S., Rajput, P., Agarwal, A., Springer, Singapore. https://doi.org/10.1007/978-981-15-0540-9_1

Nakagawa, F. and Yoshida, N., 2002. Production of methane from alasses in eastern Siberia: Implications from its ^{14}C and stable isotopic compositions. *Global Biogeochemical Cycles*, 16: 1041. doi: 10.1029/2000GB001384.

NEA, 2002. The release, dispersion and deposition of radionuclides. In *Chernobyl: Assessment of Radiological and Health Impact 2002 Update of Chernobyl: Ten Years On*. Nuclear Energy Agency. https://www.oecd-nea.org/rp/chernobyl/c02.html.

Neal, C. and Stanger, G., 1983. Hydrogen generation from mantle source rocks in Oman. *Earth and Planetary Science Letters*, 66: 315–320.

Nemani, R. R., Keeling, C. D., Hashimoto, H., Jolly, W. M., Piper, S. C., Tucker, C. J., Myneni, R. B. and Running, S. W., 2003. Climate-driven increases in global terrestrial net primary production from 1982 to 1999. *Science*, 300: 1560–1563.

Neumann, R. B., Ashfaque, K. N., Badruzzaman, A. B. M., Ashraf Ali, M., Shoemaker, J. K. and Harvey, C. F., 2010. Anthropogenic influences on groundwater arsenic concentrations in Bangladesh. *Nature Geoscience*, 3: 46–52.

Nickson, R., McArthur, J., Burgess, W., Ahmed, K. M., Ravenscroft, P. and Rahmann, M., 1998. Arsenic poisoning of Bangladesh groundwater. *Nature*, 395: 338.

NOAA, 2014. U.S. National Oceanographic and Atmospheric Administration. Earth Systems Research Laboratory, Global Monitoring Division. http://www.esrl.noaa.gov/

Nordstrom, D. K. and Alpers, C. N., 1999. Negative pH, efflorescent mineralogy, and consequences for environmental restoration at the Iron Mountain Superfund site, California. *Proceedings of the National Academy of Sciences of the United States of America*, 96: 3455–3462.

Nordstrom, D. K., Alpers, C. N., Ptacek, C. J. and Blowes, D. W., 2000. Negative pH and extremely acidic mine waters from Iron Mountain, California. *Environmental Science & Technology*, 34: 254–258.

Nordstrom, D. K. and Archer, D. G., 2003. Arsenic thermodynamic data and environmental geochemistry: An evaluation of thermodynamic data and environmental geochemistry. In *Arsenic in Ground Water: Geochemistry and Occurrence*, edited by A.H. Welch and K.G. Stollenwerk, Springer, New York, 475 pp.

Noxon, J. F., 1976. Atmospheric nitrogen fixation by lightning. *Geophysical Research Letters*, 3: 463–465.

Nydal, R. and Lövseth, K., 1983. Tracing bomb ^{14}C in the atmosphere 1962–1980. *Journal of Geophysical Research*, 88: 3621–3642.

O'Leary, M. H., 1988. Carbon isotopes in photosynthesis. *Bioscience*, 38: 328–336.

O'Hannesin, S. F. and Gillham, R. W., 1998. Long-term performance of an in situ "iron wall" for remediation of VOCs. *Ground Water*, 36: 164.

O'Neil, J. R., 1968. Hydrogen and oxygen isotope fractionation between ice and water. *Journal of Physical Chemistry*, 72: 3683–3684.

O'Neil, J. R., Clayton, R. N. and Mayeda, T. K., 1969. Oxygen isotope fractionation in divalent metal carbonates. *Journal of Chemical Physics*, 51: 5547–5558.

Osborn, S. G., Vengosh, A., Warner, N. A. and Jackson, R. B., 2011. Methane contamination of drinking water accompanying gas-well drilling and hydraulic fracturing. *Proceeding of the National Academy of Sciences*, 108: 8172–8176.

Palmer, C. D. and Wittbrodt, P. R., 1991. Processes affecting the remediation of chromium-contaminated sites. *Environmental Health Perspectives*, 92: 25–40.

Panno, S. V., Hackey, K. C., Hwang, H. H., Greenberg, S. E., Krapac, I. G., Landsberger, S. and O'Kelly, D. J., 2006. Characterization and identification of NaCl sources in ground water. *Ground Water*, 144: 176–187.

Parkhurst, D. L., 1995. User's guide to PHREEQC—A computer program for speciation, -path, advective-transport, and inverse geochemical calculations: U.S. Geological Survey Water-Resources Investigations Report 95-4227, USGS, Washington, DC, 143 pp.

Parkhurst, D. L. and Appelo, C. A. J., 1999. User's guide to PHREEQC (Version 2): A Computer Program for Speciation, Batch-Reaction, One-Dimensional Transport, and Inverse Geochemical Calculations. Water-Resources Investigations Report 99-4259, U. S. Geological Survey, pp. 312.

Patterson, L. J., Sturchio, N. C., Kennedy, B. M., van Soest, M. C., Sultan, M., Lu, Z.-T. and Lehmann, B. et al., 2005. Cosmogenic, radiogenic, and stable isotopic constraints on groundwater residence time in the Nubian Aquifer, Western Desert of Egypt. *Geochemistry, Geophysics, Geosystems G3*, 6: Q01005. doi: 10.1029/2004GC000779.

Pauling, L., 1960. *The Nature of the Chemical Bond*, 3rd ed., Cornell University Press, Ithaca, NY.

Pearson, F. J., 1965. Use of C-13/C-12 ratios to correct radiocarbon ages of material initially diluted by limestone. In *Proceedings of the 6th International Conference on Radiocarbon and Tritium Dating*, Pulman, Washington, p. 357.

Pearson, F. J., 1987, Models of mineral controls on the composition of saline groundwaters of the Canadian Shield. In *Saline Water and Gases in Crystalline Rocks: Geological Association of Canada Special Paper 33*, edited by Fritz, P., and Frape, S. K., pp. 39–51.

Pearson, F. J. and Hanshaw, B. B., 1970. Sources of dissolved carbonate species in groundwater and their effects on carbon-14 dating. In *Isotope Hydrology 1970*, IAEA Symposium 129, March 1970, Vienna, pp. 271–286.

Peterman, Z. E., Thamke, J. and Futa, K., 2010. Use of strontium isotopes to detect produced-water contamination in surface water and groundwater in the Williston Basin, Northeastern Montana. In: Proceedings, GeoCanada, Calgary, Alberta, May 10–14, 2010.

Philip, J., Dunn, H., Malinovsky, D., Ogrinc, N., Potočnik, D., Flierl, L., Dipayan, O., Harro, P. and Meijer, A. J., 2024. Re-determination of R(13C/12C) for Vienna Peedee belemnite (VPDB) *Rapid Communications in Mass Spectrometry*, 38, Issue 16 e9773.

Phillips, F. M., 2000. Chlorine-36. In: *Environmental Tracers in Subsurface Hydrology*, edited by P. Cook and A. L. Herczeg, Kluwer Academic, Norwell, Mass, pp. 299–348.

Phillips, F. M. and Castro, M. C., 2003. Groundwater dating and residence-time measurements. In *Treatise on Geochemistry*, vol. 5, edited by James I. Drever, Heinrich D. Holland, and Karl K. Turekian, Elsevier, Amsterdam, the Netherlands, pp. 451–497.

Phillips, F. M., Peeters, L., Tansey, M. K. and Davis, S. N., 1986. Paleoclimate inferences from an isotopic investigation of groundwater in the central San Juan Basin, New Mexico. *Quaternary Research*, 26: 179–193.

Piper, A. M., 1944. A graphic procedure in the geochemical interpretation of water-analyses. *Eos, Transactions American Geophysical Union*, 25: 914–928. doi: 10.1029/TR025i006p00914.

Pitzer, K. S., 1973. Thermodynamics of electrolytes. I. Theoretical basis and general equations. *Journal Physical Chemistry*, 77: 268–277.

Plummer, L. N., Eggleston, J. R., Andreasen, D. C., Raffensperger, J. P., Hunt, A. G. and Casile, G. C., 2012. Old groundwater in parts of the upper Patapsco aquifer, Atlantic Coastal Plain, Maryland, USA: Evidence from radiocarbon, chlorine-36 and helium-4. *Hydrogeology Journal*, 20: 1269–1294.

Plummer, L. N., Parkhurst, D. L., Fleming, G. W. and Dunkle, S. A., 1988. A Computer Program Incorporating Pitzer's Equations for Calculation of Geochemical Reactions in Brines. U. S. Geological Survey Water-Resources Investigations Report 88-4153, p. 310.

Plummer, N. L., Prestemon, E. C. and Parkhurst, D. L., 1994. An Interactive Code (NETPATH) for Modeling Net Geochemical Reactions along a Flow Path, Version 2.0. U. S. Geological Survey Water Resources Investigations Report 94-4169, U. S. G. S., Reston, VA.

Poole, J. C., McNeill, G. W., Langman, S. R. and Dennis, F., 1997. Analysis of noble gases in water using a quadrupole mass spectrometer in static mode. *Applied Geochemistry*, 12: 707–714.

Price, C., Penner, J. and Prather, M., 1997. NO_x from lightning, 1. Global distribution based on lightning physics. *Journal of Geophysical Research*, 102: 5929–5941.

Prinzhofer, A. A. and Huc, A. Y., 1995. Genetic and post-genetic molecular and isotopic fractionations in natural gases. *Chemical Geology*, 126: 281–290.

Proemse, B. C., Mayer, B., Fenn, M. E. and Ross, C. S., 2013. A multi-isotope approach for estimating industrial contributions to atmospheric nitrogen deposition in the Athabasca oil sands region in Alberta, Canada. *Environmental Pollution*, 182: 80–91.

Quay, P. D., King, S. L., Stutsman, J., Wilbur, D. O., Steele, L. P., Fung, I. and Gammon, R. et al., 1991. Carbon isotopic composition of atmospheric CH_4: Fossil fuel and biomass burning source strengths. *Global Biogeochemical Cycles*, 5: 25–47.

Radloff, K. A., Zheng, Y., Michael, H. A., Stute, M., Bostick, B. C., Mihajlov, I. and Bounds, M. et al., 2011. Arsenic migration to deep groundwater in Bangladesh influenced by adsorption and water demand. *Nature Geoscience*, 4: 793–798.

Rai, D., Sass, B. M. and Moore, D. A., 1987. Chromium (III) hydrolysis constants and solubility of chromium (III) hydroxide. *Inorganic Chemistry*, 26: 345–349.

Rayleigh, J. W. S., 1896. Theoretical considerations respecting the separation of gases by diffusion and similar processes. *Philosophical Magazine*, 42: 493–498.

Reardon, E. J., Allison, G. B. and Fritz, P., 1979. Seasonal chemical and isotopic variations of soil CO_2 at Trout Creek, Ontario. *Journal of Hydrology*, 43: 355–371.

Reardon, E. J. and Dewaele, P., 1990. Chemical model for the carbonation of a grout/water slurry. *Journal of the American Ceramic Society*, 73: 1681–1690.

Reichel, T., Kersting, A., Ritterbusch, F., Ebser, S., Bender, K., Purtschert, R. and Oberthaler, M., Aeschbach-Hertig, W., 2013. First Dating of Groundwater with Atom Trap Trace Analysis of ^{39}Ar: Application. EGU General Assembly 2013, held on April 7–12, 2013, Vienna, Austria, id. EGU2013-10901.

Reimer, P. J., Baillie, M. G. L., Bard, E., Bayliss, A., Beck, J. W., Blackwell, P. G. and Bronk Ramsey, C. et al., 2009. IntCal09 and Marine09 radiocarbon age calibration curves, 0–50,000 years cal BP. *Radiocarbon*, 51: 1111–1150.

Renaud, R., Clark, I. D., Kotzer, T. G., Milton, G. M. and Bottomley, D. J., 2005. The mobility of anthropogenic ^{129}I in a shallow sand aquifer at Sturgeon Falls, Ontario, Canada. *Radiochimica Acta*, 93: 363–371.

Robertson, D. E. and Cherry, J. A., 1995. In situ denitrification of septic system nitrate using reactive porous media barriers: Field trials. *Groundwater*, 33: 99–111.

Robertson, D. E., Perkins, R. W., Lepel, E. L. and Thomas, C. W., 1992. Radionuclide concentrations in environmental samples collected around Chernobyl during the International Chernobyl Project—analyses conducted by Battelle, Pacific Northwest Laboratory. *Journal of Environmental Radioactivity*, 17: 159–182.

Robertson, W. D., Blowes, D. W., Ptacek, C. J. and Cherry, J. A., 2000. Long-term performance of in-situ reactive barriers for nitrate remediation. *Ground Water*, 38: 689–695.

Robertson, W. D. and Cherry, J. A., 1989. Tritium as an indicator of recharge and dispersion in a groundwater system in Central Ontario. *Water Resource Research*, 25: 1097–1109.

Robertson, W. D., Moore, T. A., Spoelstra, J., Li, L., Elgood, R. J., Clark, I. D., Schiff, S. L., Aravena, R. and Neufeld, J. D., 2012. Natural attenuation of septic system nitrogen by anammox. *Ground Water*, 50: 541–553. doi: 10.1111/j.1745-6584.2011.00857.x.

Rosman, K. J. R. and Taylor, P. D., 1998. Isotopic compositions of the Elements 1997. *Pure and Applied Chemistry*, 70: 217–235.

Rozanski, K., Araguás-Araguás, L. and Gonfiantini, R., 1993. Isotopic patterns in modern global precipitation. In *Continental Isotope Indicators of Climate*, Monograph 78, edited by P.K. Swart, K.C. Lohmann, J. McKenzie, and S. Savin, American Geophysical Union, pp. 1–36.

Rudolph, D. L., Barry, D. A. J. and Goss, M. J., 1998. Contamination in Ontario farmstead domestic wells and its association with agriculture: 2. Results from multilevel monitoring well installations. *Journal of Contaminant Hydrology*, 32: 295–311.

Sander, R., 1999. Compilation of Henry's Law Constants for Inorganic and Organic Species of Potential Importance in Environmental Chemistry. Max-Planck Institute of Chemistry, Mainz, Germany. http://www.mpch-mainz.mpg.de/~sander/res/henry.html.

Sanford, W. E., Shropshire, R. G. and Solomon, D. K., 1996. Dissolved gas tracers: Simplified injection, sampling and analysis. *Water Resources Research*, 32: 1635–1642.

Santella, N., Ho, D. T., Schlosser, P. and Stute, M., 2008. Widespread elevated atmospheric SF6 mixing ratios in the Northeastern United States: Implications for groundwater dating. *Journal of Hydrology*, 349: 139–146.

Sarmiento, J. L. and Gruber, N., 2002. Sinks for anthropogenic carbon. *Physics Today*, August 2002, pp. 30–36.

Savin, S. M. and Epstein, S., 1970. The oxygen and hydrogen isotope geochemistry of clay minerals. *Geochimica et Cosmochimica Acta*, 34: 25–42.

Schleppi, P., Bucher-Wallin, I., Saurer, M., Jäggi, M. and Landolt, W., 2006. Citric acid traps to replace sulphuric acid in the ammonia diffusion of dilute water samples for ^{15}N analysis. *Rapid Communications in Mass Spectrometry*, 20: 629–634.

Schoell, M., 1980. The hydrogen and carbon isotopic composition of methane from natural gas of various origins. *Geochimica et Cosmochimica Acta*, 44: 649–662.

Schubert, I. M., Kopitz, J. and Knöller, K., 2019. Improved approach for LSC detection of ^{35}S aiming at its application as tracer for short groundwater residence times. *Journal of Environmental Radioactivity*, 208–209: 106022.

Seal, R. R. II, 2006. Sulfur isotope geochemistry of sulfide minerals. *Reviews in Mineralogy and Geochemistry*, 61: 633–677.

Sebilo, M., Mayer, B., Grably, M., Billious, D. and Mariotti, A., 2004. The use of the ammonium diffusion method for $\delta^{15}N$-NH_4^+ and measurements comparison with other techniques. *Environmental Chemistry*, 1: 99–103.

Séby, F., Potin-Gautier, M., Giffaut, E., Borge, G. and Donard, O. F. X., 2001. A critical review of thermodynamic data for selenium species at 25°C. *Chemical Geology*, 171: 173–194.

Seefeldt, L. C., Hoffman, B. M. and Dean, D. R., 2009. Mechanism of Mo-dependent nitrogenase. *Annual Review of Biochemistry*, 78: 701–722. doi: 10.1146/annurev.biochem.78.070907.103812.

Self, S., Widdowson, M., Thordarson, T. and Jay, A. E., 2005. Volatile fluxes during flood basalt eruptions and potential effects on the global environment: A Deccan perspective. *Earth and Planetary Science Letters*, 248: 518–532.

Shannon, R. D., 1976. Revised effective ionic radii and systematic studies of interatomic distances in halides and chalcogenides. *Acta Crystallographica*, A32: 751–767.

Sherwood Lollar, B., Weisse, S. M., Frape, S. K. and Barker, J. F., 1994. Isotopic constraints on the migration of hydrocarbon and helium gases of southwestern Ontario. *Bulletin of Canadian Petroleum Geology*, 42: 283–295.

Shkarupin, A., Saurette, E., Sharp, K., Kim, S. B. and Clark, I. D., 2025. Influence of clay-water interactions on organically bound tritium characterization in soil and sediment samples. *Journal of Environmental Radioactivity*. In review.

Shouakar-Stash, O., Frape, S. K. and Drimmie, R. J., 2005. Determination of bromine stable isotopes using continuous-flow isotope ratio mass spectrometry. *Analytical Chemistry*, 77: 4027–4033.

Shukla, S. and Saxena, A. (2019). Global Status of Nitrate Contamination in Groundwater: Its Occurrence, Health Impacts, and Mitigation Measures. In *Handbook of Environmental Materials Management*, edited by C. Hussain, Springer, Cham. https://doi.org/10.1007/978-3-319-73645-7_20

Siegel, D. I. and Mandle, R. J., 1984. Isotopic evidence for glacial meltwater recharge to the Cambrian-Ordovician aquifer, north-central United States. *Quaternary Research*, 22: 328–335.

Sigman, D. M., Casciotti, K. L., Andreani, C., Barford, C., Galanter, M. and Böhlke, J. K., 2001. A bacterial method for the nitrogen isotopic analysis of nitrate in seawater and freshwater. *Analytical Chemistry*, 73: 4145–4153.

Sivan, O., Yechieli, Y., Herut, B. and Lazar, B., 2005. Geochemical evolution and timescale of seawater intrusion into the coastal aquifer of Israel. *Geochimica et Cosmochimica Acta*, 69: 579–592.

Skuce, M., Longstaffe, F. J., Carter, T. R. and Potter, J., 2015. Isotopic fingerprinting of groundwaters in southwestern Ontario: Applications to abandoned well remediation. *Applied Geochemistry*, 58: 1–13.

Smith, S. P. and Kennedy, B. M., 1983. The solubility of noble gases in water and in NaCl brine. *Geochimica et Cosmochimica Acta*, 47: 503–515.

Solanki, S. K., Usoskin, I. G., Kromer, B., Schussler, M. and Beer, J., 2004. Unusual activity of the Sun during recent decades compared to the previous 11,000 years. *Nature*, 431: 1048–1087.

Spoelstra, J., Leal, K. A., Senger, N. D., Schiff, S. L. and Post, R., 2021. Isotopic characterization of sulfate in a shallow aquifer impacted by agricultural fertilizer. *Groundwater*, 59: 658–670.

Squeo, F. A., Aravena, R., Aguirre, E., Pollastri, A., Jorquera, C. B. and Ehleringer, J. R., 2006. Groundwater dynamics in a coastal aquifer in north-central Chile: Implications for groundwater recharge in an arid ecosystem. *Journal of Arid Environments*, 67: 240–254.

SRB Technologies (Canada) Inc., 2016. Annual Compliance Report to the Canadian Nuclear Safety Commission.

Stevenson, F. J., 1985. Geochemistry of soil humic substances. In *Humic Substances in Soil, Sediment, and Water*, edited by G. R. Aiken, D. M. McKnight, R. L. Wershaw, and P. MacCarthy, Wiley, New York, pp. 13–52.

St-Jean, G., 2003. Automated quantitative and isotopic (^{13}C) analysis of dissolved inorganic carbon and dissolved organic carbon in continuous-flow using a total organic carbon analyser. *Rapid Communications in Mass Spectrometry*, 17: 419–428.

Stotler, R. L., Frape, S. K. and Shouakar-Stash, O., 2010. An isotopic survey of $\delta^{81}Br$ and $\delta^{37}Cl$ of dissolved halides in the Canadian and Fennoscandian Shields. *Chemical Geology*, 274: 38–55.

Sturchio, N. C., Caffee, M., Beloso, A. D., Jr., Heraty, L. J., Böhlke, J. K., Hatzinger, P. B., Jackson, W. A., Gu, B., Heikoop, J. M. and Dale, M., 2009. Chlorine-36 as a tracer of perchlorate origin. *Environmental Science & Technology*, 43: 6934–6938.

Sturchio, N. C., Du, X., Purtschert, R., Lehmann, B. E., Sultan, M., Patterson, M. J. and Lu, Z.-T. et al., 2004. One million year old groundwater in the Sahara revealed by krypton-81 and chlorine-36. *Geophysical Research Letters*, 31: L05503. doi: 10.1029/2003GL019234.

Stute, M., Clark, J. F., Schlosser, P., Broecker, W. S. and Bonani, G., 1995. A 30,000 yr continental paleotemperature record derived from noble gases dissolved in groundwater from the San Juan Basin, New Mexico. *Quaternary Research*, 43: 209–220.

Stute, M., Forster, M., Frischkorn, A., Serejo, A., Clark, J. F., Schlosser, P., Broecker, W. S. and Bonani, G., 1995. Cooling of tropical Brazil (5 degrees C) during the last glacial maximum. *Science*, 269: 379–383.

Suchy, M., 2008. Nitrate and ammonium transformation and fate in groundwater of an agricultural watershed: An isotope and geochemistry approach. M.Sc. thesis, Department of Earth Sciences, University of Ottawa, Ontario, Canada. 256 pp.

Suecker, J. K., Turk, J. T. and Michel, R. L., 1999. Use of cosmogenic sulfur-35 for comparing ages of water from three alpine–subalpine basins in the Colorado Front Range. *Geomorphology*, 27: 61–74.

Suzuoki, T. and Kumura, T., 1973. D/H and $^{18}O/^{16}O$ fractionation in ice-water systems. *Mass Spectroscopy*, 21: 229–233.

Svec, H. J. and Anderson, A. R. Jr., 1965. The absolute abundance of the lithium isotopes in natural sources. *Geochimica et Cosmochimica Acta*, 29: 633–641.

Synal, H. A., Beer, J., Bonani, G., Suter, M. and Wölfli, W., 1990. Atmospheric transport of bomb-produced ^{36}Cl. *Nuclear Instruments and Methods in Physics Research Section B: Beam Interactions with Materials and Atoms*, 52: 483–488.

Teck, 2019. Elk Valley Water Quality Plan. Annex C: Mitigration Inputs to 2017 Regional Water Quality Model. July 2019. Unpublished report.

Tesoriero, A. and Puckett, L. J., 2011. O_2 reduction and denitrification rates in shallow aquifers. *Water Resources Research*, 47: 1–62.

Thomas, J. M., Hershey, R. M., Fereday, W. and Burr, G., 2021. Using Carbon-14 of dissolved organic carbon to determine groundwater ages and travel times in aquifers with low organic carbon. *Applied Geochemistry*, 124: 104842.

Thornton, S. F., Bottrell, S. H., Spence, K. H., Pickup, R., Spence, M. J., Shah, N., Mallinson, H. E. H. and Richnow, H. H., 2011. Assessment of MTBE biodegradation in contaminated groundwater using ^{13}C and ^{14}C analysis: Field and laboratory microcosm studies. *Applied Geochemistry*, 26: 828–837.

Ti, C., Ma, S., Peng, L., Tao, L., Wang, X., Dong, W., Wang, L. and Yan, X., 2021. Changes of $\delta^{15}N$ values during the volatilization process after applying urea on soil. *Environmental Pollution*, 270: 116204. doi: 10.1016/j.envpol.2020.116204.

Tsunogai, S. and Henmi, T., 1971. Iodine in the surface water of the ocean. *Journal of the Oceanographical Society of Japan*, 27: 67–72.

Turchyn, A. V. and Schrag, D. P., 2006. Cenozoic evolution of the sulfur cycle: Insight from oxygen isotopes in marine sulfate. *Earth and Planetary Science Letters*, 241: 763–779.

Urey, H. C., 1947. The thermodynamic properties of isotopic substances. *Journal of Chemical Society*, 1947: 562–581.

Utting, N., Clark, I. D., Lauriol, B., Wieser, M. and Aeschbach-Hertig, W., 2012. Origin and flow dynamics of perennial groundwater in continuous Permafrost Terrain using isotopes and noble gases: Case study of the fishing branch river, Northern Yukon, Canada. *Permafrost and Periglacial Processes*, 23: 91–106.

Van Stempvoort, D., Maathuis, H., Jaworski, E., Mayer, B. and Rich, K., 2005. Oxidation of fugitive methane in ground water linked to bacterial sulfate reduction. *Ground Water*, 43: 187–199.

Veizer, J., 1989. Strontium isotopes in seawater through time. *Annual Review of Earth and Planetary Science*, 17: 141–167.

Veizer, J., Holser, T. and Wilgus, C. K., 1980. Correlation of $^{13}C/^{12}C$ and $^{34}S/^{32}S$ secular variations. *Geochimica et Cosmochimica Acta*, 44: 579–587.

Velderman, B.-J., 1993. Groundwater Recharge and Contamination: Sensitivity Analysis for Carbonate Aquifers in South-Eastern Ontario — The Jock River Basin Study. Unpublished M.Sc. Thesis, University of Ottawa, Ottawa, Canada, 126 p.

Vitoria, L., Otero, N., Soler, A. and Canals, A., 2004. Fertilizer Characterization: Isotopic Data (N, S, O, C and Sr). *Environmental Science and Technology*, 38: 3254–3262.

Vogel, J. C., 1970. Carbon-14 dating of groundwater. In *Isotope Hydrology 1970*, IAEA Symposium 129, March 1970, Vienna, pp. 225–239.

von Brömssen, M., Jakariya, M., Bhattacharya, P., Ahmed, K. M., Hasan, M. A., Sracek, O., Jonsson, L., Lundell, L. and Jacks, G., 2007. Targeting low-arsenic aquifers in Matlab Upazila, Southeastern Bangladesh. *Science of the Total Environment*, 379: 121–132.

Warner, M. J. and Weiss, R. F., 1985. Solubilities of chlorofluorocarbons 11 and 12 in water and seawater. *Deep Sea Research*, 32: 485–497.

Wassenaar, L., Aravena, R. and Fritz, P., 1989. The geochemistry and evolution of natural organic solutes in groundwater. *Radiocarbon*, 31: 865–876.

Wassenaar, L., Aravena, R., Hendry, J. and Fritz, P., 1991. Radiocarbon in dissolved organic carbon, a possible groundwater dating method: Case studies from western Canada. *Water Resources Research*, 27: 1975–1986.

Wassenaar, L. I., 1995. Evaluation of the origin and fate of nitrate in the Abbotsford Aquifer using the isotopes of ^{15}N and ^{18}O in. *Applied Geochemistry*, 10: 391–405.

Wassenaar, L. I., Hendry, M. J., Chostner, L. V. and Lis, G. P., 2008. High resolution pore water δ^2H and $\delta^{18}O$ measurements by H_2O (liquid)-H_2O (vapor) equilibration laser spectroscopy. *Environmental Science and Technology*, 42: 9262–9267.

Webb, E. and Longstaffe, F. J., 2000. The oxygen isotopic compositions of silica phytoliths and plant water in grasses: Implications for the study of paleoclimate. *Geochimica et Cosmochimica Acta*, 64: 767–780.

Welte, J., Ritterbusch, F., Steinke, I., Henrich, M., Aeschbach-Hertig, W. and Oberthaler, M. K., 2010. Towards the realization of atom trap trace analysis for ^{39}Ar. *New Journal of Physics*, 12(6): 065031.

Whiticar, M. J., Faber, E. and Schoell, M., 1986. Biogenic methane formation in marine and freshwater environments: CO_2 reduction vs. acetate fermentation—isotope evidence. *Geochimica et Cosmochimica Acta*, 50: 693–709.

Wickens, K., 2003. Determination of isotope enrichment factors for ^{15}N-NH_4 during volatilization and oxidation. Unpublished Honours Degree Thesis, Department of Earth Sciences, University of Ottawa, Ottawa, Canada, K1N 6N5. pp. 21.

Widory, D., Kloppmann, W., Chery, L., Bonnin, J., Rochdi, H. and Guinamant, J.-L., 2004. Nitrate in groundwater: An isotopic multi-tracer approach. *Journal of Contaminant Hydrology*, 72: 165–188.

Widory, D., Petelet-Giraud, E., Bregnot, A., Bronders, J., Tirez, K. and Boeckx, P., 2013. Improving the management of nitrate pollution in water by the use of isotope monitoring: The $\delta^{15}N$, $\delta^{18}O$ & $\delta^{11}B$ triptych. *Isotopes in Environmental and Health Studies*, 49: 29–47.

Wiles, D. R., 2002. *The Chemistry of Nuclear Fuel Waste Disposal*, Polytechnic International Press, Montreal, Canada, p. 182.

Williams, P. R. D., Cushing, C. A. and Sheehan, P. J., 2003. Data available for evaluating the risks and benefits of MTBE and ethanol as alternative fuel oxygenates. *Risk Analysis*, 23: 1085–1115.

Winograd, I. J., Coplen, T. B., Landwehr, J. M., Riggs, A. C., Ludwig, K. R., Szabo, B. J., Kolesar, P. T. and Revesz, K. M., 1992. Continuous 500,000 year climate record from vein calcite in Devils Hole, Nevada. *Science*, 258: 255–260.

Wood, S. and Samson, I. M., 2006. The aqueous geochemistry of gallium, germanium, indium and scandium. *Ore Geology Reviews*, 28: 57–102.

Wu, D., Sui, Y., He, S., Wang, X., Li, C. and Kong, H., 2008. Removal of trivalent chromium from aqueous solution by zeolite synthesized from coal fly ash. *Journal of Hazardous Materials*, 155: 415–423.

Xiao, Y. K., Zhou, Y., Wang, Q., Wei, H., Liu, W. and Eastoe, C. J., 2002. A secondary isotopic reference material of chlorine from selected seawater. *Chemical Geology*, 182: 655–661.

Yi, L. M., Zhou, W.-J., Hou, X.-L., Chen, N., Zhang, L.-Y., Liu, Q. and He, C.-H. et al., 2011. Determination of low level ^{129}I in soil samples using coprecipitation separation of carrier-free iodine and accelerator mass spectrometry measurement. *Chinese Journal of Analytical Chemistry*, 39: 193–197.

Yokochi, R., Sturchio, N. C. and Purtschert, R., 2012. Determination of crustal fluid residence times using nucleogenic ^{39}Ar. *Geochimica et Cosmochimica Acta*, 88: 19–26.

Yu, H., Kim, B. J. and Rittmann, B. E., 2001. A two-step model for the kinetics of BTX degradation and intermediate formation by *Pseudomonas putida* F1. *Biodegradation*, 12: 465–475.

Zal, D. 2023. Carbon dating of agricultural soils and further understanding the transport of CO_2 gas using isotopes. Unpublished M.Sc. Thesis, University of Ottawa, 47 pp.

Zhang, J. 2018. Origins of porewater sulfate in an Ordovician aquiclude of the Michigan Basin, Ontario: Insight from stable isotopes. MSc Thesis, University of Ottawa, Ottawa, Canada

Zuo, E., Lapp, A., Jautzy, J. J. and Clark, I. D. 2022. Crustal Noble Gas Isotopic Characteristics in Low-Permeability Ordovician Sedimentary Rock, Eastern Flank of the Michigan Basin. *ACS Earth and Space Chemistry*, 6, 1: 189–196 https://doi.org/10.1021/acsearthspacechem.1c00346

Index

ISOTOPE INDEX

Note: *Italics* and **bold** page numbers refer to figures and **tables**, respectively.

ELEMENTS INDEX

Note: *Italics* and **bold** page numbers refer to figures and tables, respectively.

A

B

C

SUBJECT INDEX

Note: *Italics* and **bold** page numbers refer to figures and **tables**, respectively.

H

I

U

V

W

Z

For Product Safety Concerns and Information please contact our EU representative GPSR@taylorandfrancis.com Taylor & Francis Verlag GmbH, Kaufingerstraße 24, 80331 München, Germany